Song of the Earth

Song of the Earth

Understanding Geology and Why It Matters

ELISABETH ERVIN-BLANKENHEIM

OXFORD
UNIVERSITY PRESS

OXFORD
UNIVERSITY PRESS

Oxford University Press is a department of the University of Oxford. It furthers the University's objective of excellence in research, scholarship, and education by publishing worldwide. Oxford is a registered trade mark of Oxford University Press in the UK and certain other countries.

Published in the United States of America by Oxford University Press
198 Madison Avenue, New York, NY 10016, United States of America.

Library of Congress Control Number: 2021933128
ISBN 978-0-19-750246-4

DOI: 10.1093/oso/9780197502464.001.0001

3 5 7 9 8 6 4 2

Printed by Integrated Books International, United States of America

For those of all ages
who want to deepen their knowledge of the Earth,
and in particular for my students present and past, you inspire me.

Contents

Acknowledgments

Writing *Song of the Earth* has been a journey, one that began relatively recently in geologic time, when a friend and mentor said to me, "You should write a book." This is the book that emerged, as I thought about my teaching experiences in geology and realized my students and others wanted to know more about the Earth and why an understanding of geology matters.

Thank you to Oxford University Press, especially Jeremy Lewis, my publisher, and his capable assistant, Bronwyn Geyer. Without you, *Song of the Earth* would not have come forward to be heard. Special thanks to the Alfred P. Sloan Foundation for its grant to support this work. The grant allowed editing and fine-tuning of the manuscript and provided funds for an illustrator. I am honored to be a Sloan Foundation grant recipient.

My deepest appreciation goes to my agent, Jeff Ourvan, of the Jennifer Lyons Literary Agency, who believed in the work and added melody and harmony to the piece. Thanks to my general editors, who asked questions and provided insight and clarity, especially the ministrations of Jeff Ourvan and Sara Lippincott, science book editor and formerly nonfiction editor of *The New Yorker* magazine. Thank you to Michael Signorelli, literary agent with Aevitas Creative Management, who looked over the initial chapters and provided his input. Thank you to the biologists and paleontologists who reviewed the manuscript, Louis Taylor, Denver Museum of Nature and Science, and Chris Romero, Front Range Community College. To Heather Hollis, a colleague in my doctoral cohort in educational studies, Nova Scotia, I appreciate your fact-checking of the text and figures and your attention to detail. Sincere thanks to natural science illustrator and graphic artist R. Gary Raham, of Biostration, who took my illustration concepts and brought them to life, and to Nobumichi (Nobu) Tamura, paleoartist, who kindly permitted me to use his superb life drawings of fossil animals.

As I drafted the manuscript, colleagues and friends provided support and input as early readers; I thank, in particular, Emily N. Carey, Nancy E. Driver, Diana Perfect, Sharon Moon, Kristy Wumkes, Joan Letizio, and Mary E. Davis. To my family who encouraged and cheered me on throughout the process, Robert L. and Elisabeth T. Ervin, I am very appreciative, and especially to my husband, Richard E. Blankenheim, for allowing me space and time to research and write and for having an enduring belief in me and *Song of the Earth*. I am eternally grateful. And finally, to the Earth itself, from whence it all arose, I am thankful to be the one to compose your song through the lens of geology. Indeed, as philosopher David Hume proposes, "the past is the key to the present."

Introduction

Why Geology?

Geology is the language by which the Earth speaks. It is also a philosophy, and a poetry, offering a perspective on time—past, present, and future—through almost endless cycles of cataclysm, abundance, and decay. Many readers may have never given a second thought to geology—beyond, perhaps, looking at a peculiarly shaped rock or a dramatic landscape and wondering how the feature came to be—but the field extends beyond rocks and minerals to a holistic perspective of the Earth. It illuminates the extensive chronicle of the planet and speaks to the processes in operation, the interrelatedness of all systems, and the history of life. *Song of the Earth: Understanding Geology and Why It Matters* explores the architecture of the Earth in a way that is comprehensible and meaningful. This book will illustrate the dynamic interplay of the planet and life on it through the eons of the Earth.

From the present view, the face of the Earth appears enduring—continents are in their places, mountain ranges rise for all time, and vast oceans separate the Americas from Europe and Africa and Asia from the Americas. However, that picture differs markedly from the snapshots of the planet's past. Processes on the Earth can seem slow and steady, but transformation does not always occur gradually, and time reveals changes in sudden bursts, with dire consequences.

An asteroid the size of Mount Everest struck the Earth 66 million years ago and not only drastically altered the face of the planet but also caused a mass extinction of multiple species, including the non-bird (non-avian) dinosaurs[1] that had dominated the planet for nearly 200 million years. more recently, in 2004, tectonic plates shifted in the Indian Ocean, dropping the ocean floor and powering a wall of water, a tsunami that killed a quarter of a million people. Today, ice sheets and glaciers that were formed up to hundreds of thousands of years ago are melting and may disappear within the next generation. Unless addressed, and somehow slowed, this will lead to sea-level rise, flooding of coastal cities, the salinization of freshwater, and the destruction of agricultural areas, including low-lying deltas which provide food for a significant portion of the world.

The forces of geology are not those that have occurred in primeval times and are now complete; those processes continue today and will do so into the future. It is tempting to forget this fact because of the substantial difference between the

Song of the Earth. Elisabeth Ervin-Blankenheim, Oxford University Press. © Oxford University Press 2021.
DOI: 10.1093/oso/9780197502464.003.0001

span of human life and the life of the Earth—geologic time—but the human time frame is an illusion. The discipline of geology offers a unique understanding of the world drawn from the deep well of geologic time, the longest planetary view, going back 4.6 billion years. It provides a distinct perspective of life on Earth and of the Earth itself. Geologic conditions evident today represent just one page in a book more than 4 billion pages long—the number of years the planet has been in existence.

The study of the Earth stands in its own right, even as it owes a debt to astronomy, biology, physics, and chemistry. Geology is a science that lends itself to the narrative approach. Science educators, such as Jeff Dodick and Shlomo Argamon, state that the science of the Earth examines "ultimate causes that often lie very deep in the past, and the effects of which are observed only after very long and complex causal chains of intervening events."[2]

As a discipline, geology is relatively young, discovered and formalized during the scientific revolution in the seventeenth century in the Western world. Other societies, particularly indigenous cultures, have profound knowledge of the Earth, through induction (based on observations) and reasoning, of geologic forces and ways of working with the environment. These practices and understandings should be honored but are not covered herein.

My geology students inspired me to write this book, and it is to them and future students, those in school and those who are not but love learning, that I dedicate it. My journey began with a keen interest in the natural world, leading me to study archaeology and later geology. After a career as a U.S. Geological Survey hydrologist and geologist, I became a licensed professional geologist and started to teach as a second profession and calling. This path led me to realize that my students required more grounding in the basics of geology and why geology matters, and thus, this book came about. One methodology of historical science and geology, in particular, is the use of language and storytelling as a way of understanding geologic processes and timelines through a narrative lens. When I speak to people about geology, I tell them stories of the Earth, and often they are amazed to think that even their backyards reflect millions of years of the biography of the planet. Likewise, when we walk through the woods and encounter outcroppings of long-buried rock layers, we truly are communing across eons of time, if we learn how to read them.

Song of the Earth begins in chapters 1 and 2 by tracing the discovery of geology through the lives of individual scientists during and after the European age of Enlightenment. Some readers might wonder, why not start with the dawn of the planet's birth? One reason to begin with these snapshots is that it took centuries for generations of geologists to grasp the story of deep time and the biography of the Earth; and another is that it becomes more critical that we recognize the role of humans in both interpreting and affecting the Earth's destiny.

Some critical questions these first chapters address include: How did geologists unravel clues they observed to figure out what happened to the planet through time? How did they piece together the arrangement of the continents over the long history of the Earth? By what methods did geologists begin to understand the processes that resulted in the way the planet appears today? How did they differentiate the eons, eras, and periods of the Earth's history? The biographies of the scientists presented are distilled to highlight the immense effort and visionary thinking it took to understand the Earth in a formalized scientific manner. Thus, the stories preserve the historical record and emphasize the power of both individual lives and collaboration among those working on similar questions.

The text then delves into the history and principles of the three overarching concepts in geology: geologic time in chapters 3 and 4; the most recently revealed, plate tectonics, in chapters 5 and 6; and change in organisms through evolution in chapter 7. These theories comprehensively explain the Earth's processes. The following chapters, 8–11, elaborate the concepts for each geologic era—the Precambrian Supereon and the Phanerozoic Eon, consisting of the Paleozoic, Mesozoic, and Cenozoic—from the birth of the Earth through the present day. These chapters discuss the biography of the Earth. Finally, chapter 12 wraps up the previous ideas and examines the intersection of geology and humanity, how the Earth impacts life, and how life impacts the Earth, with an emphasis on climate change and global processes. Geology can provide a way forward in thinking about the planet and the challenges presented by the massive changes in the climate affecting all parts of the Earth.

One might assume that events on the geologic time scale are separate from individual human experience. However, geology is everywhere; its processes create the very soil, water, and air from which life springs. Understanding the Earth is essential on many levels. It is critical in determining the location and extent of natural resources, the odds of experiencing geologic and related natural hazards, and, in turn, what the human impact on these is. Understanding the origins and cycles of economic minerals can lead to better management, more protection, and sustainable utilization of those resources. Most earthquakes, tsunamis, and volcanic eruptions occur not randomly but in association with the edges of the Earth's lithospheric plates, which slip and strike and fault and fold through tectonic forces. Sinkholes and subsidence, problematic soils, floods, radon emissions, avalanches, landslides—all are understandable through the lens of geology. Such knowledge can guide the best selection of business or farm property, where to live, and where to put a highway or a dam, and it can inform many other decisions, including preparations to avert natural disasters. Learning about local, regional, national, and world-scale processes deepens people's relationship with the planet.

Additionally, a tsunami of climate-change impacts is heading directly toward humans and other life on the Earth. These challenges, based on international assessments,[3] entail severe consequences which will directly impact every element that supports life. Such factors include the air we all breathe, the clean water that all life needs, the arable land for the food supply that will feed the growing number of people on the planet (nearly 10 billion by 2050, according to the United Nations).[4] In short, the present state and the future of life itself are and will be affected by current and future climate change and the actions people choose to take today.

Geology offers a path of navigation through this coming climate crisis. Through hard-won work, geologists have developed a narrative of the planet throughout the Earth's long history. Climate events that occurred in the past exist within the rock record, including warm and hot periods when the amount of atmospheric carbon dioxide and other gases soared, and periods of great cold when nearly the entire Earth's surface froze. Both types of events, especially the former, have shaped which species survived and which did not. Absorbing the lessons from geology, in many ways, is a passage back to humanity's essential nature, revealing the substrate from which life originated and suggesting frameworks by which to manage contemporary climate changes.

Hidden within the rocks and the history of the Earth, after all, are the secrets and keys to surmounting the present challenges to every sphere—air, water, soil, and life—resulting from fluctuations in global cycles because of changes to the climate. Furthermore, our world, if not the universe, not only reflects such cycles but also adapts to them: the Earth revolves, and so the sun rises and sets, only to rise again; volcanoes erupt, and so land is created and subsided, only to return to magma to be reconstituted. This cyclic perspective, or what might be referred to over stretches of geologic time as astronomical or planetary impermanence, can foster a new, more profound relationship with the planet, effecting perhaps a keener consideration, respect, and care for our shared environment.

Song of the Earth employs the analogy of melody and harmony to explain and expand on the interconnection of life and the Earth, now and through geologic time. Recent evidence shows that this analogy may have a physical grounding in harmonics of the Earth, which can be thought of as the "music of the sphere." The planet's magnetosphere, generated in the interior of the solid Earth, not only shields organisms on the surface of the Earth from harmful rays by the protective magnetic field but also appears to act as a musical instrument.

Scientists discovered that the magnetic field responds to the influx of plasma to the Earth from the solar system with vibrations, much like a drum.[5] Not only does the magnetosphere respond to vibrations, but the Earth's atmosphere also has vibrational frequencies called Schumann resonances. These resonances were theorized in 1952 by German physicist Winfried Otto Schumann.[6] However, it was not until the mid-1960s that scientists were able to measure these low-frequency radio waves.[7] The waves range from the surface of the planet to the ionosphere, 96 kilometers (60 miles) high. Some researchers postulate that organisms respond to the frequencies.[8]

Our very atoms consist of stardust, and as the renowned planetary scientist Carl Sagan famously said, "The nitrogen in our DNA, the calcium in our teeth, the iron in our blood, the carbon in our apple pies were made in the interiors of collapsing stars. We are made of starstuff."[9] Stardust makes up the planet itself as the aggregation of molten material coalesced when the Earth first formed 4,600 million years ago. Thus, based on the origin of our atoms and molecules, the substance of life lies far back in the annals of deep time. Planet Earth (color insert I.1) has been integral to life, including that of humans, as life has been integral to the Earth in a dynamic interaction punctuated by disasters and quiescent times, all of which have shaped how the Earth and life appear today. *Song of the Earth* is the story of that journey.

1

Geology Emerges as a Science

European Roots

Our story begins with the history of geology as it was discovered by the Western scientists who discerned its foundational ideas before the dawn of the twentieth century. Most early researchers were not geologists, as there was no such formal discipline, but instead came to ponder the strange workings of the Earth from a wide variety of backgrounds in science, including medicine, anatomy, and chemistry. Even lawyers, ministers, and laypeople contributed to the burgeoning field. All had in common, however, inquiring minds and a keen interest in the natural world. Their biographies demonstrate the scope of what one person can accomplish, sometimes with quirks, and also how their work overlapped, influenced, and augmented one another's findings, particularly during the time when rationalism, through the application of modern scientific thought, replaced superstitious notions and verbatim interpretations of the Bible. The science of geology was built up stone by stone, one might say, through collaboration, argument, discussion, and the resolution of vital questions. This effort extended from the birthplace of geology in Europe to the Americas as trade, colonization, and ultimately communications evolved.

The age of exploration and the ensuing Enlightenment of the seventeenth century in Europe touched on all scientific endeavors, in particular the desire to understand the world and nature as a whole. Geology is a young discipline, not only in obvious terms within the scope of geologic time but also in the context of how it applies to human applications. It was not until the end of the nineteenth century, for example, that most of the basic tenets of geology had been established. Some fundamental geological advances—such as determining the age of rocks by using radiocarbon isotopes which decay at a known and fixed rate or discerning the movement of plate tectonics—were discovered only as recently as the twentieth century.

Queries about the physical nature of the Earth and the remains of organic life in rocks led to the beginnings of the scientific knowledge of geology. These two investigative lines merged and split, overlapped and separated, like braided stream channels that come together and then divide but which share the same source: how individual layers of rock, known as strata, get deposited and what fossils indicate about life and rocks.

Song of the Earth. Elisabeth Ervin-Blankenheim, Oxford University Press. © Oxford University Press 2021.
DOI: 10.1093/oso/9780197502464.003.0002

Such luminaries as Leonardo da Vinci and the Danish scientist and bishop Nicolaus Steno were intrigued by what they observed in the natural environment. Curiosity about rocks, their placement, and their forms stimulated questions about what different-looking layers could imply. Scientists such as these bandied about hypotheses and proposed ideas on the origins of rocks and sediments. Thinking progressed to interpreting the rocks, based on their positions, and the relationship of one rock unit to another. Steno studied strata and unraveled their layering, which led to the fundamental principles of stratigraphy—the branch of geology concerned with the sequencing of rock units and their deposition—still taught today. This work formed the basis of the early geologic time scale. Another line of interrogation centered on how remains of ocean creatures could be found in places far removed from the sea, as first noted by Leonardo da Vinci.[1]

These earliest inquiries preceded James Hutton, who was a Scottish naturalist and scientist, considered the father of geology. Hutton questioned how quickly and under what conditions rock layers are deposited, putting together a broader picture of how the various pieces on the Earth's surface fit together. He speculated that heat in the Earth was a significant driver of geologic forces. The next question centered on further sequencing of rocks, relating rocks to one another—the bailiwick of Abraham Werner, whose studies, a now-superseded theory known as Neptunism, in part refined the early geologic timetable. Around the same time as Werner was examining rocks in Germany, early paleontologists (those who study fossils), such as Etheldred Benett, began to organize these remains and put them in a logical order based on early classification systems.

Other scientists of the late eighteenth and early nineteenth centuries increasing numbers of whom were women pursued more complex and intricate fossils. Using fossils to correlate rock layers across large areas of England, Wales, and Scotland, William "Strata" Smith produced the first detailed geologic map of all of Britain. William Buckland promoted a catastrophic Noachian flood as responsible for many strata and the placement of fossils. He theorized, for example, that the biblical flood was responsible for the presence of bones in an English cave. In response to such catastrophic theories of geology, Charles Lyell, following the insights of Hutton, his fellow Scot, into geologic processes, developed "uniformitarianism," a gradualism theory that suggested to an increasingly intrigued British public that the Earth was shaped by the force and intensity of the same natural processes observable today.

Animals unknown to the world at the time, swimming marine vertebrates, were found preserved in the ocean sediments of the Jurassic cliffs of Dorset by paleontologist Mary Anning. She was an expert at not only locating these treasures but also mounting and drawing them. Louis Agassiz furthered understanding of Earth processes when he observed the presence of immense boulders from far-flung places and theorized that glaciers deposited them, intuiting that an ice age must have covered much of northern Europe. Surficial geological studies were carried forward by Archibald Geikie, who founded the discipline of geomorphology, the study of the deposits and processes that operate on the Earth's surface. French naturalists, in particular Georges Cuvier, made critical contributions to paleontology, the field of geology that studies past life forms and fossils, and along with Alexandre Brongniart, he drafted the first geologic map of the Paris Basin.

As the stories of the early geologists emerge, they carry interwoven ideas fundamental to geology, which in turn are primarily categorized by unfathomable stretches of time—deep time. How geology rose as a science is a remarkable feat, and the stories of its pioneers humanize a field that can seem dry and unremarkable to the uninitiated. Moreover, telling the history of geology through the lives of its founders is a way to appreciate how a time of significant scientific advancement intersected with the lives of fascinating individuals to create the bedrock concepts of the field. We live on the Earth, and it sustains us. Could any other science intimately affect life and, in turn, be affected by it?

European Scientific Geological Developments

Nicolaus Steno

Nicolaus Steno (1638–1686, also known as Niels Stensen, the name given to him at birth, or alternatively by his Latinized name, Nicolaus Stenonius) was born in 1638 in Copenhagen, son of a Lutheran goldsmith. He was in poor health in his early years and thus developed strong friendships with older friends of his family, rather than the traditional childhood companions, and these older friends encouraged in him a serious cast of mind. Steno studied medicine and in 1660 went to Amsterdam to further his training with anatomist Gerard Blaes (Blasius). There he discovered the parotid duct, which conveys saliva to the mouth from the parotid gland known to this day as *ductus stenonianus* (he was given full credit after his mentor tried to claim the find as his own). After a stint in Leiden and in Paris, Steno was summoned to Italy by Grand Duke Ferdinand II to become a member of the Accademia del Cimento, an early scientific society.

In 1666, a giant great white shark was caught by a fisherman in the Mediterranean and came to the attention of the grand duke. Steno was asked to dissect the shark and became enthralled with its thirteen rows of teeth. He intuited that the teeth resembled tongue stones (*glossopetrae*) (figure 1.1) that had been discovered high above sea level in layers of rock.[2] Steno was not the first to relate the *glossopetrae* fossils to the teeth of living sharks, but he confirmed

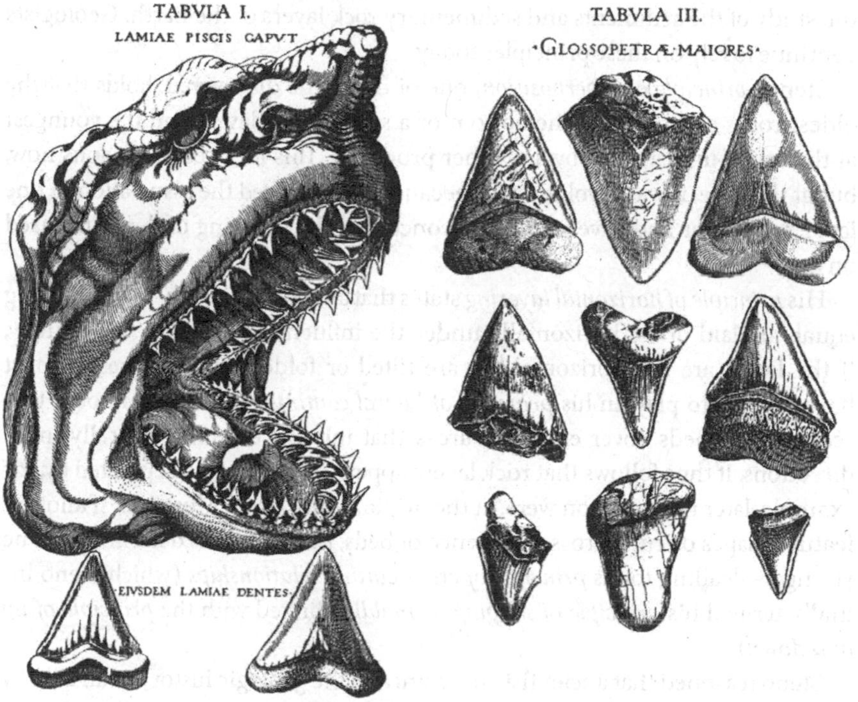

Figure 1.1 Steno's drawing of shark dissection and fossil teeth. (Steno, 1667, tab. IV, fol. 90)

the idea with his anatomical study, publishing his results in 1667[3] along with the illustration of the dissection and examples of the *glossopetrae*.[4] After that, scientists viewed fossils as the remains of ancient animals preserved in layers of sediment. Steno hypothesized that the ancient teeth found in sediments had been transformed into stone by the process of "corpusculization," whereby the "corpuscles," or molecules, of the original material were replaced by those of minerals. Why these stone teeth were found so far above the ocean was another question. Steno reasoned that either the sea long ago must have risen, spreading across the land, or the ocean floor had been uplifted to form a new land surface. If the latter was true, then environmental conditions must have changed drastically.

Steno made other observations critical to the development of the field of geology. In 1669, following his report on the shark teeth, he wrote *De solido intra solidum naturaliter contento dissertationis prodromus* (*The Prodromus to a Dissertation Concerning Solids Naturally Contained within Solids*),[5] a study of rocks around the area of Tuscany. He noted the relationships of strata, or rock layers, to one another and started to develop the basic principles of stratigraphy, the study of the sediments and sedimentary rock layers of the Earth. Geologists continue to rely on these principles today.

Steno's *principle of superposition*, one of the six he discovered, holds that the oldest rocks are found on the bottom of a sequence of layers and the youngest at the top, if undisturbed by any other processes. This idea seems obvious now, but at the time, it was revolutionary because it introduced the association of one layer to another and gave rise to the concept of the ordering of deposits based on age.

His *principle of horizontal layering* states that all sediments, other factors being equal, are laid down horizontally under the influence of gravity. Accordingly, if the layers are not horizontal but are tilted or folded, other processes must have come into play. In his *principle of lateral continuity*, Steno also noted that sedimentary beds cover extensive areas that initially extended laterally in all directions. It thus follows that rock layers appearing similar but separated by, for example, later river erosion were, at the origin, continuous. Moreover, if another feature shapes or cuts across a sequence of beds, that feature is determined to be younger—leading to his *principle of cross-cutting relationships* (which Steno initially termed his *principle of shaping or molding* linked with the *principle of up and down*).

Steno reasoned that a scientist could unravel the geologic history of an area by examining and theoretically removing one layer of strata at a time from the top of the stack of rocks (the youngest) to the bottom of the stack (the oldest layer), resulting in his *principle of reconstruction* (or *back stripping*). This concept (figure 1.2) shows the most recent condition (20 in the figure). By peeling off one layer and then the next one through the entire sequence, a scientist can arrive at the arrangement of the layers (25 in the figure) at an earlier time and thus understand the geologic processes that have occurred.

Steno, among his many talents, was a mineralogist, and he developed one other stratigraphic principle, the *principle of included fragments* (alternatively called the *principle of inclusions*). The idea is that if a fragment of another mineral or rock is found within a sedimentary rock, it must be older than the rock it is contained within. As he developed his principles, Steno also theorized how fossils came to be located within rocks, speculating that at one time, the rock material was liquid and settled out of the ocean in layers, thereby trapping animals to create fossils.

Steno's theories sometimes are called laws, but it is more accurate to refer to them as principles. Laws are statements or rules mathematically expressed that appear to be without exception at the time and have become consolidated by repeated testing, for example, Isaac Newton's laws of motion. Principles, on the other hand, are guiding ideas and theories about phenomena demonstrated with

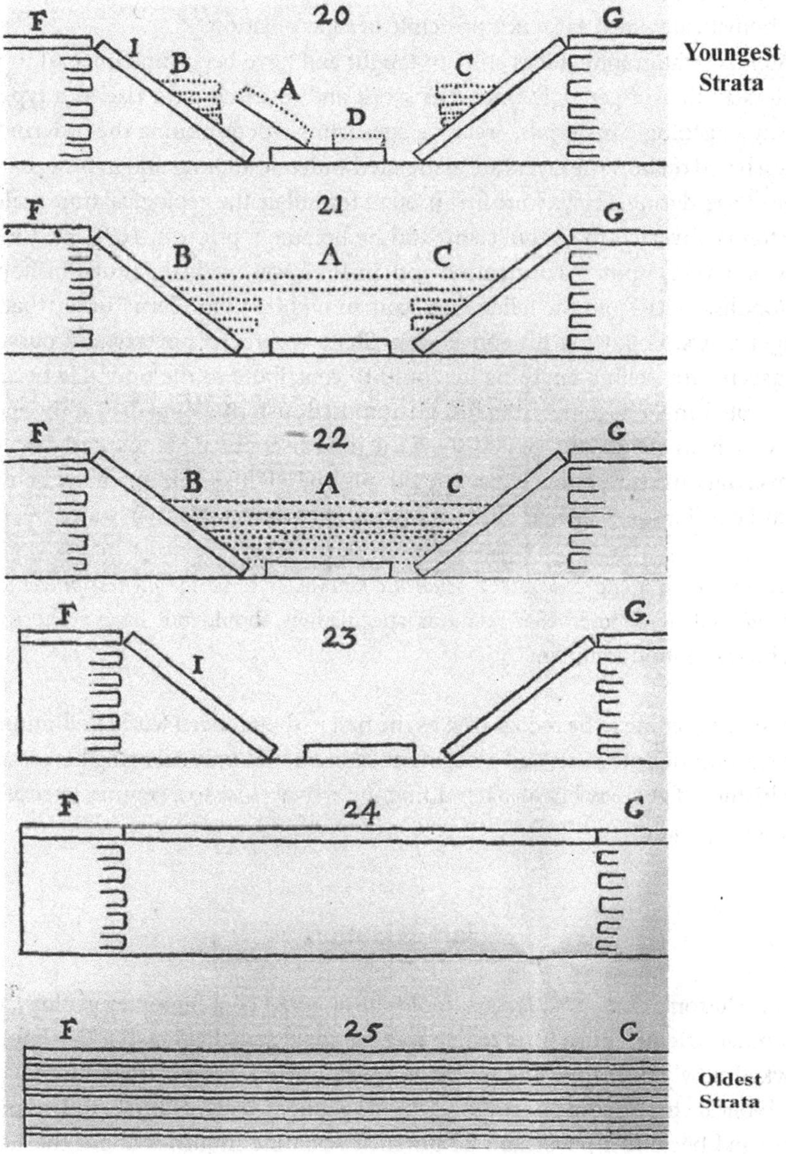

Figure 1.2 Steno's principle of reconstruction. (Steno, 1916 [1669], p. 226)

a preponderance of the evidence. They may or may not be able to be expressed mathematically, such as Steno's principle of superposition.

Steno's stratigraphic ideas still are taught and have been fundamental to the understanding of geologic time. His work and research gave rise to a type of dating of geologic materials—relative age-dating—determining the ordering of strata based on how the layers are associated with one another and in time. In this type of age-dating, fossils were first used to formulate the geological time scale.

Steno converted to Catholicism, and he became a priest in 1675. He took a vow of poverty upon his ordination and, in 1677, was made the titular bishop of Titiopolis, a city from the fallen East Roman Empire in modern Turkey that no longer exists. Following his conversion, Steno embraced poverty and pursued an ascetic life, selling anything he could to contribute to the poor. He became debilitated and emaciated from his self-imposed austerities and died at the age of forty-eight in 1686. Other scientists of the time discounted Steno's work because of his religious conversion. However, although he left his scientific career behind, he did not abandon science. His biographer, Jens Morten Hansen, states:

> To his death Steno considered *scientific knowledge to be the highest praise to God*, and he claimed that religious speculations should not have authority above scientific arguments.[6]

Steno has come to be recognized as the first in the modern world to illuminate the significance of fossils and placement of strata, not only securing his place in the history of geology but also heralding the arrival close to a century later of the "father of geology."

James Hutton

James Hutton (1726–1797), considered the father of contemporary geology, was a Scottish scientist born in 1726. He was a gradualist, believing that Earth forces worked slowly over time and in cycles, as contrasted with catastrophes, such as the biblical story of the Great Flood. In his younger years, Hutton studied medicine and became a physician; he pursued scientific inquiries about the Earth his entire life but did not publish his findings until the age of sixty. Hutton lived and worked during the Scottish Enlightenment of the late eighteenth century and was a contemporary of Benjamin Franklin, David Hume, and Adam Smith. He, along with Franklin, Hume, Smith, John Playfair, and other notables, was a founding fellow of the Royal Society of Edinburgh (RSE) in 1783, created by royal charter for "the advancement of learning and useful knowledge," which met in its early days at the Edinburgh University Library.

Along with the noble houses and salons of Edinburgh and the surrounding areas, the RSE was a place in which ideas and scientific findings were presented, discussed, and debated. Franklin visited Edinburg in 1759 and again in 1771, meeting fellow thinkers, and he corresponded with Hutton and others for years after. Like many of his colleagues, Hutton was a deist, believing the creator put into place the natural laws and created the world for humans, who, through rationality and thought, could uncover the Earth's mysteries. Moreover, as revealed throughout his, at times, complicated writings, he struggled to frame his epiphanies about the natural world within this deist context.

In 1785, Hutton presented a talk to the RSE on his geological findings, *Concerning the System of the Earth, Its Duration, and Stability,*[7] transcribed and printed three years later as *Theory of the Earth.* A full two volumes of this work, approximately 1,200 pages, were later published in 1795, two years before his death.[8] In *Theory,* Hutton elaborated on his earlier ideas and addressed criticisms raised by opponents of these thoughts. His third, posthumous volume set forth the origin of granite, which he theorized was from an igneous (molten magma) origin rather than deposited through sedimentation. Hutton ascribed the formation of these rocks to the heat of the Earth and was responsible for promoting Plutonism (named for Pluto, the god of the underworld and heat), a school of thought correctly theorizing that igneous rocks created in the Earth's fiery magma eventually were eroded and deposited in sedimentary layers. In Hutton's Plutonist view, primordial—or primitive—rocks were the result of magmatic intrusion or eruptions from deep in the Earth's interior.

Hutton recognized that sediments are deposited over long periods, mostly through the action of water (in the form of rivers and streams and some by oceans), and laid down horizontally (as Steno observed) with the oldest materials on the bottom of the stack and the youngest on top. Over time, the deposits are lithified (made into rock) by burial and temperature changes and then uplifted by mountain-building processes that distort and deform the beds. Later, the exposed rocks are eroded, and with more time, new sediments settle on top of the eroded beds. Even without modern numerical age-dating methods to go by, it was evident to Hutton that one of those natural sequences would take millions of years, giving rise to what became known as the *theory of uniformitarianism.*

Hutton also examined and discussed breaks in time in the rock record, called unconformities (a period in geologic time where no rocks remain to indicate massive uplift, erosional periods, or the forces by which old ones were destroyed by weathering and erosion, as shown by missing strata). These "irregular junctions of primary and secondary strata"[9] are critical to an understanding of the great age of the Earth and the geologic time scale, because they illustrate how rock strata form over a vast period. Hutton was not the first to notice changes in angles and orientation of rock layers, but he was the first to interpret the phenomenon correctly and recognize the implications. Steno had noted angular variations in layers of rock in his *Prodromus*, published in 1669.[10] Swiss and French geologists, among them Horace Bénédict de Saussure, also a mountaineer, had reported changes in the orientation of rock layers in the Alps.[11] Hutton was familiar with these earlier findings and searched for junctions where different strata were in contact. Initially, he examined inclined rocks on the northern end of the island of Arran but knew he needed further evidence. Friends informed him of outcroppings along the eroded banks of the river Jed where flat-lying beds of marl (a sedimentary rock made of limestone and clay) and sandstone lay over other layers with a pronounced dip. Hutton reasoned that the near-vertical beds had been uplifted and were older than the overlying horizontal beds. He postulated that some event had eroded the top surface of the lower beds, but over a long period, conditions changed, and sediments in a horizontal position settled on top. He documented what he termed this "great unconformity" of Inchbonny at Jedburgh, Scotland (figure 1.3) in 1787. His ideas on the heat of the Earth in the formation of rocks, cycles, and age of the Earth had more support from his excursions in the field, but he needed further proof. Hutton began looking for other examples where younger strata were in contact with much older layers with missing rock units indicating a significant discontinuity in time.

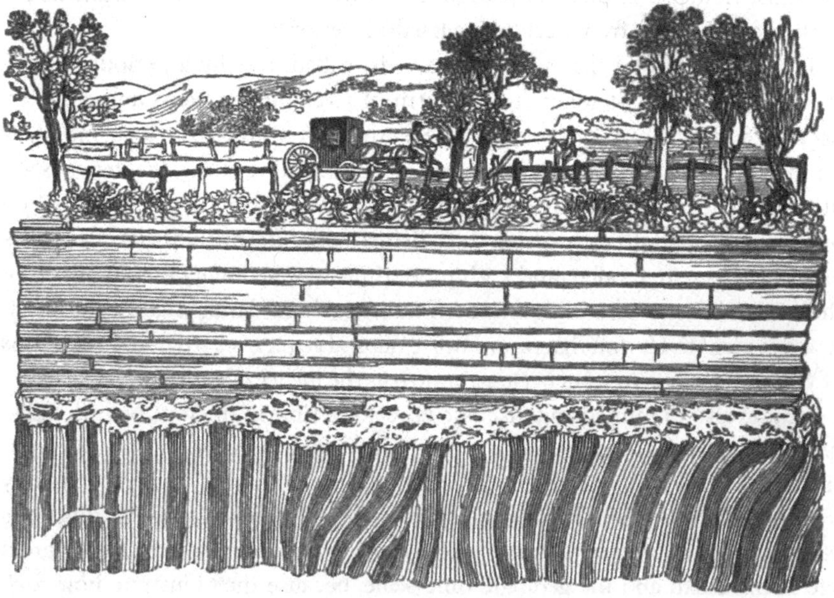

Figure 1.3 The great unconformity of Inchbonny at Jedburgh. (Hutton, 1795, Vol. 1, plate III; engraving of a drawing by John Clerk of Eldin)

Further verification came in the rocks at Siccar Point (color insert 1.1). Upon hearing of the sea-washed exposure from Sir James Hall, a politician, landowner, and chemist, Hutton took his fellow RSE member Playfair to observe the angular unconformity (adjacent strata at sharp angles to one another) in a small boat, risking life and limb. A tilted red sandstone layer overlies the spectacular near-vertical beds of schistus (defined as shale and limestone), in what is now termed "Hutton's unconformity." This feature was confirmation of a break in geologic time. Hutton realized that significant events must have occurred between the older schistus and sandstone layers at an angle to them, which upended the older rocks and then eroded or destroyed intervening strata, leaving a flat surface on which later sand built up and which over time coalesced into sandstone. Forces that would cause the loss of strata and the apparent missing time in the rock record can be episodes of heat, uplift, the action of rivers, or eroded strata, all of which represent an extensive period of millions of years.

Related to the idea that geologic change occurs slowly over time, Hutton suggested the pre-Darwinian *principle of variation*, which, though stopping short of a theory of evolution, held that life forms also change gradually over time. He disputed the findings of Simon Pallas, also known as Peter Simon Pallas (1741–1811), a German zoologist and botanist, that land-dwelling animal and marine fossils found together in Siberia were placed there by the great Noachian flood. Hutton said no such flood was needed; the presence of marine fossils among the land-mammal bones could be explained by geologic processes, water moving them, and that indeed "formerly the Elephant and Rhinoceros had lived in Siberia."[12] He never adopted the idea of extinction, the dying out of a species, genus, or family, and instead argued that life forms remained unchanged, although they could vary within specified limits.[13]

Hutton died in 1797, but his third volume of *Theory of the Earth: With Proofs and Illustrations*, which contained not only his thoughts on the origin of granite but also an account of his travels in the field, was not published until 1899.[14] This 102-year delay was precipitated by the Geological Society in London because, according to Sir Archibald Geikie, the editor of Hutton's last volume, some of Hutton's illustrations were incomplete.

To be sure, the peregrinations in Hutton's somewhat tortuous writings at times obscured his ideas, including his attempt to explain his deism and how he reconciled it with his geologic understanding of the world. In a draft preface to his 1788 *Theory*, titled "Memorial justifying the present theory of the Earth from suspicion of impiety," Hutton writes:

> It belongs to religion to teach, that God made all things with creative power; that perfect wisdom had then presided in the election of ends & means, and that nothing is done without the most benevolent intention. But it belongs not

> to religion to give a history of nature, or to inform mankind of those things which actually are; it belongs not to religion, to teach that natural order of events, which man, in his science, may be able to unfold, and, in the wise system of intellect, find means to ascertain.[15]

Hutton went on to discuss the authority of religion versus science. He was attempting to forestall religious criticism of his geological theory, but his argument became unclear as he delved deeper into the subject. A friend and colleague, William Robertson, principal of the University of Edinburgh, made tactful changes in an attempt to shield him from religious controversy and to mitigate his verbosity; however, despite these ministrations, Hutton decided against publishing the preface, and it first appeared almost two centuries later, in a 1975 paper by Dennis R. Dean called "James Hutton on Religion and Geology: The Unpublished Preface to His Theory of the Earth."[16]

Upon Hutton's death, Playfair set out to write a biography of his friend and mentor, devoting his time to isolating the vital kernels of geology hidden in Hutton's long-winded descriptions, lack of clarity, and digressions into theological matters. Ultimately, Playfair temporarily shelved the biography and concentrated instead on clarifying Hutton's ideas about geology. On July 1, 1799, he read the first part of his manuscript to the RSE and followed with the second part at the next meeting.[17] These two parts ultimately became Playfair's *Illustrations of the Huttonian Theory of the Earth*, published in 1802. Unlike Hutton himself, Playfair concisely explained the Huttonian concepts of geothermal heat, unconformities, the gradualist view of the rock cycle (how rocks transform from one type to another), and periods of erosion and uplift as repeated events in Earth's history. Thus, Playfair illuminated Hutton's theory of uniformitarianism and his statement that "The present is key to the past."[18] Natural forces that occur on the Earth at present have operated in the past in the same way and over similar lengths of time.

Playfair finally read his biographical sketch of Hutton to the RSE on January 10, 1803, five years after Hutton's death, and published it in the RSE *Transactions*[19]—an elegy and tribute to his longtime friend and mentor. Playfair's work in recasting Hutton's hypotheses and theories guaranteed the latter's preeminent place as the father of geology. Thirty years later, through the agency of Playfair, Lyell's writing furthered Hutton's ideas of uniformitarianism, gradualism, and the igneous origins of rocks.

The geologic research and writings by Hutton, with Playfair's interpretations, have stood the test of time. Principles gleaned from Hutton are the ancient age of the Earth, cycles in geology—including uplift and erosive factors—the effects of rivers, the heat of the Earth's core as contributing to the development of rocks, shaping of the crust and subsurface, and gradual changes over long periods whose processes continue today. In the 1788 *Theory of the Earth*, Hutton famously said:

> The result, therefore, of our present inquiry is, that we find no vestige of a beginning,—no prospect of an end.[20]

Abraham Werner

Abraham Werner (1750–1817) was born more than a century later than Steno and a quarter of a century after Hutton, in 1750 (some reports cite 1749 as the year of his birth), to a family of mining engineers. His work was fundamental to understanding the age order of rocks and their sources, building on the tenets that Steno established.[21] Werner's sequencing of rocks formed the basis of the early geologic time scale and provided a way to categorize layers of the Earth.

Werner's father was the inspector of ironworks for the duke of Solm, Germany.[22] As a child, he was frail, and sickness plagued him throughout his life. His father instructed him, and though he collected minerals as a boy, his ill health made him unable to continue fieldwork as he developed an interest in geology. After that, Werner was educated at the Mining Academy (Bergakademie) of Freiberg and the University of Leipzig and then appointed as a teacher of mining and mineralogy as well as curator of the mineral collection at the Mining Academy of Freiberg at age twenty-six. He remained at the university teaching for the next forty years and was a dynamic and engaging lecturer, making minerals exciting and relevant to his students.

Werner contributed significantly in the area of mineral identification, classification, and color analysis. He wrote several volumes on these subjects. His major work, *Von den äusserlichen Kennzeichen der Fossilien* (*A Treatise on the External Characteristics of Fossils*), was first published in 1774 (before his teaching appointment) and was translated into English in 1805.[23] The term "fossil," introduced by Georgius Agricola (1494–1555) in his 1546 work *De natura fossilium* (*On the Nature of Fossils*, known as his *Textbook of Mineralogy*), was used in Germany for both "earths" (minerals) and "figured stones" (fossils)[24] until 1820, after a gradual realization that the two categories were different. Most minerals are formed from inorganic sources, while organic life processes form most fossils. In his preface, Werner states his definition of mineralogy as the natural history of "fossils." He classified minerals systematically and logically, the basis for how they are studied today.

Along with his research on minerals, Werner studied rocks, their formations, and how they relate in time, which he called geognosy. He was mainly influenced by the works of earlier German geologists Johann Gottlob Lehmann (1719–1767) and Georg Christian Füchsel (1722–1773).[25] Lehmann and Füchsel, publishing separately, discussed the positioning of rock units in relationship to one another in the Saxony area of northern Germany, now identified as Triassic and Permian rocks. Werner wrote up his ideas on rock formations in his 1787 *Kurze Klassifikation und Beschreibung der verschiedenen Gebirgsarten*.[26] He divided formations into four categories: *Primitive*, *Secondary*, *Alluvial*, and *Volcanic*. He added a fifth, *Transitional* rocks, between the Primitive and Secondary layers.[27] These divisions were time-based: Primitive rocks were the oldest, and the youngest rocks were the Alluvial and Volcanic.

Lehmann had further proposed that a universal ocean had once covered the world.[28] Werner carried these ideas forward and hypothesized that the formations were originated from particles and sediments settling out of a vast sea. Werner thus became the founder of the Neptunist school (named for Neptune, the Roman god of the sea, which held that rocks had formed by sedimentation in water). His hypothesis on the origin of rocks sometimes was transmuted by others into deposition by the global Noachian flood, or the Great Flood in the Bible. Werner himself never referred to the Bible or the stories in it as a source of geology, and he is not considered a biblical geologist by those who have studied his writings.[29] Igneous rocks, such as basalt (a dark, fine-grained igneous rock), were thought by Werner to be unusual, extraneous, and somehow deposited in the same manner—through settling out of sediments in the worldwide ocean, contrasting the theory of the Plutonists like Hutton who held that heat and magma produced basalt.

Given the state of his health, Werner was unable to travel, but his reputation was such that students and scientists from all over Europe came to take his classes and to visit him at the Mining Academy of Freiburg. A member of the Institute of France and the Wernerian Natural History Society in England, Werner, on his deathbed, learned with great happiness of the professorships in mineralogy at Cambridge, Oxford, London, Glasgow, Dublin, and Belfast founded on his principles.[30] He died in 1817 and is renowned for leaving a legacy of inspiring many young geologists.

Part of the difficulty with Werner's Neptunist hypothesis was that he was unable to study rocks firsthand in other parts of Europe; he therefore projected information from his local settings to the rest of the geological world. Additionally, Werner's theory could not account for what happened to all of the water that had made up the ancient worldwide ocean. Werner nevertheless retains a prominent place in the history of geology, credited for several theories. He concluded that rocks should be classified by their geologic ages rather than traditional methods of categorizing them by the minerals they contained.[31] His investigations of layers relative to each other led to one of the earliest versions of the geologic time scale, and he provided a unifying theory of geology for that time.[32]

Etheldred Benett

Critical work on locating and classifying fossils continued in England with Etheldred Benett (1776–1845), who was born in Wiltshire. (Some sources list her birth year as 1775.) Few, if any, actual images remain of Benett. She was encouraged to study natural science by her sister-in-law's half brother, a botanist and fellow of the Geological Society of London, and she developed a keen interest in fossils and stratigraphy.

Benett amassed an extensive collection of fossils by 1809 from quarries in the Wiltshire area. She was a contemporary of, and was influenced by, William "Strata" Smith, the famous mapmaker of England, and followed his stratigraphic methods (tools to relate one geological bed to another), though she did not always agree with them. She collected her specimens by herself and was proficient in taxonomy, finding and naming new taxa (from taxonomy, the branch of science concerning classification). She made a stratigraphic section (a way of portraying rock units in a vertical sense arranged from oldest at the bottom to youngest at the top, as seen in the field) of the Upper Chicksgrove Quarry in Wiltshire, which she first sent to the Geological Society in 1815. It was published without her knowledge and failed to credit her work.[33]

Many prominent geologists of the time came to see and study the renowned fossils Benett found and curated, including William Buckland, Charles Lyell, and Louis Agassiz. She was generous in offering her specimens to these and other scientists and sent some from her collection to museums. She published "A Catalog of the Organic Remains of the County of Wiltshire" in 1831, with local stratigraphy, an index of her collection, and illustrations by E. D. Smith. In the preface, she states:

> I have endeavored to render this catalog as correct as possible; and when I mention that it has been approved by Mr. [George Bellas] Greenough, it will run no risk of being despised in the Geological world.[34]

Among the material she collected were rare mollusks with preserved soft parts—the first ever to be identified (figure 1.4).[35] Many of her finds were "type" fossils (representative specimens utilized to define a species), and the majority of her collection was made up of invertebrates (animals with external shells and soft body parts), such as the shells shown (figure 1.4). However, a critical bone of a vertebrate (an animal with an internal skeleton), *Ichthyosaurus trigonus* (a

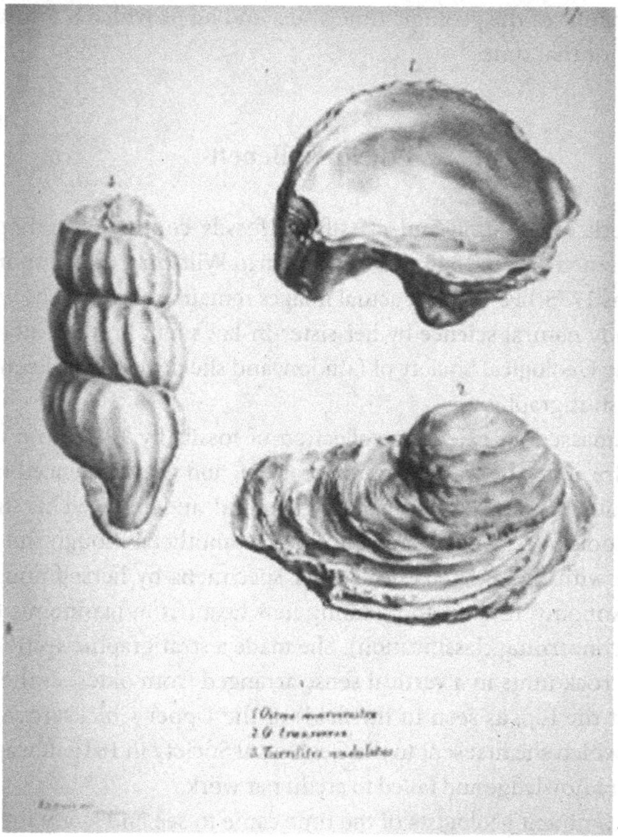

Figure 1.4 Illustrations of (1) *Ostrea recurvirostra*, (2) *Ostrea transversa*, and (3) *Turrilites undulates* from the collection of E. Benett (Benett, 1831, figure 17; illustration by E.D. Smith)

giant, carnivorous marine reptile, now extinct), was so important an example of its type that when the materials were rediscovered, it replaced another type fossil per the International Commission on Zoological Nomenclature.

Because she was a woman, Benett was not permitted to join any of the geological societies of the time. She sent some of her fossils to the Museum of Saint Petersburg in Russia, and in appreciation, the Imperial Natural-History Society of Moscow under the impression that she was a man, awarded her an honorary degree. In a letter to Samuel Woodward April 12, 1836, she noted:

> it is provoking that no-one will believe that a Lady could write such a trifling thing—in this Diploma I am called Dominum Etheldredum Benett & Mr Lyell told me that he had been written to by foreigners to know if Miss Benett was

> not a gentleman . . . so you see that scientific people in general have a very low opinion of the abilities of my sex.[36]

The majority of Benett's collection was sold after her death to Thomas Bellerby Wilson (1807–1865), a naturalist and philanthropist, in Newark, Delaware. He donated the material to the Academy of Natural Sciences of Philadelphia (ANSP) in successive allotments between 1848 and 1852.[37] However, the fossils were not cataloged by the museum until well into the 1980s, in part because the focus of the ANSP was on Neogene (younger) fossil mollusks. Further, there were difficulties transcribing the notes and nomenclature of the various species. Occasional references to the collection over the years even misidentified the name of the museum, such that until 1989, curators considered the Benett collection lost.[38] Now that it has been rediscovered, museums and researchers recognize Benett's collection as pivotal to the history and field of paleontology.

William "Strata" Smith

No discussion of present-day geology would be complete without recognizing the contributions of William "Strata" Smith (1769–1839). Smith was born in Oxfordshire in 1769, the son of a blacksmith. His father died when Smith was eight, and his uncle raised him in the limestone hills known as the Cotswolds. Jurassic fossils were abundant in the area, and young Smith collected them intently. At eighteen, he went to work for Edward Webb, a mapmaker and surveyor, and was sent to survey wealthy estates, learning his trade while on the job.

In studying the rock formations of southern Britain, Smith first realized that specific, well-defined beds of sediments held particular sets of fossils, which were unlike those in overlying and underlying beds. This finding resulted in his development of the *principle of faunal succession*, which holds that geological strata can be identified and categorized by the fossils found within them.

A series of sedimentary formations consisting of Carboniferous Period coal, Triassic Period sandstones, Jurassic Period limestone, and Cretaceous Period chalk (the last make up the 300- to 400-meters-thick white cliffs of Dover) underlies southern Britain. These formations were laid down horizontally in a marine setting, then gently folded. This area, the Weald, encompasses from Tunbridge Wells in the north to Brighton in the south. Younger Cenozoic Era deposits, such as the London Clay, were deposited to the north of the Weald.

Two canals were commissioned in 1795 for barges to deliver coal from Bristol and Bath to London. In overseeing the digging of trenches, Smith kept an extensive record of the fossils found in the excavations and collected them in their stratigraphic (age and rock unit) order. He further managed to correlate layers from one canal location to another based on the fossils found in them. Over time, he devised a stratigraphic column of rock units around Bath[39] based on the organic remains in the layers and the dip of the beds—the first biostratigraphic study (relating strata based on the fossils founds within) done on this scale. A keen observer, he examined and correlated many outcrops of rock across all of England, Wales, and southern Scotland.

In 1815, Smith published *A Delineation of the Strata in England and Wales, with Part of Scotland*, the first geologic map of Britain (color insert 1.2). Geologic units on the map were hand-painted in watercolor, and most important, the map showed in its legend the stratigraphic units in order, from older at the bottom to younger at the top.[40] Two years later, Smith produced the *Stratigraphical System of Organized Fossils*.[41] His work revealed that fossils within the rock units allow the formations to be distinguished from one another and classified by the life forms they contain.

A competing map published in 1820 by George Bellas Greenough, president of the Geological Society of London, undercut sales of Smith's map. Smith was not a member of this prestigious society,[42] presumably only because of his humble origins. While Greenough rejected Smith's correlation of biostratigraphy and delineation of specific geologic units, notes on his map[43] showed that he relied in large part on Smith's map, without crediting him.[44]

Smith ended up in debtors' prison for ten weeks; nevertheless, the venerable Geological Society finally recognized his contributions to geology, and in 1831, he received its first Wollaston Medal.[45]

Smith remained active in geology, lecturing on the strata of Yorkshire and making local geological maps. He eventually moved to the seaside town of Scarborough, where he assisted in recommending sources of suitable building stone for the Houses of Parliament. In 1838, he applied for a position with the Geological Survey of London but did not receive it. The following year, on his way to the annual meeting of the British Association for the Advancement of Science in Birmingham, he stayed with a friend and colleague in Northampton. They embarked on a series of trips in the area to view the local geology, but Smith became gravely ill after catching a chill and died that same month, August 1839, engaged in doing what he most loved.

William Buckland

William Buckland (1784–1856) was the son of a rector, who took him along on his fossil-collecting trips to local quarries. He was awarded his BA from Oxford in 1804 and his MA, along with his ordination, in 1808. Remaining at Oxford, he was appointed reader of mineralogy in 1813 and reader of geology in 1819.

Among his students were Lyell and, curiously, the Victorian art critic, philanthropist, and writer John Ruskin. At the inauguration of his geology readership, Buckland gave a speech titled "*Vindiciae geologiciae*; or the Connexion of Geology with Religion Explained," which interpreted geology according to biblical accounts, especially that of the Great Flood.

In 1818, French naturalist and paleontologist Georges Cuvier (1769–1832), considered by paleontologists to be the father of vertebrate paleontology, visited Buckland, who showed him unusually large bones from Stonesfield Quarry, Oxfordshire. Through the study of the animal's teeth, jaw, and bones, Buckland described the first complete skeleton of a sizable carnivorous reptile, which he named *Megalosaurus* (which current paleontological studies have determined to be an extinct, land-dwelling, carnivorous theropod—a dinosaur with hollow bones and three-toed feet).

He read his findings to an 1824 meeting of the Geological Society and published them as "Notice on the Megalosaurus or Great Fossil Lizard of Stonesfield,"[46] with drawings by Mary Moreland, a skilled illustrator of fossils[47] for Cuvier (figure 1.5). Buckland became president of the Society that year and in the following year married his illustrator. Mary Buckland also was a fossil collector who did fieldwork, prepared fossils for display, and illustrated her husband's papers. *Megalosaurus* was eventually reclassified along with several other reptiles in 1841 by British paleontologist Sir Richard Owen, the premier paleontologist at the British Museum of Natural History in the mid-1800s. Owen was first to introduce the term *Dinosauria* ("fearfully great lizards" or "terrible lizards").[48] Thus, Buckland's *Megalosaurus* has the distinction of being the earliest dinosaur to be named as such.

Buckland was influential in promoting his views of the Noachian flood through his studies of the Kirkdale Cave deposits in Yorkshire. These deposits consisted of a significant number of elephant, rhino, hippo, horse, and hyena fossil bones. Some force had shattered the herbivore bones, leading Buckland to conclude that while the Great Flood covered some of them up, most of the bone material was brought in by the hyenas. He published his findings in an 1823 paper with the unwieldy title of Reliquiae Diluvianae: *Or Observations on the Organic Remains Contained in Caves, Fissures, and Diluvial Gravel, and other Geological Phenomena, Attesting to the Action of the Universal Deluge.*[49] Buckland also initially thought floods created the deep grooves found in rock pavements, which the Swiss glaciologist Louis Agassiz later determined were caused by the action of glaciers dragging boulders over the surfaces. Buckland's views on glaciers were modified in 1838, when he traveled to Switzerland to meet Agassiz. Two years later, Agassiz would visit Buckland at Oxford, and they spent time in Scotland looking for evidence of striations, moraines, and other glacial features, including the transportation of erratic boulders.

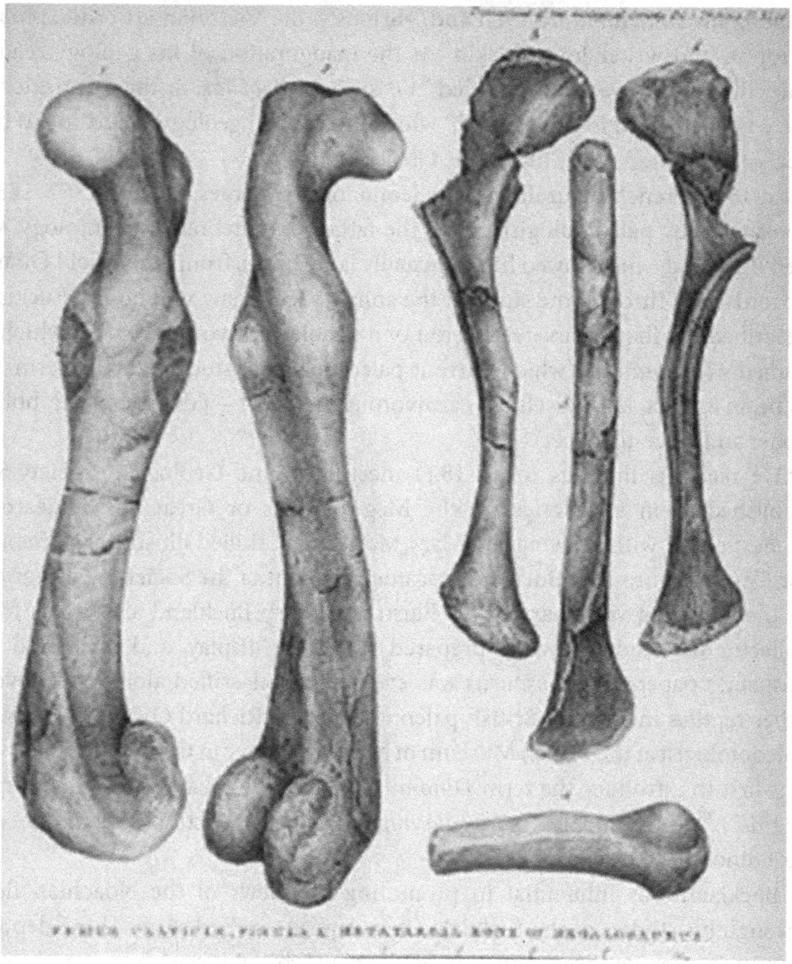

Figure 1.5 *Megalosaurus* femur, clavicle, fibula, and metatarsals. (Buckland, 1824, plate XLIV; drawn by Mary Moreland Buckland)

The Geological Society of London named Buckland its president in 1824, and he was selected as canon of Christ Church, Oxford, the following year. In 1836, Buckland published a sizable two-volume set: *The Bridgewater Treatises on the Power, Wisdom and Goodness of God as Manifested in the Creation: Animal and Vegetable Physiology, Considered with Reference to Natural Theology*. Helpfully, he included copious diagrams, some in color, for those readers unfamiliar with geology. The work was philosophical and included all of his paleontological and mineralogical findings. It took five years to compile.

Buckland was a prankster and an eccentric. His son, Francis T. Buckland, notes that they both had "eaten their way through the animal kingdom." [50]He kept all manner of animals at his house, and among those he dined on were mice, served on toast for breakfast. At a dinner party at the Harcourts of Nuneham House, Oxfordshire, he was shown a precious relic, the preserved heart of King Louis XIV of France, and said, "I have eaten many strange things, but I have never eaten the heart of a king before," upon which he popped the item into his mouth and swallowed it.[51]

Among the honors accorded to Buckland in his later years was his appointment in 1845 as dean of Westminster Abbey. Nevertheless, he became ill with "depression and lethargy" in 1849 and died seven years later, in 1856. When gravediggers proceeded to excavate the plot Buckland had selected, they found solid bedrock—consisting of Jurassic-aged limestone—several inches below the surface. The site had to be blasted to make the grave. Surely, Buckland knew of this fact when he designated the site, and some consider this his final jest.

Charles Lyell

Charles Lyell (1797–1875) was born in Scotland to a wealthy family, the eldest of ten children. Following on Hutton's heels, Lyell is best known as the seminal author of *Principles of Geology* (1830–1833).[52] When he was quite young, he moved with his family to Hampshire in southern England, where he luxuriated in nature and collected moths. At Exeter College, Oxford, he studied under Buckland but became disenchanted with his mentor's biblical interpretations of geology. After earning his bachelor's degree in the classics, Lyell pursued law, but his eyesight was weak, and law studies were difficult. Although he was called to the bar in 1822 at Lincoln's Inn, his father's financing allowed him to pursue his childhood love of nature and geology.

He joined the Geological Society in 1823 and served as its president for two terms, becoming an ardent proponent of Hutton's ideas, traveling widely, and studying geology not only in England but in France, Sicily, the Alps, the Canary Islands, and North America (from Nova Scotia to the Mississippi Delta).[53]

Lyell was influenced profoundly by Hutton's research and writings but was more interested in the Earth's history than in its origin, noting that one did not need to know the origin of humankind to write about the history of a nation.[54] He observed no difference between the present and past processes, recognizing that the Earth was a stable system, and thus he advanced Hutton's theory of uniformitarianism and the gradual change of the Earth over vast periods. He restated the argument and strengthened it by supporting Hutton's ideas with copious examples and observations from the field, discussing water-related (aqueous) and igneous processes as sculptors of the Earth's surface. Aqueous agents denuded, eroded, and leveled the surface, while igneous agents worked to create uplift, unevenness, and new sources of materials to be acted upon by water. The power of water was central to his thesis. An example from his fieldwork was the study of delta systems, notably that of the Mississippi River. The land was created and destroyed in these areas by the action of waves and tides over geologic time. This information was gleaned on one of the four trips Lyell and his wife, Mary, made to the United States between 1841 and 1853.[55] To further bolster uniformitarianism, Lyell added detail to the cycles discussed by Hutton, such as the role of earthquakes in creating valleys. He also recognized that processes deep in the Earth were at work, producing heat, lava, and gas.

Lyell published the first volume of his *Principles of Geology* in 1830, the second in 1832, and the third in 1833. After Volume 1, opposition to his views on uniformitarianism mounted from religious quarters and also from William D. Conybeare (1787–1857), a well-known geologist and paleontologist, and Adam Sedgwick (1785–1873), president of the Geological Society.[56] The "catastrophists," as they were known, held that the Earth underwent cataclysmic changes in the past and was not in the state of steadiness advocated by Hutton and Lyell. They considered volcanic and igneous activity to be diminishing in the present and cited natural theology—the thought that geology supported the biblical account of the history of the world—to describe the non-cyclical, unidirectional course of the Earth.[57]

Volume 3 (first edition) of *Principles* is notable for its discussion of Werner's classification of rocks and Lyell's own thoughts on their origin. Lyell argued, based on his studies of granitic masses intruding into sedimentary rocks and associated igneous veins and dikes from Glen Tilt in Scotland, Cornwall, South Africa's Table Mountain, and other localities that Werner's Primitive rocks could have formed at any time. In a nod to Hutton's Plutonism, Lyell proposed the term "plutonic" (rocks originating from magma) for granites, holding that they were produced by magma and should not be age-dated by placement relative to other rocks. He also suggested abandoning Werner's category of Transitional rocks as too confusing and not supported by what was seen in the field. Volume 3 also contains his interpretation of the rock cycle—how rocks transform from one type to another. He noticed that granitic bodies of rock seemed to change sedimentary rocks around them to a crystalline form because of the heat they generated; he termed these transformed rocks "metamorphic," altered by temperature and pressure. He further developed his ideas on metamorphic theory in *Elements of Geology* in 1838. These arguments were a culmination of much of his work and his understanding of geology. He associated the same origin of volcanic rocks with granites since both were formed by magma. He also noted changes in other rocks brought about by metamorphism. The framework for understanding surface and deep processes, cycles of erosion and rock formation, and changes in the Earth over time was laid out by Lyell's third volume of *Principles*.[58]

Lyell's publications not only revolutionized the nineteenth-century study of geology but also shook the foundations of other natural sciences, especially biology. Eleven editions of *Principles* came out during Lyell's lifetime, and a twelfth edition appeared after his death. These latter books incorporated recent advances in the field and his rebuttals to critics. Lyell was a formidable influence on Charles Darwin, particularly concerning geologic time. Indeed, Darwin traveled with Lyell's Volume 3 on his journey on the HMS *Beagle*.[59]

However, when it came to changes in organisms, Lyell's views largely mirrored those of Hutton. Lyell thought that species remained virtually unchanged and did not represent a progression of the complexity of life throughout geologic history, in line with uniformitarianism (the only exception to which was the special creation of man, which holds that humans were made directly by divine forces). The idea was at odds with the fossil record—Hutton and Lyell were never able to rationalize the lack of fossils in the oldest sediments fully—and ran counter to Darwin's developing theory of evolution by natural selection.

Several possible reasons exist for Lyell's nonacceptance of evolution. He held steadfastly to the ideas in *Principles* opposing the progression of life forms because of religious scruples, partly promoted by the catastrophist school of thought. He also may have resisted the idea that human beings descended from animals.[60] In his later work, *Antiquity of Man* (1863), Lyell summarized Darwin's arguments for natural selection without embracing them. However, in a limited sense, he begrudgingly accepted Darwinian evolution in the tenth edition of *Principles*, 1866–1868, after reading the preponderance of evidence for natural selection. He maintained, however, that Darwin's work did not "explain creation" and held fast to his view on the uniqueness of humans relative to the animal world.

In any event, critics shifted in the mid-1830s from dismissing Hutton's works to debating Lyell's.[61] Ultimately, scientists who believed that catastrophic events had shaped the Earth modified their theories in response to the evidence Lyell produced for uniformitarianism. The Wernerian school of thought gradually diminished in prominence as geology became a more unified science—mainly based on the works of Hutton and Playfair and the extensive fieldwork and meticulous writing of Lyell.

Mary Anning

In a discipline and in a time dominated by men, Mary Anning (1799–1847), like Benett a generation before her, was a renowned female paleontologist who, despite social and educational obstacles, contributed profoundly to geology's development. Born in Lyme Regis, Dorset, on the coast and in one of the most historically rich fossil locations in the south of England, Anning was one of ten children, only two of whom survived to adulthood.

Her family was not well-to-do. Her father was a cabinetmaker and carpenter who collected "curiosities" (fossils) from the cliffs and sold them to supplement their income. He died in 1810, leaving the family in dire financial straits. Thereafter, the Annings subsisted by selling fossils they found in the cliffs. In 1817, Lieutenant-Colonel Thomas Birch, a wealthy fossil collector, met the family, was moved by their plight, and sold his own fossil collection to assist them financially.

Fossil collecting in the Lyme Regis cliffs was fraught with danger, in part because of the steepness of the slopes, the weakness of the rocks, and the action of the waves that rendered the coastline unstable, prone to landslides and rockfalls. In 1833, for example, Anning was severely injured, and her dog Tray was killed, when blocks of chalk from the cliff collapsed onto the beach.

Here the nearly vertical cliffs of Jurassic Period marine clays of the Blue Lias Formation are topped unconformably (with layers separated by a break in geologic time) by hills of Cretaceous greensand formations (marine silts and clays colored green by the mineral glauconite) containing chert (rocks made of the shells of silica-rich shells) and chalk.[62] The Blue Lias strata record the rise and fall of sea level during the Mesozoic and have the most wide-ranging numbers, types, and varieties of ammonite species, extinct spiral-shelled mollusks, in the world. Within these Liassic clays are sea fossils, such as ammonites, belemnites, and the rare *Ichthyosaurus* and *Plesiosaurus*. Ammonoidea (a subclass of mollusks, class Cephalopoda), represented by ammonites, ranged in size from 1 centimeter to 2 meters (6 feet). Their fossils are prevalent in the rock record, and variations in types of ammonites helped sort out part of the geologic time scale. Belemnites (another marine cephalopod, order Belemnitida, now extinct, resembling a squid with a straight shell) varied from 1 to 46 centimeters (18 inches), but the living creature would have been 3 meters (10 feet) in length.

Swimming reptiles of considerable size consisted of the *Ichthyosaurs* (from 3 meters, or 10 feet) and *Plesiosaurs* (from 1.5 meters, or 5 feet, up to 15 meters, or 49 feet). *Ichthyosaurs*, classified as neither fish nor mammals nor dinosaurs, were fish-shaped, with large eyes to see predators, and they swam in an undulating manner. *Plesiosaurs* had long necks, rounded bodies, and flipper-like appendages.

Along with her brother Joseph, Anning discovered in 1811–1812 the first complete *Ichthyosaur*. Her brother noticed the head first in the winter months when erosion was highest along the coast, and she found the rest of the *Ichthyosaur*'s body the following winter. The fossil was not the first of its kind to be located in Lyme Regis, but it was the best example to date and was sold to a "gentleman scientist." In 1823, Anning found a complete, articulated *Plesiosaur* skeleton (figure 1.6) high in the cliffs and hired local workers to help her extract it.[63] It was immediately recognized by the scientific community as one of the most significant finds ever uncovered and had no living equivalent. The animal had a relatively small head and an extremely long neck. The English paleontologist Henry De la Beche presented information about the skeleton, which he classified as an *enaliosaurian* (Greek for a marine lizard, representing the extinct group of saurian fossil dinosaurs including the *Plesiosaur* and the *Ichthyosaur*) at the Geological Society meeting in February 1824. Conybeare sent Cuvier a letter describing the *Plesiosaur*, but because he had never seen a neck of that length, Cuvier declared it a fake. He quickly retracted this view when he received detailed drawings from Buckland and Conybeare, made by Buckland's wife, Mary, based on Anning's sketches. Cuvier acknowledged Anning from this time

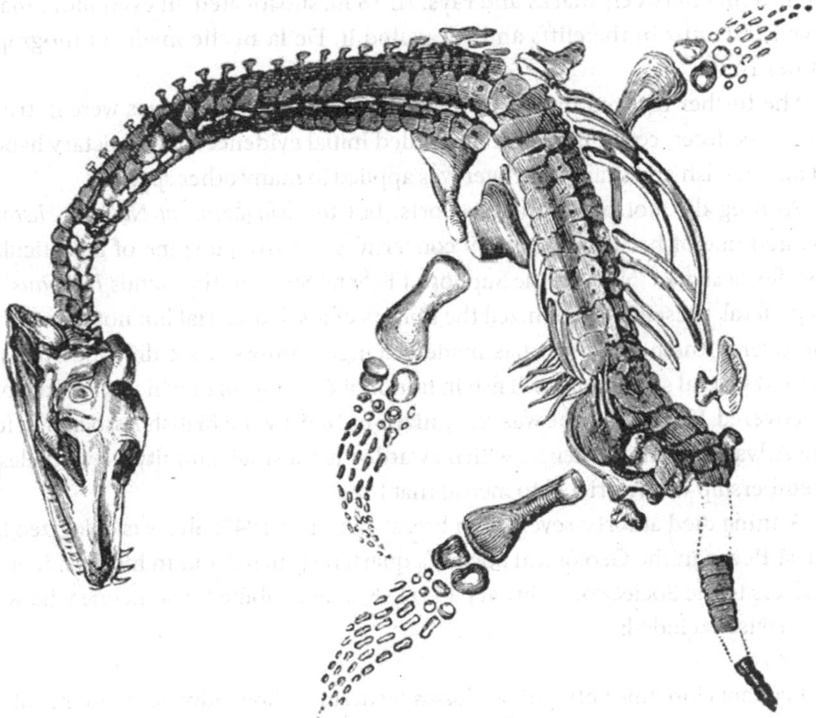

Figure 1.6 *Plesiosaurus macrocephalus*, found by Mary Anning, named by William Buckland in 1836, and later described by Richard Owen. (Owen, 1840, plate 43; lithograph by George Scharf)

forward for her work as a respectable "fossilist" (paleontologist) and gave her credit for her work.[64]

Increasingly, scientists came to Lyme Regis to seek out Anning for her knowledge and expertise in the developing field of paleontology. They included De la Beche, Buckland, Conybeare, Lyell, and Smith.[65] Though not formally trained, Anning read papers written by well-known scientists of the day and transcribed them, copying them word for word and also replicating the illustrations in order to learn them thoroughly.[66] She also dissected fish and rays to understand their anatomy. She was a frequent correspondent with Buckland and other geologists. Anning not only found and excavated fossils but also prepared them for display, which involved great knowledge and skill to affix the bones in proper anatomical order. A third famous discovery by Anning, in 1828, was of a *Pterosaur* (an extinct flying reptile that lived during the Jurassic and Cretaceous Periods and had a lengthened fourth finger that supported the wing), the first flying reptile found in Britain. The following year, she discovered the fossil of a *Squaloraja* (an extinct fish), which proved to be a link between sharks and rays. In 1830, she located an even more massive *Plesiosaur* in the cliffs and excavated it. De la Beche made a lithograph of her finds.

She further realized that the round stones called bezoar stones were in truth fossilized feces, coprolites, which provided initial evidence for the dietary habits of ancient fish and sharks and later was applied to many other species.

Anning did not publish any reports, but the *Magazine of Natural History* printed one of her letters in 1839 concerning the frontal spine of a particular species of shark, "Note on the Supposed Frontal Spine in the Genus *Hybodus*."[67] In general, museums recognized the donors of fossil material but not those who discovered them. This fact has made tracing Anning's work difficult. Agassiz named several species of fossil fish in honor of Anning, all of which she probably discovered. Ultimately, she was recognized in 1838 by the British Association for the Advancement of Science, which awarded her a small annuity. Nevertheless, membership was restricted to men at that time.

Anning died at forty-seven from breast cancer in 1847. She was eulogized by De la Beche in the Geological Society's quarterly journal and in his presidential address to the Society on February 14, 1848, a rare tribute for someone who was otherwise excluded:

> I cannot close this notice of our losses by death without advertising to that of one, who though not placed among even the easier classes of society, but one who had to earn her daily bread by her labour, yet contributed by her talents and untiring research in no small degree to our knowledge of the great Enalio-Saurians [*sic*], and other forms of organic life entombed in the vicinity of Lyme

> Regis . . . there are those among us who know well how to appreciate the skill she employed (from her knowledge of the various works as they appeared on the subject), in developing the remains of many fine skeletons of *Ichthyosauri* and *Plesiosauri*, which without her care would never have been represented to the comparative anatomists in the uninjured form so desirable for their examination.[68]

Through Anning's short life, fossils went from being seen as "curios" to undergoing full scientific inquiry. In finding and preserving the specimens, she made significant contributions to the field of paleontology and understanding former life.

Louis Agassiz in Europe

Louis Agassiz (1807–1873) in 1837 became the first geologist to formally propose the existence of ice ages and glacial epochs marked by the spreading of vast ice sheets. The son of a minister, he grew up in the Jura Mountains of the Swiss Alps. Like other early geologists, including Hutton, he studied different fields; in 1829, he received a doctorate in philosophy, graduating from the University of Erlangen, Germany, and a year later, he earned his medical degree at Munich.

The following year, he went to Paris to study comparative anatomy with Cuvier, and though Cuvier died several months later in 1832, the paleontologist profoundly impacted Agassiz. He subscribed to Cuvier's philosophy of classification and defended his views. Agassiz taught for the next fourteen years at the College of Neuchâtel, Switzerland, studying paleontology, specifically the classification of fishes. In 1836, he began research under Swiss-German geologist Jean de Charpentier (1786–1855) on the glaciers of Savoy near Lake Geneva, realizing with Charpentier that large boulders from the mountains found far away in valleys—erratics—had been carried there by glaciers. He was the founder of early glacial theory.[69]

Agassiz presented a paper at Neuchâtel in 1837 proposing an "ice age" that spanned the European continent. He published the talk in 1840 with illustrator and author Joseph Bettannier *Études sur les Glaciers* (*Studies on Glaciers*); in it, he built on the work of Charpentier but without attributing early glacial theory to him. Agassiz dismissed the Great Flood as responsible for erratics, and as more data were collected, he would later claim that there had been several periods when ice covered parts of the Earth, not just one ice age. Agassiz's glacial theory influenced Lyell's thoughts, resulting in Lyell not only acknowledging the import of glacial theory but also discounting an earlier hypothesis that icebergs, not glacial movements, were responsible for erratics.

Agassiz faced financial and personal challenges in 1845 with the failure of a scientific publishing firm he had founded and the departure of his wife from their home.[70] They subsequently divorced. Then, in 1846, with little incentive to stay in Europe, he left for a lecture tour of the United States funded by the king of Prussia. He spent the rest of his life in America, lecturing on geology and zoology at Harvard University (see chapter 2).

Early French Geologists

After the French Revolution, science in France was reorganized and moved from the royal academies to more egalitarian scientific societies. Geology flourished during this time, appearing in the work of such naturalists as Jean-Baptiste Lamarck (1744–1829), Cuvier, and Alexandre Brongniart (1770–1847).[71]

Britain and continental Europe have similar geology, which is present in sedimentary strata deposited in continuous layers that were joined by a land bridge as recently as the Quaternary Epoch, 450,000 years ago. Before that time, Britain and the European continent had been on the same landmass, which periodically underwent incursions by the ocean responsible for laying down sedimentary units that underlie the Weald in Britain and the northwest portion of France in the Paris Basin. The large fold of sedimentary strata stretches across the English Channel to France and is called the Weald-Artois Anticline (in an anticline, the rocks have bent into a structure that looks like an upside-down U). About 450,000 years ago, a mega-flood, the first of two from a glacial lake, carved through the relatively soft rock of the Dover Straits.[72] The flood left deep scour marks and plunge basins in the modern-day English Channel, recently confirmed through remote sensing. Similar to the early British geologists, these French geologists inquired about the nature of sedimentary deposits and the fossils associated with them.

Within the Paris Basin, miners quarried the Lutetian limestone and the Montmartre gypsum strata for building stones as far back as when France was a province of Rome. Lutetian limestone (also called *calcaire grossier*) makes up the outer structure of many buildings in Paris. Other deposits, such as Fontainebleau sandstone, from the surrounding areas form Parisian sidewalks, buildings, and squares. Because of the extensive mineworks and quarries around Paris, there was abundant opportunity for those interested to examine the sedimentary strata and fossils without the need for far-flung excursions. Moreover, quarrymen had a side business of extracting fossils as they came upon them during excavation and selling them to collectors and scientists.

Lamarck studied and classified invertebrate fossils of shells in the sedimentary beds around Paris. Cuvier became famous for identifying and describing fossil vertebrates and, along with Brongniart, made the first geologic map of the Paris Basin in 1810 (color insert 1.3);[73] they republished the color map in 1811 in a series of twelve copper plates.[74] Although influenced by Smith's work and his map of England, Cuvier and Brongniart did not display their geologic units in stratigraphic order from youngest to oldest formations until the 1822 edition.

Cuvier was famous for his studies of comparative anatomy, fossil remains, and the idea of the extinction of species. He was the first to document an extinct species related to living elephants in a 1796 talk given to the National Institute of Sciences and Arts in Paris, later published as *Memoires sur les espéces d'eléphans vivantes et fossiles* (*Study on the Species on Living and Fossil Elephants*).[75] Like Hutton, however, he did not subscribe to a theory of evolution and believed that species remained fixed. This view was in contrast to Lamarck, who thought species changed over time and passed on their successful attributes to the next generation. Lamarck's idea was called the theory of inheritance of acquired characteristics, published in 1801.[76] Cuvier continued publication of reports on the natural history of vertebrates from 1815 to 1822.[77]

In summary, the Western understanding of geology originated in Europe during the time of the Enlightenment. America was soon to follow, and not only did ideas cross the pond, but European geologists also brought their knowledge to the vast North American continent. The study of geology expanded to North America as that continent was explored and colonized. New challenges were presented because of the New World's vast spaces, but adventurous geologists applied their understanding of how the Earth worked according to methods developed earlier in England and in France, which, as will be seen, served them and the study of geology exceptionally well.

2

Geology Emerges as a Science

On the Other Side of the Pond

American Scientific Geological Developments

The story of geology in the Americas is itself metamorphic: a transformative synthesis of existing and new ideas, bursting into discoveries of the "New World" and its geological resources. The Western understanding of the nature of the Earth, these new theories of geology, arrived late to the North America when the lands were colonized by Europeans. Moreover, geological studies in America lagged far behind European research, because of the War of Independence and other issues to be resolved as the United States became a nation. As with many things American in these early days, the challenges that geology presented pertained to the vast, and still wild, expanse of the country.

Perhaps unsurprisingly to historians of the era, an early advocate of science and applied geology was Thomas Jefferson, the third president of the United States, who was born in 1743 in Virginia. Jefferson did not particularly care about theoretical debates in geology, such as the origins of rocks, but he was an adherent of fossils and paleontology and is said to have had carried vertebrate fossil bones in his pocket, which he wanted to be identified, as he took the oath of office as the nation's second vice president.[1] Besides fossils, Jefferson was interested in minerals and mineral exploration. During his presidency, he sponsored several famous expeditions. These included the one headed by Meriwether Lewis and William Clark in 1804–1806 to explore the lands of the Louisiana Purchase west of the Mississippi in pursuit of a northwest water passage to the Pacific Ocean. Jefferson also backed the expedition in 1806–1807 led by Zebulon Pike, the eventual namesake of Colorado's Pikes Peak, whose mission was to locate the sources of the Mississippi and Arkansas Rivers, and although he failed to reach the summit of his own mountain, the journey led him to explore the American Southwest, the Plains, and parts of the Rocky Mountain region.

In 1802, Yale University appointed a lawyer, Benjamin Silliman, to the new position of professor of chemistry and natural history—covering the fields of geology, mineralogy, botany, and zoology—none of which he had extensively studied. Nevertheless, Silliman, who would help to formalize geology in the

Song of the Earth. Elisabeth Ervin-Blankenheim, Oxford University Press. © Oxford University Press 2021.
DOI: 10.1093/oso/9780197502464.003.0003

American academy of the early nineteenth century, enthusiastically studied chemistry at the University of Pennsylvania, then geology and the natural sciences in England and in Scotland at the University of Edinburgh in 1805–1806. He initiated the purchases of mineral collections for Yale, such as the Perkins Cabinet in 1807, consisting of two thousand minerals from England and the European continent, and the Gibbs Cabinet in 1811–1812, made up of European specimens collected by American George Gibbs III (1776–1833) of Newport, Rhode Island. These collections later constituted the foundational materials for the Peabody Museum.[2]

American colleges and universities did not teach geology as a subject until 1840; that year also saw the founding of the first geological society, the Association of American Geologists. This group would eventually publish the *American Geologist* journal in 1888, the earliest periodical devoted entirely to geology in the United States.[3]

Geology became fashionable in the 1830s and 1840s as lectures were given on the topic along with other natural sciences, in the form of the "lyceum" or lecture series. It was such a set of lectures that drew Louis Agassiz to the United States in 1846.

Louis Agassiz in the United States

After his successful 1846 lecture tour, Louis Agassiz, the geologist and glaciologist (see chapter 1), decided to make the United States his home. The Lowell Institute in Boston appointed him to teach that same year. Two years later, he became a professor of geology and zoology at Harvard University, where he founded Harvard's Museum of Comparative Zoology. In 1864, he studied the glacial deposits and drifts in Maine, examining them from Mount Katahdin to Bangor, reporting his results for the public in the *Atlantic Monthly* and later in his book *Geological Sketches*, published in 1866.[4]

In America, and despite his later-proven insights into the occurrence of ice ages, Agassiz was no proponent of Darwin's theory of evolution. In an 1860 article in the *American Journal of Science*, he responded to Darwin's *On the Origin of Species*, which had been published the preceding year:

> Until the facts of Nature are shown to have been mistaken by those who have collected them, and that they have a different meaning from that now generally assigned to them, I shall therefore consider the transmutation theory as a scientific mistake, untrue in its facts, unscientific in its method, and mischievous in its tendency.[5]

Agassiz believed species did not change over time, but, like Cuvier, he was a proponent of extinctions and held that species were recreated anew after each extinction. He never varied in his objections to evolution. Some authors cite his unwillingness to be seen as wrong on any matter,[6] and perhaps, despite urging from colleagues to give Darwin's theory another look, he fell into dogmatism and tarnished his legacy based not on logical but on personal biases and beliefs.

After Agassiz arrived in the United States, he encountered Samuel George Morton, whose study of skulls of various races supposedly showed differences between them (a conclusion later disproved by Harvard paleontologist Stephen J. Gould and others). Buoyed also by work of Josiah C. Nott, a physician, surgeon, and author who believed that the races represented different species altogether, Agassiz taught such notions in his classes, and his racist views strengthened over time, along with a conviction that America should belong to whites only.[7] He believed that the "races should not interbreed," that they should remain "pure,"[8] and other offensive, repugnant ideas.

Curiously, in 1906, during the Great San Francisco Earthquake, a marble statue of Agassiz took a tumble from Stanford University's Zoology Building and ended up head first in concrete (figure 2.1), an iconic image of that historic earthquake and perhaps a fitting end for the professor.

Figure 2.1 The Agassiz statue, Stanford University, April 1906, after the San Francisco earthquake. (USGS, 1906; photo by Walter Curran Mendenhall)

James Dwight Dana

Another early and indispensable American geologist was James Dwight Dana (1813–1895), born in Utica, New York. Dana, a student of Silliman, was notable for his study of mineralogy and also for writing one of the earliest textbooks on geology. In 1836, Silliman offered him a position as his assistant in the Yale chemistry laboratory, where he examined and studied the Yale mineral specimens, which had a profound influence on mineralogy in the United States. Dana wrote up his findings based on chemistry, crystallography, and the Linnaean classification system first promulgated for minerals in 1837 by German mineralogist Friedrich Mohs in *A System of Mineralogy*, a seminal work. By 1844, Dana had married Henrietta Silliman—Professor Silliman's daughter—a union that produced another leading American geologist, his son, Edward Salisbury Dana (1849–1935), who became a famous mineralogist. It was during time in the elder Dana's life that he produced a torrent of critically acclaimed geology works.

In 1848, Dana wrote *The Manual of Mineralogy*, a work so comprehensive that it is available today in its twenty-first edition. After Silliman retired from Yale in 1853, Dana was appointed to the professorship and took over the curation of the mineral cabinets and the collection. The following year, he published the fourth edition of *A System of Mineralogy* and detailed, for the first time, the still-used chemical classification of minerals (native elements, oxides, carbonates, phosphates, sulfides, sulfates, halides, and silicates). After teaching for a number of years, Dana realized he should publish a newer work to help his students understand geology and mineralogy, and in 1863, his *Manual of Geology* was released. Two years later, he wrote an extraction for less advanced audiences called *Text-Book of Geology*.

New editions of *A System of Mineralogy* were carried forward with the assistance of Dana's son (and Silliman's grandson), mineralogist Edward Salisbury Dana, and others until its further publication was interrupted by World War II. Remarkably, the tome, in an eighth edition, is still available, with modifications, updates, and new minerals. Dana, Silliman's disciple and son-in-law, died in 1895 at the age of eighty-two after a long, active career, having carried forward the knowledge from his professor to his own son, making for three generations of mineralogists in the family.

Sir John William Dawson

One of the most remarkable yet little-known geological features the world over is found in Nova Scotia. Known colloquially as the "coal measures" (an older term for geologic units of the Upper Carboniferous Period enriched in coal), these sedimentary deposits of Paleozoic Carboniferous age are the source of coal deposits and remnants of fossil forests. In particular, a UNESCO World Heritage Site in Joggins, Nova Scotia, uniquely contains coal deposits with fossil trees and plants growing in life position (in their original locations in beds, which have subsequently been tilted) (figure 2.2). The coast of the Bay of Fundy famously boasts the largest interval between high and low tides in the world, and these massive weathering influences have shaped cliffs that erode, decay, and slump over time to reveal the fossil plants trapped within. It was near here that Canada's preeminent paleobotanist Sir John William Dawson (1820–1899) grew up and where he studied fossils for decades.

Based on knowledge gleaned when Dawson studied at the University of Edinburgh in Scotland, he was able to identify an ancient forest ecosystem that over time, and with burial, led to formations of the coal measures. Unlike some of his contemporaries, who only studied specimens in museums, far removed from their original context, Dawson's work was based on experience in the field (going out to examine rocks, fossils, or other features as they are found in nature), examining the outcrops (rock exposures).[9] As could be said for virtually every groundbreaking scientist, Dawson's success was enhanced not only by

Figure 2.2 Fossil lycopsid tree, rooted in place, Joggins Fossil Cliffs, Nova Scotia. (Michael C. Rygel, 2010)

his work in the field but also by his attention to detail, careful observation, and persistence.

Dawson was born in Pictou, Nova Scotia, to a family with strong Presbyterian ties and beliefs. Like many early geologists, Dawson was deeply religious and active in the Presbyterian Church, which shaped his outlook on the natural world and, ultimately, like Agassiz, his objection to Darwinism. As a boy, Dawson was intrigued by nature, exploring areas around Pictou, including the surface mines of the Albion Coalfield. He was familiar with and read Lyell's 1830 *Principles of Geology*. The family suffered economic hardship when their import-export fortunes were reversed during the economic depression of the mid-1820s and with the loss of several of their merchant ships at sea, causing Dawson's father to owe others a great amount of money.[10] This debt was a motivator for the younger Dawson, who worked for decades to pay off the amount owed by his father until it was finally discharged around 1850. Despite this hardship, Dawson attended local schools and was sent to the University of Edinburgh in 1840, where he studied chemistry, botany, and geology. The following year, he was called back to his family to help financially. This setback turned out to be fortuitous, however, because he happened to meet his hero, Lyell, in 1842, when the latter came to Pictou to see the coal measures at the Albion Mines.[11] Dawson served as Lyell's guide for the area, and the two developed a lifelong friendship and mentoring relationship. Dawson eventually returned to Scotland in 1847 and completed his studies at Edinburgh, after which he returned to the province in 1852 and, along with Dawson, explored the Joggins Fossil Cliffs. Here they not only examined the Carboniferous fossil forest but also found a unique reptilian fossil located within some of the tree trunks, *Hylonomus lyelli* (the first land-dwelling reptile) (color insert 2.1).[12] There was a great debate about whether these creatures had fallen into the trees, had been washed in by floods, or had made their homes among the roots. The debate was not settled until years later, when evidence indicated that the reptiles denned within the old stumps and roots of the trees.

Dawson also was an educator, both in the province of Nova Scotia and later as principal of McGill University until 1893, where he transformed the institution into a leading international university. Dawson, a prolific writer, counts among his most well-known works *Acadian Geology*, first published in 1855, with four total editions. Later in his career, Dawson became embroiled in a number of controversies, which set him apart from the dominant thinking in geology. He did not subscribe to Agassiz's glacial theory that ice sheets covered the land, instead proposing that ice was rafted in on drifting pieces of ice; this was later determined to be impossible. Although his thoughts on glaciation and evolution were ultimately debunked, his life's work with Lyell and his training of future geologists formed a critical bridge linking European geologists to those of the New World.

John Wesley Powell

Dawson's American contemporary John Wesley Powell (1834–1902) was born the fourth of nine children in western New York, son of a Methodist minister. The family moved to Ohio, then to Wisconsin, and ultimately settled in Illinois. In school, he was bullied by classmates because his father was a staunch abolitionist. A tutor, George Crookham, a self-taught naturalist, took charge of the boy's schooling, and it is due to Crookham that Powell became exposed to natural history and botany and spent time in outdoor studies.

In Illinois, with his father away much of the time, young Powell ran the family farm and taught elementary-school students in Macon County, at the age of sixteen. He pursued a college degree, first at the Illinois Institute at Wheaton in 1855 and later at Oberlin College, where he took classes in botany, Latin, and Greek. In 1861, Powell had been teaching in Hennepin, Illinois, for several years and had become the Illinois superintendent of schools when the Civil War broke out.[13] He enlisted in the 20th Illinois Infantry Regiment and was wounded a year into the war, at the Battle of Shiloh in April 1862. He lost his right arm below the elbow, an amputation that plagued him on and off until, toward the end of his life, he had surgery to sever the affected nerves.[14] Powell continued to serve in the Union Army as a major for three more years, working closely with General Ulysses S. Grant and becoming chief of the artillery. He was known for collecting fossils from the trenches on the battlefield at Vicksburg.

Powell had amassed an extensive collection of fossil Mollusca (mollusks, the second-largest phylum of invertebrates) and joined the State of Illinois Society of Natural History in 1854. Prior to the war, in 1856, he had journeyed alone down the Mississippi River and later descended the Ohio, Illinois, and Des Moines Rivers. These trips and his enthusiasm for nature set the stage for his more famous explorations of the West.

Powell, of course, is famous for his expeditions down the Grand Canyon of the Colorado River in 1869 and 1871–1872 to study geology and natural landscape, but he had made several earlier ventures to scope out the area with students before his great expeditions.[15] In 1867, for example, Powell, along with his wife and a group of students funded in part by the Illinois Society of Natural History and supplied through Army contacts, made a reconnaissance trip to the high mountain parks, South Park and Middle Park, of Colorado. The following year, he went back to Colorado and developed his plan to follow the course of the Green River to the Colorado River. The latter was known to drop in elevation 10 to 15 feet per mile, a substantial change in slope. A colleague, William H. Brewer, a botany professor at Yale, later wrote that with the degree of gradient on the river, Powell should encounter falls and rapids, and he expressed concern at the undertaking of such a daring project.[16] Powell, however, had observed that the river carried a considerable amount of sediment; thus, he expected that the river, seeking its base level, had eroded any falls, and while he anticipated rapids, he correctly predicted that there would not be many falls.

The 1869 expedition included ten men, four small wooden boats, and enough supplies for ten months.[17] The trip through the canyon proved complicated, with many rapids, a shortage of food, exhaustion, loss of valuable scientific equipment when boats overturned, and discontent among the men, four of whom abandoned the journey, the first early in the expedition. The other three climbed out of the canyon at what came to be known Separation Rapid and were never heard from again, presumed to have been killed or died The latter three left two days, unknown to them, before the journey ended at the confluence of Virgin River. The expedition yielded little in terms of scientific results. The team produced no map—or if a map existed, it was lost. The mapmaker of the crew, was among the three that had left the expedition and climbed out of the canyon at Separation Rapid. Still, it was a critical first step in understanding the geology and natural history of the region. Powell's notes and journal recorded the river's course and described how it had carved the canyon, laying the groundwork for his second expedition, in 1871–1872, which would go more smoothly, although it ended early, downriver of Marble Canyon near Kanab Wash.[18]

Powell published several reports of these and other journeys. In 1875, he wrote up the information he had gleaned from the two Colorado River expeditions.[19] The following year, he reported on the eastern Uinta Mountains and surrounding areas,[20] in 1877 on the geology and geography of the Rocky Mountain region as a whole,[21] and in 1895 again on the Colorado River and its canyons.[22] His geologic studies detailed the slow development of the Grand Canyon by downcutting of the river and the gradual evolution of the landscape. He was a follower of Darwin, and he hypothesized that evolutionary processes shaped not only species but also rocks and geological settings over long periods. Out of the experience, Powell made illustrations that are still stunning today (figure 2.3).[23]

Powell also ventured into ethnographic studies of the western Native American tribes. He became a well-respected student and author of anthropological works and published several ethnographic reports and dictionaries of tribal languages, working cooperatively with the tribes of the areas he visited, who respected him in turn.

Gradually, the Wild West would be mapped and measured through four major surveys led, respectively, by Powell, the Yellowstone explorer F. V. Hayden, Sierra Nevada explorer and Dana's student Clarence King, and cartographer George Montague Wheeler. Inevitably, competition and disagreement ensued among the four studies and their leaders.[24] Accordingly, U.S. Senator Abram Hewitt of New York proposed a consolidation bill to clarify the matter, which resulted in the establishment of the U.S. Geological Survey (USGS) and the Smithsonian

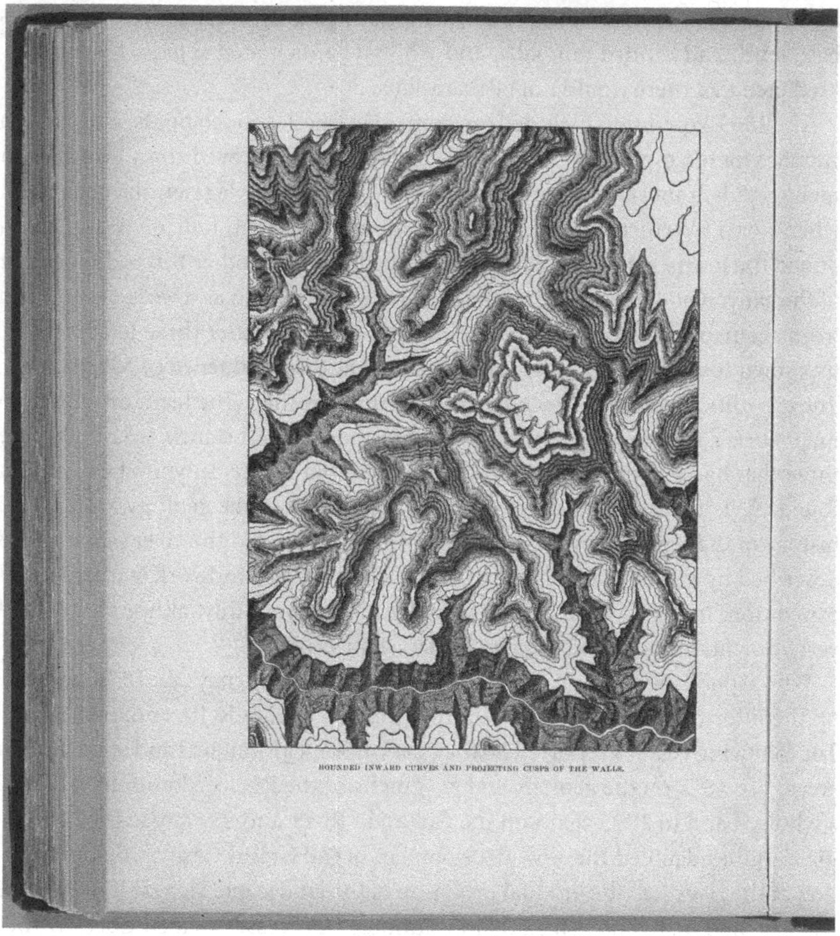

ROUNDED INWARD CURVES AND PROJECTING CUSPS OF THE WALLS.

Figure 2.3 Grand Canyon curves, projecting cusps of the walls, illustration by John Wesley Powell, 1895. (Courtesy of David Rumsey Map Collection, David Rumsey Map Center, Stanford Libraries, https://purl.stanford.edu/dy316pc6530)

Bureau of Ethnology in 1879. Powell was appointed head of the Smithsonian Bureau of Ethnology, and King became the first director of the USGS, much to Hayden's dismay. A little more than a year later, however, King resigned from the USGS to pursue mining interests, appointing Powell as its second director. Under his leadership, from 1881 to 1894, the USGS became the premier scientific institution in the world. Publications and reports based on geological research and field studies poured out of the Survey. Powell directed USGS geologists to make a topographic map of the contiguous United States, and he formulated standardized map symbols that are still in use today. He also worked on irrigation projects and issues of farming and dams in the arid West. Powell resigned from the USGS in 1894 but remained the director of the Bureau of Ethnology. His 1895 book, *The Exploration of the Colorado River and Its Canyons*, is still in print. He died in 1902, at age sixty-eight, from a stroke.

Clarence King

As noted, the founding USGS director, Clarence King (1842–1901), was a contemporary of Powell's and one of American geology's more colorful rogues. King was born in Rhode Island to a well-off religious family. His father died when he was young, but his mother supported his interest in fossils and natural science.[25] King's stepfather underwrote his studies at Yale, and King was one of the first enrollees in Yale's Sheffield Scientific School, graduating in 1862. He took Agassiz's glacial geology class at Harvard one winter during his college years, which in part contributed to his passion for pursuing work in the field.

Enthralled by the work of Josiah Whitney, who was creating a geological survey of California, King traveled west to find him in 1863. He persuaded Whitney to let him work for the California Geological Survey as a volunteer assistant geologist and, along with others, made first ascents of many of the Sierra Nevada peaks, naming a number of them as they went, including Mount Dana, Mount Brewer, Mount Lyell, and Mount Whitney. He published his memoirs of the expeditions and climbs, beginning in 1871, in the *Atlantic Monthly*. King wrote up the series as a book in 1872, *Mountaineering in the Sierra Nevada*, and later republished an edition with several maps along with a note stating that he had climbed and described the wrong mountain.[26]

King's contribution to the four great geologic surveys of the West was the conception for a survey of a 100,000-square-mile area along the Central and Union Pacific transcontinental railway line, following the fortieth parallel, a 100-mile swath of Wyoming, Colorado, Utah, and Nevada, to connect two of the three other surveys being completed at the time, those led by Hayden and Whitney. He convinced Congress to fund the plan, thanks in part to endorsements from his former professors and California Congressman John Conness, for whom he had named one of the Sierra Nevada peaks. His *Report of the Geological Exploration of the Fortieth Parallel* in seven volumes was published in 1870–1880. Volume 1, *Systematic Geology*, was and still is considered an essential discussion of mountain-building processes. King and his crew were excellent mapmakers, using a scale of 1 inch per 4 miles that many others adopted. He also pioneered the use of topography as a base for geologic maps, a practice that holds today for geologic mapping. The plates for the report (color insert 2.2) consisted of large-format atlas maps in color, rock types listed in stratigraphic order (youngest on top and oldest at the bottom), and detailed cross-sections at the bottom of the map collar. King and members of his survey completed six additional volumes of the series.[27]

King adhered to the catastrophic school of geology. He did not think that features he observed in the areas he mapped—in particular, the volcanic structures of the late Cenozoic on the Columbia River Plateau and the glacial drainages—could have been produced by gradual processes. He also believed that biological evolution did not proceed gradually.[28] He was not alone in his thinking. Many other geologists and scientists of the time had modified their thoughts on the gradualist school of geology and proposed more significant variations in rates and intensity of geologic processes than Lyell had.

As the first director of the USGS, King instituted high standards for mapping and mineral exploration, and he established a laboratory to investigate the impact of temperature and pressure on the melting point of various rock samples. He resigned from the post, however, in 1881 to pursue a fortune in mining. King purportedly had a taste for extravagance, high living, art collections, and travel.[29] He also suffered reversals of finances and retired from scientific inquiry, but he remained a member of society, was best friend of Abraham Lincoln's personal secretary and future secretary of state John Hay, and was received by the king of England during his tour of Europe in 1882–1884. He spent most of his money on expensive art and dabbled in mining interests and cattle sales, ending up in Cuba in 1894, supporting the overthrow of Spanish colonial rule.

King led a double life for many years. He had an African-American common-law wife, Ada Copeland, who had been born a slave in 1864, and because interracial marriage was illegal, he passed himself off (even to her) as a light-skinned black man. During the day, he was Clarence King, the Manhattan-based prospector and geologist; at night, he crossed the Brooklyn Bridge to be with Copeland and their five children in their home, posing as James Todd, Pullman porter.[30] He only revealed his true identity in a letter he wrote to Copeland from Arizona as he lay dying of tuberculosis in 1901. Copeland, incidentally, lived to be 103, long enough to watch another King express his dream from the Georgia marble steps of the Lincoln Memorial.

Florence Bascom

Florence Bascom (1862–1945), only the second woman in the United States to earn a PhD in geology, was the daughter of a Massachusetts suffragist and schoolteacher and a professor at Williams College. While studying for her doctorate at Johns Hopkins University,[31] she was required to sit behind a screen so that her presence would not disturb the male students.[32] Her father, who would later be president of the University of Wisconsin, encouraged her to persevere, and in 1893, Bascom received her PhD. She taught at several colleges before being hired by Bryn Mawr College and soon after founded the geology department there.

Bascom was the first female geologist hired by the USGS (in 1896). She edited the *American Geologist* magazine from 1896 to 1905 and in 1930 became the first woman vice president of the Geological Society of America. She taught, mentored, and led students on field trips for the next thirty-three years, at the same time mapping the Piedmont and other regions in Pennsylvania and surrounding areas.

Bascom specialized in crystallography and mineralogy, in addition to being a field geologist and petrologist (a scientist who studies the formation and constituents of rocks). As important, Bascom supported and encouraged women to conduct fieldwork as their male counterparts had. She produced several USGS folio series maps for the *Geologic Atlas of the United States*, at the 1:62,500 scale, including the Philadelphia Quadrangle in 1909.[33] Bascom also authored eleven reports and folios during her time at the USGS. Her geologic mapping was of such precisely fine quality that her maps are still used to this day.

The founders of geology are best understood in the context of their times. The sixteenth and seventeenth centuries, for example, were ages of discovery: the Scientific Revolution, which gained momentum and evolved into the Enlightenment, leading the way for the development of rationalism, scientific method, and the rise of individual rights. These shifts impacted democratic ideals and were ultimately poured into the foundations of current Western civilization. Scientists and early geologists viewed nature in new ways through radically different spatial and temporal relationships.[34]

Many in the nascent field of geology were experts in other disciplines—medicine, biology, chemistry, philosophy, law, and logic. Such a range of experience is the very definition of a polymath. Yet as science continues to progress in the twenty-first century, there has been a trend toward specialization, perhaps not to the overall benefit of the field. The beginnings of the contemporary science of geology and the varied disciplines from which those early geologists arose might be a reminder that a comprehensive grounding in the sciences and philosophy is as valuable as specialization, and maybe even more so.

Early theories of catastrophism, uniformitarianism, the age of the Earth, and how species were defined and changed through time—and how religion fits in—not only were passionately debated but also informed notions of both geological time and physical transformation. William Robertson, in his rewriting of James Hutton's preface, stated:

> It is not the end of Revelation to instruct mankind in speculative science, to communicate to them a history of Nature, or to explain the true system of the Universe; Intent upon inculcating the religious doctrines which we ought to believe, & the moral virtues which we are required to practice, it rests satisfied with describing the phenomena of nature not according to philosophic truth, but as they present themselves to our view.[35]

The lives of the earliest geologists serve to illustrate the intersection of individual achievement and collective knowledge building. The modern world owes the founders of geology an enormous debt for their hard-won knowledge and contributions. If anything, the matrix running through the lives and works of the scientists and early geologists discussed in these chapters springs from a love of nature, a keen eye, and a desire to understand and learn.

The theory of uniformitarianism prevailed, but also included in the present-day understanding of geology are elements of early theories, such as the importance of cataclysmic events, where supported by evidence. The catastrophists proposed floods, for example, which have been observed over the past hundred years to precipitate large-scale geologic change. Indeed, during the seventeenth to the nineteenth centuries, the foundational principles were intuited by which geology would emerge as a science in its own right. This groundwork laid down deep roots for the field, and those core ideas developed into the trunk of the geological tree. Three dominant limbs branched out in the twentieth century that make up the tenets of geology: geologic time, plate tectonics, and evolution.

These three basic principles define the Earth temporally and spatially, and they frame both the unifying concepts and the defining precepts that explain many phenomena on the Earth. And yet, as we have seen and will continue to learn, the philosophical, scientific, and technical breakthroughs that inform geology today are interwoven with the life stories and philosophical outlooks of the geologists and other scientists who discover and advance those theories.

3

Geologic Time

From an Early Geologic Time Scale

The Age of the Earth: Geologic Time

One of contemporary civilization's most critical philosophical principles is the concept of geologic time. Also called deep time, geologic time is the evidentiary scaffolding upon which the entirety of geology is built. It provides order to the seeming jumble of rocks on the Earth's surface and below, and it reveals the long life story of our planet. Deep-time chronicle both gradual changes and catastrophic ones within the rock layers. Some strata document slow processes, such as the layering of sediments over millions of years; others contain stories of devastation, such as an asteroid strike on the Earth 66 million years ago that ended the reign of the non-avian dinosaurs.

The time scale of the Earth is immense compared to the human reference, and this fact makes it difficult to grasp. Even geologists find themselves checking details and requiring multiple immersions in the time scale to fully comprehend its vast span. Scientists, nonscientists, students, and the public can begin to learn the geologic time scale in myriad ways, from memorization to analogy and visual metaphors. Getting out in the field to see the strata in their natural settings is an additional way to start to grasp the geologic time scale. The multidisciplinary stories of various animals and events on the Earth present another gateway to understanding deep time.

The concept of geologic time inspires profound philosophical insights. Deep time gives context to life, illustrating the interconnectedness of all parts of the planet: air, water, soil, rocks, and life itself. Deep time enables a look back into the Earth's past when severe climate events occurred, and it provides lessons, warnings, and analogies with respect to humanity's current environmental stewardship. Alarmingly, in view of our contemporary treatment of the natural world and the use of the Earth's resources, it may seem that we have not yet absorbed the implications of deep time. Are we, in fact, living at the very end of time? Or are there still lessons to be gleaned, including ways to frame the climate crisis the planet faces? Let's start with an orientation to deep time and examine how early geologists figured out the ordering of events of the time scale, ways in which

Song of the Earth. Elisabeth Ervin-Blankenheim, Oxford University Press.
DOI: 10.1093/oso/9780197502464.003.0004

life forms fit in, and the stories of those discoveries. And then let's consider the implications of those discoveries for us today.

About the Time Scale of the Earth: Ordering of Events through Relative Age-Dating

The fact that the Earth is 4,600 million years old is a challenge to assimilate and comprehend. A frequent analogy employed to illustrate geologic time is a twenty-four-hour clock with hands for minutes and seconds (color insert 3.1). The birth of the Earth is recorded starting at midnight on the clock, as the planet coalesced from dust and accretion of particles. Humans emerged just within the last two minutes of the twenty-four hours. All the rest of geologic time to the present occurs within the preceding twenty-three hours and fifty-eight minutes on the clock.

Three conventions for modifiers of geologic time units are employed in *Song of the Earth*. In the first several chapters the terms "upper, middle and lower" are used in a historical sense as the geologic time scale was developed and are not capitalized. Later in the book, use of the terms "Upper, Middle and Lower" refer to specific stratigraphy, based on the International Chronostratigraphic Chart (International Commission on Stratigraphy)[1] to discuss geologic time and rock units. These terms are chronostratigraphic (time-rock) units and are capitalized. Finally, "early, middle and late" are utilized, in a broader sense, to describe events, such as when certain plants or animals lived, or extinctions occurred through geologic time. These terms are geochronological terms and are not capitalized.

At first, geologists developed the geologic time scale through *relative dating* techniques, locating the position of strata and beds in relation to one another and based on changes in life forms over time (see chapter 1). In a sequence of undisturbed rocks, older beds are at greater depths, with younger strata toward the top, consistent with the principle of superposition first identified by Nicolaus Steno. Conclusions about the relative dating of strata also rely on Steno's other principles, as discussed in chapter 1, such as original horizontality and cross-cutting relations. Geologists, moreover, examine fossils to identify and correlate layers for relative dating.

William Smith pioneered these ideas and techniques in the early 1800s, following Steno's work on the order of geological units. Smith linked unique fossils to rock layers in central England and developed the principle of faunal and fossil succession. As he surveyed the outcroppings revealed by the construction of canals for coal transportation through the central part of England, he observed that specific fossilized shells appeared in particular layers. Smith compared the strata between the two channels, applied his concepts to the geologic layers in the area, ultimately used the same methods throughout the country, and created the first geologic map of England in 1815. Later in the nineteenth century, James Hutton realized that certain rocks, which should have continuously recorded the Earth's history, in actuality did not appear over otherwise expected strata, causing what he deduced to be missing layers. He called these breaks in the geologic record unconformities, resulting from erosion and other rock-destroying events. Hutton's unconformities gave rise to his idea of cycles, the great age of the Earth, and the uniformity of processes. Charles Lyell developed these ideas about unconformities and the use of present processes to understand past events, a concept called uniformitarianism. These techniques—the principles of stratigraphy (Steno); fossil identification, location, and correlation (Smith); and gaps in time and rates of change (Hutton and Lyell)—led geologists in the eighteenth and nineteenth centuries to piece together a cogent order of events, including the gaps, resulting in the construction of an early geologic time scale.

There are many methods to correlate strata from one location to other areas, whether in the same country or other parts of the world. Geologists can relate rocks by fossils, fossil groupings, evolutionary changes in organisms, and extinction events; this is called biostratigraphy. Biostratigraphy works in sedimentary rocks because these layers were made up of loose deposits containing animals and plants. Geologists also correlate strata by lithostratigraphy, which is looking at characteristics of the rocks—grain size, cement type, the environment of deposition, presence of unconformities, and chemistry. Further, geologists use climatostratigraphy to connect rock units through similar climates that laid down the various strata.

Starting in the late eighteenth century, geologists utilized biostratigraphic methods and, to a lesser extent, lithostratigraphic techniques to assign rock units to geologic periods. Smith's principle of faunal succession, along with the work of Georges Cuvier, Alexandre Brongniart, and Louis Agassiz on fossils and extinctions, was critical to understanding how specific fossils related to particular strata. The geologic time scale was at first built on these principles.

However, matching fossil succession and change in organisms to chronostratigraphic events is not an easy task. The original environments where the organisms lived differ from place to place, containing completely different species. Fossilization is a relatively rare process and requires special conditions such that the body does not decay or get swept away; moreover, preservation is enhanced if the material undergoes rapid burial with little or no oxygen. And then, after many millions of years, the deposits have to be located and excavated.

Index fossils are particular types of fossils that stratigraphers and geologists use as keys, so to speak, to unlock the sequence of rocks through time. Index fossils consist of particular animals or plants that lived across a wide geographic area for a relatively short time, were distinct from other species or subspecies, and were abundant and readily identifiable. Key characteristics of index fossils allow geologists to separate the beds based on their morphology (shape) and distribution.

In the Paleozoic Era, for example, long-extinct trilobites (a group of extinct marine arthropods, class Trilobita) are considered the primary index fossil. There were thousands of species of these marine invertebrates, and they changed in distinctive ways over time (figure 3.1a). The reign of the trilobites ended with the largest mass-extinction events ever recorded, the evidence of which marks the conclusion of the Paleozoic Era and clearly demarcates the boundary with the subsequent Permian Period. Likewise, in the Mesozoic Era, ammonites, with their coiled shells, are the primary index fossil (figure 3.1b). Ammonites were prolific breeders, leading to many species that allow geologists to identify and correlate geologic beds. And their extinction marks the end of the Mesozoic.

The work of paleontologist Mary Anning, as we have seen, along with others, including the German paleontologist Friedrich von Quenstedt and his student Albert Oppel in the 1850s, became the basis for the development of ammonite zones, mapped zones defining the age of a rock unit based on changes in ammonites. Ammonite zones have been calibrated for deposits all over the world, including within the layers of the Cretaceous Period Western Interior Seaway (WIS)—a large inland sea that once extended from Canada to Mexico and, for more than 30 million years, split the North American continent into two landmasses.

(a) (b)

Figure 3.1 Middle Cambrian Period trilobite, Mount Stephens, British Columbia. (Mark A. Wilson, 2009) (b) ammonite, Lyme Regis, Dorset Coast, United Kingdom. (Fluffybiscuit, 2010)

Ammonites scuttled through the waters of the WIS, as dinosaurs made trackways and left their bones on its edges. *Pterosaurs* flew over its vast, salty currents. Where today mile-high Denver and the Rocky Mountains rise, the one-time depths of the seaway, depending on time period, ranged from 182 meters (600 feet) with a flat bottom to a dipping basin 200–500 meters (656–1,640 feet) deep. The bottom consisted of an oxygen-poor mud interbedded with sand layers that made for excellent fossil preservation. Eventually, the materials lithified, becoming the Pierre Shale, the Niobrara Formation, and others. Those rocks were uplifted some 40–70 million years ago along with the young Rocky Mountains in a mountain-building event called the Laramide Orogeny.

Geologists have pinpointed age zones within the Pierre Shale, near Loveland, Colorado, based on the distribution and various species of ammonites (color insert 3.2); the map extent marked in blue represents the youngest age defined by one species called *Sphenodiscus*.[2] Geologists also utilize the fossils of straight-shelled ammonites, belemnites, for time distinctions.

Both ammonites and belemnites can provide additional information about the environmental conditions under which they lived. Belemnites, in particular, align themselves to the flow of currents; accordingly, their fossilized shells, called rostrums, are found recording the direction of water or current flow. Studies of ammonite shells and belemnite rostrums reveal other climatic variables, such as water temperature.

Sometimes geologists look at groupings of creatures and their fossils, as a multitude of animals lived in the same ecosystem, called fossil assemblages. An example of a fossil assemblage is one that has preserved dwellers on the seafloor and in the water column above, including the famous Cambrian-aged deposits of the Burgess Shale in British Columbia. Fossil assemblages found in these rocks include trilobites; large, odd-looking arthropods called *Anomalocaris*, also known as "strange shrimp"; early sponge-like creatures, hinged two-shelled invertebrates called *Brachiopods*; the first recorded animal with a notochord, called *Pikaia*; proto-crabs; worms; and the ancestors of horseshoe crabs, among others.[3] Scientists map these distributions and thereby correlate rocks over large areas to develop the geologic time scale. Caution is needed when interpreting the assemblages, because currents and other forces can move the animals after death, but they are handy for the correlation of rocks.

Geologists were able to order events of the time scale based on relative dating methods, but it was not until advancements in physics and chemistry, and their applications to rocks, that the boundaries between eons, eras, periods, and epochs could be definitively pinned down. These innovations ultimately transformed the conceptual theories of both evolution and geologic time into spectacularly broad evidence-based sciences.

Development of the Geologic Time Scale

In part spurred by the Industrial Revolution and the race to mine valuable minerals and other commodities such as coal, geologists and mineralogists of the late eighteenth century attempted to predict the location of such critical deposits by organizing known geologic beds in time order. In particular, the German mining geologist and professor of mineralogy Abraham Werner made what was among the earliest attempts to formulate a geologic time scale of at least the fossilized strata, those within the Phanerozoic Eon. It would not be until much more recently that the ages of the oldest, largely non-fossilized rocks, those of the Precambrian Supereon, would be determined by numerical methods.

Abraham Werner and an Early Geologic Time Scale

Werner divided rocks at the Earth's surface initially into four divisions, or periods:[4] Primitive, Secondary, Alluvial, and Volcanic; he then added a fifth division, Transitional rocks, between the Primitive and Secondary (figure 3.2).

Werner's Primitive rocks (*Urgebirge*) undergirding the base of the formations were inferred by him to make up the Earth's oldest materials and consisted of crystalline (igneous) rocks. Werner and his colleagues speculated that the

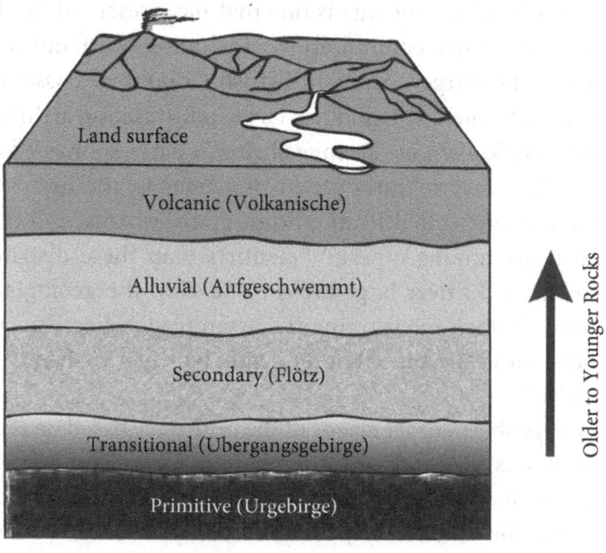

Figure 3.2 Werner's early geologic time scale. (Concept by the author; illustration by R. Gary Raham, 2020)

Primitive strata did not contain any fossils or organic remains. The German prefix *Ur-* means fundamental or original, and *Gebirge*, in current usage, means mountains or mountainous. The term was employed in a more mutable sense by Werner for any ancient formation, in the mountains or not, and he employed other terms related to mining as a source for many of the designations.[5] Transitional rocks (*Ubergangsgebirge*) were located stratigraphically above Primitive rocks, were younger, and were composed of both chemically formed rocks (those precipitated from water or other chemical reactions, e.g., some limestones, chert) and others made of detrital (weathered, e.g., sandstones) grains of material. These Transitional layers consisted of hardened limestones, as well as greywackes (in German *Grauwackes*), and occasional sills and dikes (horizontal or vertical intrusions of newer rock within preexisting layers). *Ubergang* is German for crossing. Thus, these strata spanned the Primitive and overlying Secondary deposits. Above the Transitional units, and younger yet, were those designated as Secondary (*Flötzgebirge* or *Flötz*, also spelled *Floetz*) layers. *Flöz* is a mining term used to describe strata that are flat-lying, or horizontally bedded, stratified deposits of economic importance, such as coal, containing organic material and fossils. Werner postulated that these first three layers, forming the base of his time scale, were deposited from a universal ocean and made up the bulk of the rocks on the Earth. Above Primary, Transitional, and Secondary were the Alluvial layers (*Aufgeschwemmt*), referring to "washed-out" deposits, thinly placed on top of the three foundational ones. These were created by flowing water, with mainly horizontal beds of various thickness and limited extent. In Werner's scheme, the topmost layers were the Volcanic (*Volkanische*). He theorized that these strata originated from coal seams set on fire and quenched with water to create layers of basalts under aqueous conditions.[6] Geologists later determined that volcanic rocks are cooled from magma.

Among European stratigraphers investigating rocks and strata at the same time as Werner was Peter Simon Pallas, a German zoologist and botanist who studied rock formations in the Ural Mountains of Russia. Pallas divided the strata into a (tripartite) system consisting of Primary, Secondary, and Tertiary materials. Deposited first were the Primary rocks, the oldest, made up of crystalline units; next were the Secondary strata containing fossils and mainly found in flat-lying layers; and finally, Tertiary deposits, the youngest, were composed of loose gravels and sediments.[7]

The tripartite system is also known as the Tertiary. The Tertiary was included as a period in the formal designation of the geological time scale until it was replaced by the Neogene Period in most versions of the time scale.[8] The tripartite system was applied by Johann Gottlob Lehmann, a German geologist and mineralogist, in his study of the Harz and Erz Mountains (the Ore Mountains) in Germany. He noted that the core of the mountains contained Primary rocks of crystalline nature, covered by Secondary formations of sedimentary rocks with fossils, and on top of some of the Secondary formations were loose sediments making up the Tertiary units.

Italian geologist Giovanni Arduino (1714–1795) worked in the Alps and refined the three-part classification of Pallas and Lehmann.[9] He observed that these deposits had pebbles identified as originating from Werner's older Primary and Secondary formations. He termed these strata *monti teriazi*, or Tertiary, and noted that they contained distinctive fossils with horizontal layering. Arduino distinguished these from overlying, loose alluvial valley deposits. The latter he placed in a fourth group called *quatro ordine*, eventually to be known as Quaternary.

Werner's ideas on geologic time merged with those of Pallas, Lehmann, Arduino, and others to produce an early time chart (figure 3.3) with Primary (or Primitive), Transitional, Secondary, Tertiary, and Quaternary categories, in age order from older to younger rocks.

Indeed, the remnants of early geologic time divisions are present in today's time scale, specifically in the names of the Tertiary and Quaternary Periods. Werner proposed that all rocks originated from deposition underwater. Although his theory, called Neptunism, was disproved by later geologists, Werner's ideas were vital in ordering the strata. He applied Steno's principle of superposition to formations in the field and is credited as the founder of the first geologic time scale.

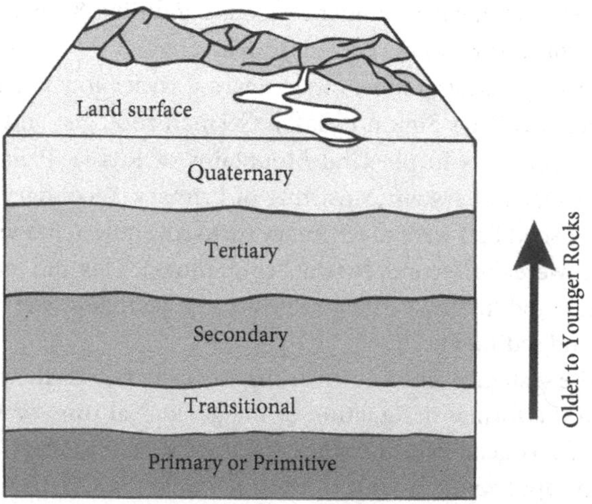

Figure 3.3 Aggregated early geological time scale. (Concept by the author; illustration by R. Gary Raham, 2020)

Expansion of the Time Scale of Geology

Geologists revised and refined the time scale of geology as the field progressed with further research into rock origins and formations. Fossil discoveries and improved classification led to a better understanding of how life developed, enabling a more precise sequencing of deep time. Ultimately, geologists made subdivisions of the five main categories proposed by Werner and others based on these discoveries. Starting in the early 1800s, as discussed below, European geologists suggested formal periods, basing their categorizations on descriptive analysis of rock types and interpretive faunal (animal life represented by fossils from a particular region at a particular time) relationships. Again, the drive to organize field data to assist mining efforts spurred the work of resolving the subdivisions within deep time.

The categorization of the next expansion of the geologic time scale (table 3.1) involved intense research, debate, and discussion—and bitter Victorian-era infighting. The relative order of fossils helped geologists determine the relational order of the geological time scale, but a great deal of discussion and debate focused on the placement of firm time boundaries and distinctions of data upon which to base these decisions.[10] Still, it was not until the twentieth century that methods for radiometric dating—the use of isotopes that decay at a known rate—were developed to determine "absolute" numerical dates.

The basic units of geologic time, which geologists first demarked, ultimately became known as periods, two or more periods are known as eras, and two or more eras are referred to as eons. For example, the Paleozoic (old life) Era was the first to be unraveled, followed by the Mesozoic (middle life) and the Cenozoic (new life). Together, these three eras make up the Phanerozoic Eon. As the study of fossils and rocks advanced, further divisions into distinctive periods gave rise, for example, to the shorter Carboniferous, Triassic, and Jurassic classifications. As will be seen, the older Transitional beds were most difficult to decipher because of their contorted natures, a lack of fossils, and a variety of rock types, including "old greywackes," below a particular marker bed from England that geologists call the Old Red Sandstone, as compared to the gently dipping, fossil-rich Secondary sedimentary beds.

Periods of the Paleozoic Era

The Cambrian, the oldest of the periods located within the Paleozoic Era and originally classified as Transitional by Werner, was first demarcated by British geologist Adam Sedgwick.[11] Within the Cambrian System, Sedgwick noted three groups of rocks with the uppermost portion of the unit containing a

Table 3.1 Phanerozoic Eon geologic periods and their history from youngest to oldest eras and periods.

Era	Period	Named For	Discovered By	Additional Information
	Quaternary	Werner and others' original time scale	Desnoyers,[a] 1829, Paris Basin, France	Alluvium and other unconsolidated deposits; glacial and interglacial cycles
Cenozoic	Tertiary or Neogene and Paleogene	Werner and others' original time scale	Tertiary: Arduino,[b] 1760, Italy; Cuvier and Brongniart, 1810, France; Lyell, 1833, Europe Paleogene: Naumann[c] Neogene: Hörnes[d]	Divided by IUGS[e] into Neogene (new life) and Paleogene (old life); rise of mammals; iridium layer[f]
	Cretaceous	*Terrain Cretacé* (Chalky Terrain)	D'Halloy,[g] 1822, Paris Basin, Belgium; Smith,[h] 1816–1819, England	Chalk and greensand; ammonites continue; based on biostratigraphy; not as sharp a boundary with Jurassic; major mass extinction at the end of Cretaceous; asteroid
Mesozoic	Jurassic	Jura Mountain Limestone and Lias, Dorset Coast, England	Smith, 1816–1819; Conybeare and Phillips,[i] 1822, England	Specific conodonts; all ammonite groups; dinosaurs
	Triassic	*Trias* (three formations in Germany: Bunter, Muschelkalk, and Keuper)	Alberti,[j] 1834, Germany	Ubiquitous ammonites; flood basalts; mass extinction at the end of Triassic; beginning of Age of Dinosaurs
	Permian	Town of Perm near the Ural Mountains	Murchison,[k] 1841, Russia	Conodonts, Foraminifera, and ammonites; appearance of Therapsids; mass extinction event at the end of Permian
	Carboniferous	Name means "carbon bearing"; subdivided in the United States into Pennsylvanian and Mississippian[l]	Conybeare and Phillips, 1822, England and Wales	"Coal measures"; megaflora plants; specific conodonts
Paleozoic	Devonian	Devonshire, England	Sedgwick and Murchison,[m] 1835, England	"Old Red Sandstone"; originally defined by the change in graptolites

Silurian	Silures tribe of Britain	Murchison,[n] 1839; Sedgwick and Murchison, 1835–1838, England	The upper Silurian System identified first; lower was problematic; graptolites
Ordovician	Ordovices tribe of Britain	Lapworth,[o] 1879, England	Later identified to solve "Silurian–Cambrian problem"; graptolites and conodonts
Cambrian	Cambria, Roman name for Northern Wales	Sedgwick and Murchison, 1835, Wales	Trilobites

[a] Desnoyers, J., 1829, Observations sur un ensemble de dépôts marins plus récents que les terrains tertiaires du bassin de la Seine, et constituant une formation géologique distincte: Précédés d'un aperçu de la nonsimultanéité des bassins tertiaires: Annales Scientifiques Naturelles, v. 16, p. 171–214, 402–419.

[b] Vacarri, E., 2006, The "classification" of mountains in eighteenth century Italy and the lithostratigraphic theory of Giovanni Arduino (1714–1795), *in* Vai, G.B., Glen, W., and Caldwell, E., eds., The Origins of Geology in Italy: Geological Society of America Special Paper 411, p. 157–177.

[c] Naumann, C.F., 1866, Lehrbuch der geognosie (Vol. 3): Leipzig, Englemann, p. 8.

[d] Hörnes, M., 1853, Mittheilung an Professor BRONN gerichtet, Vienna: Neues Jahrbuch fur Mineralogie, Geologie, Geognosie und Petrefaktenkunde, p. 806–810.

[e] International Union of Geological Sciences (IUGS) International Commission on Stratigraphy, 2005, Definition and rank of Quaternary (unpublished report to the IUGS), 9 p., available online at http://www.stratigraphy.org/.

[f] Ogg, J.G., Ogg, G.M., and Gradstein, F.M., 2016, A concise geologic time scale: Amsterdam, Elsevier, 234 p.

[g] D'Halloy, J.J.O., 1822, Observations sur un essai de carte géologique de Pay-Bas de la France, et de quelques contrées voisinés: Namur, Imprimerie de Madame Huzard, p. 23.

[h] Smith, W., 1816–1819, Strata Identified by Organized Fossils, Containing Prints on Colored Paper of the Most Characteristic Specimens in Each Stratum (in 4 parts): London, Arding.

[i] Conybeare, W.D., and Phillips, W., 1822, Outlines of the Geology of England and Wales: London, William Phillips, George Yard, 470 p.

[j] Alberti, F.A., von, 1834, Beitrag zu einer Monographie des bunten Sandsteins, Muschelkalks und Keupers: Und die Verbindung dieser gebilde zu einer Formation: Tubingen, Stuttgart University, 366 p.

[k] Murchison, R.I., 1841, First sketch of some of the principal results of a second geological survey of Russia, in a letter to M. Fischer: Philosophical Magazine and Journal of Science, series 3, no. 19, p. 417–422.

[l] Pennsylvanian and Mississippian were not recognized by the USGS until 1953. Berry, W.B.N., 1987, Growth of a Prehistoric Time Scale (revised edition): Palo Alto, Blackwell Scientific, p. 101–102.

[m] Sedgwick, A., and Murchison, R.I., 1835, On the Silurian and Cambrian Systems, exhibiting the order in which older sedimentary strata succeed each other in England and Wales: British Association for the Advancement of Science, Report 5th Meeting, p. 59–61.

[n] Murchison, R.I., 1839, The Silurian System, Founded on Geological Research, in Two Parts (Part 1): London, John Murray, p. 11.

[o] Lapworth, C., 1879, On the tripartite classification of the lower Paleozoic rocks: Geological Magazine, N series, v. 6, p. 1–15.

limited number of fossils. Sedgwick studied these outcrops in northern Wales. His Scottish colleague and friend Sir Roderick I. Murchison (1792–1871) investigated rocks in southern Wales, which he named the Silurian System (Period) for an indigenous Silures tribe later conquered by the Romans, along with twenty-six other Iron Age tribes, in 76–77 CE. The rocks Murchison examined were much more fossiliferous than the strata Sedgwick described to the north but were correlated with other formations in England and Wales using the principle of faunal succession developed by Smith.

There was a significant disagreement between Sedgwick and Murchison regarding the location of the boundary between the Cambrian and Silurian Systems. Because of the relative lack of fossils, Sedgwick did not subscribe to Smith's faunal succession principles for his research in the Cambrian. Murchison clouded the waters by saying that the organisms found in the upper portions of the Cambrian were Silurian, leading the two friends to become involved in an acrimonious dispute, the result of which was they ultimately became estranged. Indeed, the dispute became exceedingly bitter. Murchison was the director of the Geological Society in 1855 and refused to publish any of Sedgwick's papers related to the topic after that time,[12] and he denied that the Cambrian was a system in its own right.[13] The controversy raged on even after their deaths, and it was not until 1879 that Charles Lapworth (1842–1920), a geology professor from the University of Birmingham, proposed adding a period called the Ordovician—named for the Ordovices, another British tribe—between the Cambrian and the Silurian Systems to settle the argument.[14] He based his analysis on the study of graptolites (extinct marine animals, subclass Graptolithina, class Pterobranchia, which lived mainly during the Paleozoic Era) distinctive to the beds. It turned out that the three periods—Cambrian, Ordovician, and Silurian—did, in fact, have different fossil assemblages unique to each. And geological research, presumably to Sedgwick's eternal ultimate satisfaction, later proved that Murchison misinterpreted the age and location of the lower Silurian rocks.

Sedgwick had classified the Cambrian System based on descriptive factors of the rocks, as opposed to doing an interpretive analysis, which would examine faunal relationships and identify characteristic fossil assemblages and index fossils, in the style of Smith. Without an investigation of this type, it was not possible, initially, for other geologists to correlate strata that might have been Cambrian in different locations. Around 1850, trilobites were found to be associated with rocks in the Cambrian System, and the relationships of other units of this system were discerned based on the fossil's presence.[15] By 1879, six years after Sedgwick's death and with Lapworth's work clarifying the upper zone of the Cambrian and the lower part of the Silurian as Ordovician, the Cambrian was reinstated by the Geological Society as a system in its own right.

The Ordovician System proposed by Lapworth was one of the most problematic of the periods because of Sedgwick and Murchison's dispute, and it was not accepted as a system by the British Geological Survey until 1960, though British geologists started to use the term as early as 1900. The Ordovician was adopted in the United States by the USGS in 1903.[16] The Ordovician initially was arranged in six series in Britain based on shelly fossils and graptolites from Wales, Shropshire, the Lake District, and southern Scotland. When this was applied internationally, it became challenging to correlate similar strata based on graptolite fauna alone, the standard practice at the time. The information gleaned from conodonts (extinct marine animals, class Conodonta, with a backbone, thought to be eel-like, whose fossils mainly consist of tooth-like structures), brachiopods, and trilobites was added later to distinguish the series.[17] Numerical age-dating has since verified the subdivisions of the period.

The upper[18] Silurian of Murchison became the new Silurian Period after the introduction of the distinct and older Ordovician. Murchison utilized both description units for the rocks and biostratigraphic zones to delineate this period in an interpretive versus descriptive way. He published *The Silurian System* in 1839[19] and dedicated it to Sedgwick—before their falling out. He found the rocks were consistently different in color, texture, and character from those above and below, identifying four epochs. The organic remains were different as well, with crinoids (animals also known as sea lilies, related to starfish, brittle stars, and sea urchins), crustaceans, and early fishes, not found in overlying rocks.[20] The Upper Ludlow Shales,[21] of the youngest Silurian, contained a bone bed full of pieces of skin, teeth, vertebrae, scales, and other calcareous fish parts. Agassiz, the famous (and later infamous) paleontologist, who specialized in fossil fish, came to England to examine the specimens Murchison found and assisted him in their identification.

Murchison wrote that many of the species had never before been seen and were "the most ancient beings in their class."[22] Agassiz determined that some of the fragments were from an early jawed fish (*Plectrodus*), others from a jawless fish (*Spagodus*), and suggested a new species: *Thelodus parvidens*, which grew up to 1 meter (3 feet) in length and whose skin was made up of non-overlapping, tooth-like scales. *Thelodus*'s mouth faced forward, rather than down, so it was probably an open ocean fish rather than a bottom feeder. Scales and impressions in rocks are all that remain of *Thelodus*; no actual fossil bones have yet been discovered.

The Devonian System was outlined in the report on the Cambrian by Sedgwick and Murchison in 1839, again well before they parted as colleagues. And determining which strata were Devonian was not without its own debate, known as the "Great Devonian Controversy."[23] Having studied the Silurian rocks and described them as a system in Wales, Sedgwick and Murchison decided to try their hands at unraveling the Transitional rocks of Devonshire and Cornwall by examining the formations of the same age and fauna in 1836.

An area of Devon called the Great Culm Trough contained rock strata at the center of the argument surrounding the Devonian-aged layers. This formation, now known to be Carboniferous in age, lies in a valley between north and south Devon located above strata older than what Sedgwick and Murchison were interested in. The sequence of beds consists of mudstone, shale, low-ranked coal, and sandstone called the Culm Measures. These sequences of sedimentary rock, up to thousands of meters thick, contain coal and are of economic import.

Before the delineation of the rocks in the area by Sedgwick and Murchison, British geologist Henry De la Beche, director of the Ordnance Geological Survey, was hired in 1834 to create a geological map of southwestern England. Working in the Devonshire area, he found fossil plants in rocks of the Culm and presented a talk based on his findings in December of the same year.[24] He claimed the plants came from the older Transitional units below the Carboniferous (called greywacke or *Grauwacke*). This view was counter to work done by the leading paleobotanist of the time, John Lindley, who pronounced the plants to be very similar to, and of, the Carboniferous, akin to the fossil plants of the northern and western coal measures. De la Beche also was rebutted by Murchison, who did not believe fossil plants existed in the older strata.

In 1835, De la Beche produced a map of the Devonshire area that lumped the strata of the Culm together in one geologic unit. The following year, Sedgwick and Murchison, here partnering against De la Beche, presented their views to the annual meeting of the British Association in Bristol, not only portraying De la Beche as having misrepresented the Culm but calling him, for good measure, incompetent. The war of words and ideas progressed with great rancor. In February 1839, De la Beche published his work *Report on the Geology of Cornwall, Devon, and West Somerset*, claiming he still did not think the Culm was Carboniferous in age, but he reluctantly separated it from the greywacke deposits and named it the Carbonaceous Series. Additionally, he took issue with Sedgwick and Murchison's definition of the Cambrian and Silurian Systems, stating that they were local names and "impeded the progress of geology."[25]

Rushing to strike back, Sedgwick and Murchison published their report *On the Classification of the Older Stratified Rocks of Devonshire*,[26] not through the Geological Society, which had a process of review and consideration, but through the *Philosophical Magazine and Journal of Science*. They submitted the article only a week before the journal published it in April 1839. This scathing review of De la Beche's findings accused him of moral failings in using, but not citing, the work they had done in the Culm. They also stated (italics in original):

> the great *culm trough of Devon* and the settlement of its true geologic position, is *the key to the whole structure of the two counties*; and that no one was in possession of this key, until in 1836, we offered it to the British Association in Bristol.[27]

They went on to propose that an unconformity separated the Devonian System from the lower Culm rocks and also stated that the use of the term "greywacke" belonged to a "dark, undefined era" and was as absurd as using Werner's *Flötz* term. The much-maligned greywacke was added to their new period as they found it correlated to the Old Red Sandstone. The upper section of the Culm was determined to be Carboniferous based on fossils within related to like organisms in rocks of known Carboniferous age through biostratigraphic methods.

Lyell said that the Culm question was one of the most important theoretical issues ever discussed in the Geologic Society.[28] Not only did it put into action Smith's principle of faunal succession, but it also verified biostratigraphy as a vital tool in unraveling relative age relationships of strata. Thus, the use of fossils became established as essential stratigraphic markers.

The actual disagreement between the geologists was more than an academic exercise in different points of view on the formations; it had economic implications as well. Murchison strongly believed that no land plants, indicative of coal measures used for providing heat to most homes and cities in England, could be found beneath the Old Red Sandstone in the greywacke or Transition rocks. Over the next several years, new fossil finds verified the divisions between the periods, including adding the upper portion of the Culm to the Carboniferous Period.

Eventually, geologists of the day settled the dispute about the Devonian and confirmed it as a period in its own right containing the Old Red Sandstone, initially placed in the overlying deposits of the Carboniferous.[29] The Old Red Sandstone has been a critical marker bed in the stratigraphy of Britain not only because of its vivid red color but also because it had few apparent fossils and was easy to distinguish from other strata.[30]

The Carboniferous Period was one of the first of all the periods to be identified by British curate and geologist William Conybeare and geologist William Phillips, in 1822,[31] because of its economic importance. The definition of this period was much less contentious than some of the other periods, in part because it was easier to distinguish based on its dark to black color resulting from a high carbon count. Conybeare and Phillips placed it among the Transition rocks of Werner, noting that the beds dipped and were contorted in places unlike the overlying, horizontally bedded *Flötz* rocks, with an unconformable surface indicating missing geologic strata between them. The rocks of this period are coal-bearing (sometimes called the coal measures), high in carbon, and consisting of lower units of clay and "grit," limestone, and sandstone. The sandstone to which Conybeare and Phillips referred was the Old Red Sandstone that was later recognized to be older and assigned by Murchison in 1839 to the Devonian.[32] In association with the coal measures were iron ore deposits, also of commercial value. Conybeare and Phillips observed a wide variety of plant matter with few genera but more than four hundred species. They noted little in the way of animal remains.[33] Most of the coal beds were deposited under marshy conditions in lakes or near the edges of water with an abundance of plant material.

In the United States, the Carboniferous Period was further divided into the Mississippian and the Pennsylvanian sub-periods. The lower portion of the Carboniferous was recognized in 1839 by American geologist David Dale Owen, who worked for the U.S. General Land Office surveying for mineral resources, as having extensive beds of limestone and calcareous (limey) shale, but no coal, spanning several states. Owen listed these layers as the "sub-Carboniferous" separated from the coal-bearing beds of the Carboniferous. Alexander Winchell, professor of geology and paleontology at the University of Michigan, named the lower Carboniferous the Mississippi Limestone Series (Mississippi Group) in 1870. In 1891, Henry Shaler Williams, a geologist from New York working for the USGS on correlation studies designed to organize the many confusing names given to various rock units, identified the upper Carboniferous as the Pennsylvanian Series and made a minor modification to Winchell's terminology for the Carboniferous, calling it the Mississippi Series.[34] In 1906, the names were raised from the series level to periods. The USGS finally accepted the designations for these periods in 1953.[35]

Murchison named the Permian Period in 1841 for the town of Perm, after he visited Russia to study the stratigraphy and to conduct fieldwork with Edouard de Verneuil, a French palaeontologist, Alexander von Keyserling, a German–Russian mining expert, and Nikolai Koksharov, a Russian mineralogist before.[36] Prior to delineating the Permian, he searched for Devonian deposits in Russia that were correlated stratigraphically to the same aged deposits in England. He traveled with de Verneuil and during this trip identified Silurian, Devonian, and Carboniferous layers by examining fossils in the Russian rocks. The Old Red Sandstone of Russia had the same fauna as that of Britain, which confirmed Murchison's view that it belonged in the Devonian. During the 1841 trip, he identified strata that were younger than the Carboniferous and had consistent biostratigraphy with a formation in Germany named the *Zechstein* (*zech* = tough, *stein* = stone)—a limey unit made from sea level rising and falling, creating oil-rich layers. These *Zechstein* deposits were associated with a limestone unit (Magnesian Limestone) above the Carboniferous in Britain. There was a debate about whether or not the beds Murchison identified in Russia were lower

Triassic or Permian,[37] but based on the fossil evidence, he named this new formation with its unique flora and fauna Permian, for the town of Perm in the Ural Mountains.

Periods of the Mesozoic Era

The first period of the Mesozoic Era is the Triassic Period. Stratigraphically younger than those of the Permian, Triassic beds were identified by German geologis Friedrich August von Alberti in 1834[38] based on formations in Germany—the Bunter Sandstone, the *Muschelkalk* (mussel chalk) Limestone, and the *Keuper* strata (representative of an onlap, rising sea level). He named the formation Trias after the three strata. Werner and other early geologists identified these beds initially as the lower portion of the Secondary strata, or *Flötz*. Copious fossils in the upper sections of Secondary strata allowed the unit to be more readily classified. Alberti's salt deposits, rocks, and fauna near the town of Sulz, the site of the three famous Trias deposits, were dissimilar in character from the older *Zechstein* (Permian), as well as being different from the younger Lias (Jurassic). The Bunter Sandstone makes spectacular outcrops of weathered red rock. The Triassic layers were eventually traced to the corresponding marine rocks of the Alps and, in time, to other continents.

The Jurassic Period, one of the divisions defined initially, was the most studied of the Mesozoic intervals, in part because of the extensive outcrops of rock from that time along the Dorset coast, Anning's terrain.

Alexandre von Humboldt, Prussian geographer, explorer, and naturalist, was the first to note Jurassic rocks on a trip to southern France, western Switzerland, and northern Italy, calling them in 1799 Jura-Kalkstein limestone, in honor of the Jura Mountains. He based his thoughts entirely on the rock characteristics and stratigraphy, and he did not include any fossil information in his writing about them. In England, Smith did not formally identify the Jurassic strata as a system until his report of 1816–1819.[39] Conybeare and Phillips further characterized the Jurassic in 1822[40] as containing the Oolite (or Oolitic) Series (oolites are rounded sphericals making up some limestones) consisting of the Oolite as younger and the Lias as older. Early time scales contained the Oolite Series for several years after that time (figure 3.4), created by John Phillips, an English geologist from Yorkshire who had assisted Smith with surveying for the county geologic map of Yorkshire.[41]

The Cretaceous Period is the last of the Mesozoic Era and was first discovered and mapped by Jean-Baptiste Julien d'Omalius d'Halloy, a Belgian geologist working in the Paris Basin between 1817 and 1822. He named the chalk formations (a soft, fine-grained limestone made from tiny marine organisms)

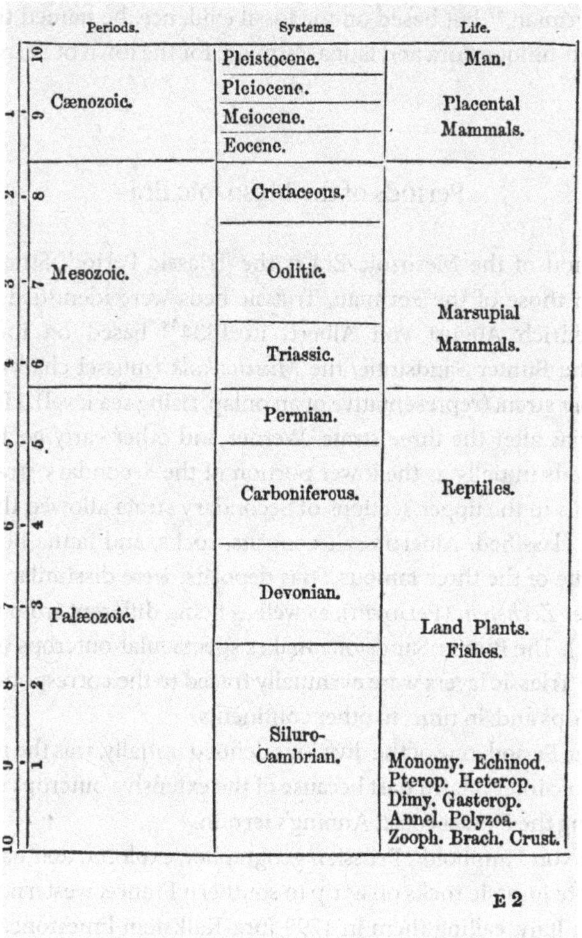
Periods.
Systems.
Life.
Pleistocene.
Man.
Cænozoic.
Pleiocene.
Meiocene.
Eocene.
Placental Mammals.
Cretaceous.
Mesozoic.
Oolitic.
Triassic.
Marsupial Mammals.
Permian.
Carboniferous.
Reptiles.
Devonian.
Palæozoic.
Land Plants.
Fishes.
Siluro-Cambrian.
Monomy. Echinod.
Pterop. Heterop.
Dimy. Gasterop.
Annel. Polyzoa.
Zooph. Brach. Crust.
E 2

Figure 3.4 Development of the time scale of geology and life forms. (Phillips, 1860, p. 51)

found there *Terrain Cretacé* (Chalk Terrain), which ultimately became the Cretaceous System. In England, Smith mapped the strata and was the first to show the Cretaceous as an areally extensive formation. Cretaceous-age chalk, built up from compressed calcite-rich foraminifera (single-celled tiny marine animals, sometimes called forams), makes up the White Cliffs of Dover.

The end of this period has garnered as much attention as the beginning and middle portion of the Cretaceous because of the Chicxulub asteroid and its role in the end-Cretaceous extinction event. Of particular note is the research on Chicxulub by the physicist Luis W. Alvarez and his son Walter Alvarez, a geologist. Luis Alvarez won the Nobel Prize in 1968 for his work in particle physics.[42] His remarkable career encompassed research during World War II at MIT on building a sophisticated detonator for the "fat boy" plutonium bomb, co-discovering tritium, working with cosmic rays, and coming up with three different types of radar systems, one of which is still used in air traffic control today. He was also known for designing and constructing a 40-foot linear particle accelerator and was a professor at the University of California (UC) Berkeley.

In 1977, Walter Alvarez, a professor of geology at UC Berkeley specializing in geoarchaeology and paleomagnetism, was in Gubbio, Italy, collecting samples for paleomagnetic studies. While examining a particular limestone bed at the juncture of the Cretaceous-Tertiary layers, he noticed a thin, red clay bed at the top of the limestone. The limestone below contained foraminifera, but the clay bed had none, and above, in the Tertiary limestone, only one species of foraminifera could be found. He took samples of the clay layer back to the UC Berkeley labs and consulted with his father, who recommended further testing by fellow nuclear scientists Frank Asaro and Helen Michel. Asaro and Michel used a technique they developed to test low levels of elements precisely. They discovered the element iridium in the clay samples at a level six hundred times greater than typically found on the Earth. By further comparing the iridium concentration ratios to those of other materials, they determined the iridium to be from an extraterrestrial source. Based on their iridium research, these four scientists published their results in 1980 detailing the theory of a tremendous asteroid impact causing the extinctions at the Cretaceous-Paleogene (K-Pg) (formerly called the Cretaceous-Tertiary, or K-T) boundary.[43] It would be ten more years until researchers found the impact location of the asteroid. It is this asteroid impact and the associated extinction event that mark the end of the Cretaceous Period and the Mesozoic Era and the beginning of the Cenozoic Era.

Periods of the Cenozoic Era

The Cenozoic Era is the current era on the Earth, and its first period, from a historical perspective, is the Tertiary. The term "Tertiary" was proposed in 1760 by Italian geologist Arduino, based on Werner's system.[44] He identified strata in the mountains of Italy by their lithology alone, not utilizing fossils, and located them stratigraphically above Werner's Secondary system, like those of the third period—the Tertiary. Geologists further identified these rocks across the continent and noticed that distinct fossil assemblages were associated with them. In 1810, for example, Cuvier and Brongniart wrote an extensive description of Tertiary rocks from fossils and strata in the Paris Basin.[45]

The Tertiary became a formal period based on the work of Lyell in 1833[46] and is composed of Paleocene, Eocene, Oligocene, Miocene, and Pliocene "Epochs," which are shorter divisions of time within certain periods. Lyell outlined the Eocene, Miocene, Pliocene, and one later stage in the Quaternary Period, the Pleistocene, in his *Principles of Geology*, Volume 3, in 1833. At first, fossils and biostratigraphy formed the basis for these divisions. Later, as numerical dating came to prominence, they were correlated into time (chronostratigraphic) units. Lyell added the Oligocene and the Paleocene in the 1854 and 1874 editions, respectively, of his book.

The Paleogene (ancient-born) has become accepted by the International Commission on Stratigraphy (ICS) for the lower part of the Tertiary, consisting of the Paleocene, Eocene, and Oligocene Epochs. Carl Friedrich Naumann, professor of mineralogy and geology at the University of Leipzig, first proposed the Paleogene in 1866. He based the determination on his work in northern Germany on rocks that exhibited different fauna and flora of neighboring Tertiary beds.[47]

The ICS reclassified the upper Tertiary to the Neogene Period, consisting of the Miocene and Pliocene Epochs. Mortiz Hörnes, director of the Natural History Museum in Vienna, first used the term "Neogene" in 1853, in referring to the work of Henrick Georg Bronn, professor of zoology and technology at the University of Heidelberg, twenty-plus years earlier.[48] Hörnes had researched fossils in the Venetian Basin and noted that the Miocene and Pliocene fauna were more like each other than older layers, leading him to place them in a new category, the Neogene. He correlated his findings to Sicily, Cyprus, and Rhodes and also included younger glacial and flood deposits, later identified as Quaternary, making this distinction in part because of Bronn's categories.

French geologist and archeologist Jules Desnoyers defined the Quaternary Period from his studies of sediments in the Loire-Touraine Basin and Languedoc, France, in 1829.[49] He found and labeled three units younger than the Tertiary and called this period the Quaternary, for the fourth period. Within the Quaternary Period, Desnoyers identified the Recent (youngest), the Diluvium, and the Faluns de Touraine (marl with shells). Not all units had fossils; some were the result of cobbles lithified into rock from the action of rivers.

In 1833, French geologist Henri-Paul-Irénée Reboul identified the Quaternary as consisting of fossils of animals still living, as opposed to the Tertiary.[50] Lyell recognized the Quaternary Period as defined by Desnoyers in the 1839 French edition of his *Elements of Geology*. Within the Quaternary, the ICS has adopted the Pleistocene and Holocene Epochs. It defined the Pleistocene as the beginning of the most recent glacial and interglacial times, when the climate cooled and massive ice sheets covered large parts of the northern hemisphere.

Controversy about the Quaternary and subdivisions within the Quaternary—how or whether to include them in the Cenozoic Era—has gone on for many decades. As far back as Lyell, for example, some geologists disagreed with associating glacial deposits with the Pleistocene; Lyell preferred to put them in his "Newer Pliocene" category.[51] Some of these challenges arose because the units initially were determined by biostratigraphy using index fossils or climatology (ice ages) and have been converted more or less imperfectly to chronostratigraphic units as age dates became available. In its most recent edition of the geologic time scale, the ICS lists, from older to younger, the periods Paleogene, Neogene, and Quaternary.

Geologic Eras

Classification of the eras, the longer divisions of the time scale, was proposed by several British geologists. In 1838, Sedgwick had suggested the term "Paleozoic Series" (later an era) for British strata from the lower Cambrian through the upper Silurian (the Ordovician had yet to be proposed).[52] John Phillips further organized the early scale in 1840 and 1860 by expanding the Paleozoic to include strata of the Devonian, Carboniferous, and Permian "Systems," those above the Cambrian and Silurian. He offered the terms "Mesozoic" and "Cenozoic" and included within the Mesozoic Era the Triassic, Oolitic, and Cretaceous Systems and, in the Cenozoic, the Eocene, Miocene, and Pleistocene Systems of Lyell.[53] Phillips studied the number of marine species over time in the fossil record of Britain and based the three eras upon the breaks caused by mass-extinction events, the first to use a statistical approach (figure 3.5).[54] The chart illustrates Paleozoic species dying out in the Permian and Mesozoic genera dying away in the Cretaceous. In addition, of the several "depressions" in the curve of species abundance, the oldest occurs during the Devonian Period and shows up in changes in species diversity in "peroxidated" sediments;[55] and the second occurs in the middle of the Mesozoic, with a loss in the number of organisms in uppermost Oolitic rocks. Later studies uncovered five significant extinctions, but the two events, the end-Permian extinction and the end-Mesozoic asteroid impact, as illustrated by Phillips, are the hallmarks of transition between the eras.

Measuring Time: The Final Frontier

The age of the Earth was first calculated from the western European viewpoint in 1650 by Anglican bishop James Ussher of Ireland, based on his detailed calculations of the number of human generations recorded in the Bible, ages in the Bible, and other historical pre-Christian dates. According to Ussher's math, the Earth was born on October 23, 4004 BCE, making it today more

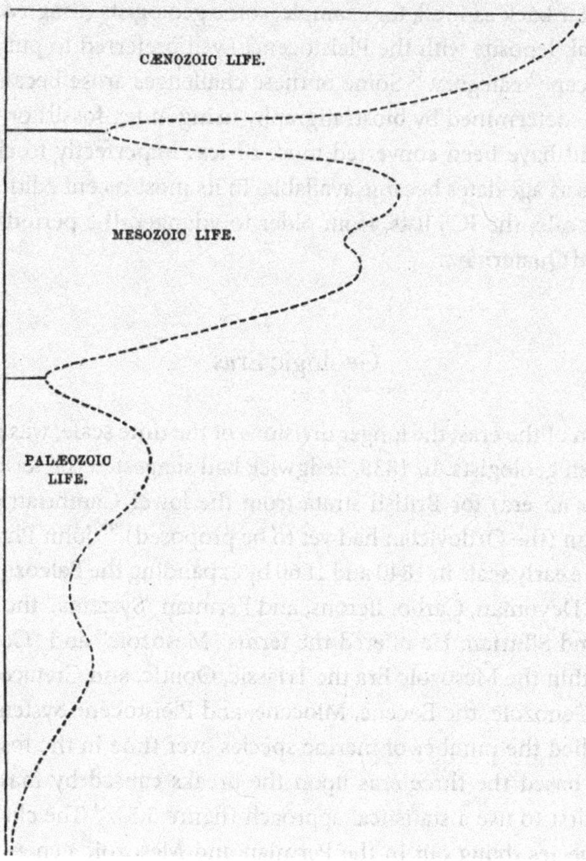

Figure 3.5 Delineation of geologic eras and the prevalence of species. (Phillips, 1860, figure 4)

than six thousand years old. As geology flourished during the beginning of the eighteenth-century Enlightenment, the research and thinking of Hutton and, later, Lyell particularly called into question Ussher's age of the Earth; they reckoned it, based on their scientific observations, more likely in the range of millions of years. This age has been pushed back as science expanded, to the point where the age of the Earth is known now to be in the billions of years.

Even though scientists are unraveling the eons and eras of the Earth, less is known about deep time, and these oldest time divisions are more generalized and broader the further back we look. These primordial times, however, are coming into closer focus through chronostratigraphic means, as will be detailed further in chapter 4.[56]

4
Geologic Time
Measuring Time and the Nature of Deep Time

Quantifying Time: Numerical Dating

The record of geologic time (see chapter 2) was at first measured by applying relational methods, as introduced by Steno. Smith later linked particular fossils to distinct rock layers in central England, developing the *principle of faunal and fossil succession.* And for some two hundred years, geologic time was calibrated in this way, until the second type of age dating—numerical dating—was developed in the mid-twentieth century. Dating strata by relative means is mainly applied to sedimentary rocks or surficial deposits, because of the critical need for fossil material, but numerical dating or geochronology applies to all rock types. This was a revolutionary development for deep time, opening up all of the Earth's rock layers to precise age-dating methods.

As a precursor to numerical dating, American geologist and professor Henry S. Williams, of Cornell University, established the term *geochronology* in 1893, when he related the geologic time scale to the age of the Earth.[1] Relying on thick deposits along the Mississippi River, he attempted to subdivide geologic time into standardized units—"geochrones"—based on known sedimentation rates and comparisons between different eras. Given the sediment thickness observed by Williams of 1700 feet, laid down over the course of an Eocene deposition, he calculated the respective geochrone to be approximately one-third of the Cenozoic Era. Age estimates using this method still are considered relative, as sedimentation rates vary depending on where the sediments accumulate in the world and other processes such as erosion. And although the concept of geochrones has not persisted, all geologists use temporal markers—called geochronology or chronometric dating—to help define deep time.

The oldest mineral found so far was a zircon crystal discovered in the Jack Hills north of Perth, Australia, 4,400 million years old.[2] Ancient rocks from the early Precambrian Supereon are scarce. One of the oldest rock outcroppings on Earth is more than 4,000 million years old, located on the coast of the Hudson Bay in Canada.[3] Today's earth scientists know the age of the Earth in numerical terms because ancient minerals and rocks have virtual clocks inside them. These "clocks" consist of radioactive isotopes—unstable atoms that change and

Song of the Earth. Elisabeth Ervin-Blankenheim, Oxford University Press. © Oxford University Press 2021.
DOI: 10.1093/oso/9780197502464.003.0005

decay at predictable rates. Numerical or actual ages (with ranges) are obtained by measuring the radioactive isotopes which decay spontaneously over time, transforming unstable atoms to more stable, lighter elements at a given rate. Earth scientists calculate the age of a respective sample based on the proportion of initial, unstable atoms to the final, more stable ones.

Numerical age-dating consists of not only of radiometric but also other chronometric methods, including electron spin resonance, which measures the number of free radicals in a sample (a number that increases over time from exposure to cosmic radiation), and recording the accumulated dose of radiation in an object since it cooled after it was formed (thermoluminescence). Other measures include the recording of the Earth's magnetic field and polar reversals as captured in rocks (magnetostratigraphy using paleomagnetism) and also the study of fossilized tree rings (dendrochronology). Each method has age ranges over which it is most useful; for example, dendrochronology can date material up to 11,000 years old; electron spin resonance dating can determine the age of teeth that cannot be dated by carbon-14 methods for samples up to hundreds of thousands of years old (with increasing uncertainty as the materials get older). Geologists use isotopes with long decay rates, such as the potassium-argon isotope series, to age-date the oldest minerals and rocks as far back as billions of years. Within particular age ranges, the various methods are used to check against each other and confirm dates.

Numerical age-dating of rocks revised the age of the Earth further back than anyone could have imagined. No longer were scientists talking about millions of years or tens of millions of years or even hundreds of millions of years. Geologists today reckon the age of the Earth in billions of years. How the science got to this point draws from other fields, especially physics and chemistry.

Radioactivity and Isotopes: History and Use in Geology

X-rays were initially observed by researchers back in the 1880s, but the first discovery of radioactivity was made by French physicist Henri Becquerel (1852–1908) in 1896. Becquerel placed a sample of uranium salts on a photographic plate (isolated from sunlight and any other radiation sources), which, when developed, showed the outline of the uranium sample. Becquerel asked himself why this would occur, and the only reasoning that made sense was that "invisible rays" were coming from the sample.[4] He published his work the same year.[5] Becquerel died at the relatively young age of fifty-five, possibly from handling radioactive material that caused numerous burns on his hands and exposure to radiation.

In 1898, the French physicists and husband-and-wife team of Pierre Curie (1859–1906) and Marie Sklodowska Curie (1867–1934) became engrossed by Becquerel's work and carried his studies further. The Curies, with Marie in the lead, investigated thorium, realizing it gave off the same "uranium rays" as Becquerel had discovered. She named this process *radioactivity*—the first use of the term.[6] Historians of science give credit to Marie Curie for recognizing that radiation came from the nucleus of the atom and not from its array of electrons. This concept was fundamental to the development of physics.[7] The Curies discovered two new elements: polonium and radium.

In 1903, the Nobel Prize in Physics was awarded jointly to Marie Curie, Pierre Curie, and Becquerel. Judges cited the Curies "in recognition of the extraordinary services they have rendered by their joint researches on the radiation phenomena discovered by Professor Henri Becquerel."[8] Marie Curie went on to receive a second Nobel in Chemistry in 1911 for her work with polonium and radium, carrying on the work after her husband tragically slipped and fell under the wheels of a carriage, sustaining a fatal skull fracture, in 1906 at the age of forty-six.

Isotopes were first discovered in 1913 when radiochemist Frederick Soddy (1877–1956), based on his work in isotopes of lead and iridium and thorium, realized that atoms of the same chemical character and atomic volume could have different weights.[9] He was awarded the Nobel Prize in 1921 but did not receive the distinction until the following year, in 1922.[10] Earlier, in 1899, another Nobelist (awarded the prize in 1908), Ernest Rutherford (1871–1937), found that thorium gave off a gas, later known as radon, producing two types of radiation: alpha and beta radiation. Two years later, Rutherford and Soddy observed that thorium (in a sample left in the lab over Christmas vacation) degraded through beta decay to a different form, which then was determined to be an entirely different element: radium. In 1902, they published a paper, "The Cause and Nature of Radioactivity," on the transmutation—changes—in elements, in two parts.[11] In 1904, Rutherford published a book titled *Radio-Activity* and had the following to say:

> The most remarkable property of radio-active bodies is their power of spontaneously and continuously radiating energy at a constant rate, without, as far as it is known, the action upon them of any external cause. . . . The phenomena appear still more remarkable when it is considered that the radio-active bodies must have been steadily radiating energy since the time of their formation in the earth's crust.[12]

Among the isotopes of hydrogen, protium (hydrogen-1) is the most stable and common form—created with one proton and one electron—and comprising more than 99 percent of all hydrogen on the Earth. Protium exists as gas or combines with other elements or another hydrogen. Deuterium (hydrogen-2) contains one proton, one electron, and one neutron. The bonds in deuterium are stronger than those of protium, and it is physically heavier. When combined with an oxygen molecule, deuterium becomes the "heavy water" used in nuclear reactors to slow uranium reactions, as well as for some medical imaging processes. Tritium (hydrogen-3) is an unstable, radioactive isotope of hydrogen, with one proton, one electron, and two neutrons. It takes 12.43 years for half the mass of tritium to decay to helium-3 (^3He), a gas. Tritium is of special interest to geology because its relatively short half-life can be used to measure the age of ground water, and the influence of surface water on ground water—one of many ways to use radioactive isotopes in earth science.

Parent isotopes are the primary isotopes in the rock or mineral that decay. Daughter isotopes are the materials to which parent isotopes decay at a specific rate called the half-life (see table 4.1). A half-life is equivalent to the decay constant for a radioactive isotope. Half-lives vary from relatively short time frames, such as in the case of tritium, to very long time frames, such as that of uranium; accordingly, their uses vary according to the length of time they cover. This characteristic makes isotopes valuable for age-dating deposits, minerals, or rocks of a variety of ages, even billions of years.

Water in the ground, ground water, is vital as one source of water, along with surface water, for life. It is part of the hydrosphere and interrelated to the other parts of the Earth. Ground water varies in age depending on how quickly it is recharged from water trickling down from the surface or its connection to lakes, streams, and rivers. Knowing the age of ground water helps determine whether it is readily replenished, which would make it a steady source

Table 4.1 Radioactive elements with parent, daughter, age range, and materials dated.

Parent	Daughter	Half-life (years)	Materials
Tritium (Hydrogen-3)	Helium-3	12.43	Surface and ground water
Carbon-14	Nitrogen-14	5,730	Shells, limestone, carbon in tissues
Uranium-235	Lead-207	713 million	Zircon crystals
Potassium-40	Argon-40	1.3 million	Zircon crystals
Uranium-238	Lead-206	4.5 billion	Zircon crystals

of drinking water. Hydrologists commonly analyze the presence of tritium to determine the age and sources of ground water. Ground water in the range of 60–100 years old is considered by hydrologists to be "young," as determined by the short half-life of tritium. Tritium is a useful isotope because, though it is relatively rare, occurring naturally in low concentrations when cosmic rays interact with the atmosphere, it is relatively easy to measure. Units are TUs (tritium units), where 1 TU corresponds to 1 atom of tritium in 10^{18} hydrogen atoms, a tiny number. Before 1950, ground-water values of tritium were low (less than 0.8 TU).[13] Most of the modern tritium comes from atmospheric thermonuclear weapons testing from the mid-1950s to the 1960s, with the peak tritium release in 1963, a period during which tritium was released into the atmosphere at a rate of 1.3×10^9 TU. The high concentrations of tritium mix with clouds and get carried into oceans, through precipitation, along with precipitating into surface and ground water. Recordings of TU readings were on the order of 5,000 units of tritium in rainfall from stations in many northern hemisphere countries after the bomb-testing peak.[14] The amount of tritium in ground water is indicative of how recent it is and if there has been replenishment of the ground water by other water from the surface.[15] The age of ground water is helpful to know for water supply and drinking-water well emplacement. If a ground-water reservoir is tens of thousands of years old, the water in it is old and will not be replenished by other sources. Such a repository would not be sustainable as a drinking-water supply because the reservoir would get depleted quickly and would not last.

Over time, depletion and attenuation of anthropogenic (human-made) tritium in surface and ground water will cause it to become a less useful in dicator of the age of water. Other radiogenic compounds used to age-date ground water, sediments, archaeological material, and anything containing carbon are carbon-14 and oxygen isotopes. More than 99 percent of all carbon atoms on the Earth have twelve neutrons in their nucleus (carbon-12) and are stable; they do not undergo radioactive decay. Other forms of carbon exist with varying numbers of neutrons. One of the most well known is carbon-14, containing the same number of protons as carbon-12 but with 14 neutrons, rendering it heavier and slightly radioactive. This form of carbon is unstable and will undergo radioactive decay with a half-life of 5,730 years. Carbon-14 data allow geologists to definitively date the age of many rocks or minerals as their isotopes decay.

Other isotopes, those in uranium, for instance, act as other "clocks in rocks." This process works like other decay series discussed: a parent material decays into a daughter product, except the half-life is much longer. The dating of zircon crystals by numerical methods, for example, has produced some of the oldest dates on the planet.

The Present Geologic Time Scale

Stephen Jay Gould, in his book *Wonderful Life: The Burgess Shale and the Nature of History*, originally published in 1989, relates that in teaching his students to memorize the periods of the geological time scale, he would try mnemonics, games, word associations—any measures for remembering the time divisions—but nothing made great sense. He wrote that he would not have minded this failing on the part of his students if the periods were random breaks in the scale, but they are not. In his words:

> The history of life is not a continuum of development, but a record punctuated by brief, sometimes geologically instantaneous, episodes of mass extinction and subsequent diversification. The geological time scale maps history, for fossils provide our chief criterion in fixing the temporal order of rocks. The divisions of the time scale are set at these major punctuations because extinctions and rapid diversification leave such clean signatures in the fossil record.[16]

Gould ultimately informed his students that they would have to learn the old-fashioned way: by memorizing these essential divisions in the geological time scale. As it turns out, this is also the most effective way to understand and assimilate the drastic changes in conditions that impacted life and caused extinction and repopulation events.

The vastness of geologic time is difficult for anyone to fathom. *The New Yorker*'s John McPhee, who rhapsodized about the geology of the United States in its beauty and many forms, wrote in 1981 in *Basin and Range*, the first of his four books on geology: "The human mind may not have evolved enough to be able to comprehend deep time. It may only be able to measure it."[17]

A simplified geologic time scale (color insert 4.1) records the biography of the Earth from its birth to the present. Time runs from the older Precambrian Supereon, 4,550 million years ago (rounded to 4,600 million years, or 4.6 billion) at the bottom of the scale, to the younger Cenozoic Era, to the present at the top of the scale.

The divisions of geologic time, proceeding from the larger to the smaller units, are eons, eras, periods, epochs, and ages. Stratigraphers refined the time scale into smaller units, epochs and ages, as more became known about the Earth's biography. The time scale is shown vertically from the oldest ages on the bottom to youngest ages on the top, according to the *principle of superposition* (see chapter 1). Segments within the time scale are related to major geologic events and form a basis for the understanding of the Earth. It places the length of human life in a different perspective and is a reminder that humans are among the many and varied types of life in the history of the planet. This figure is not proportional;

the time before the Phanerozoic Eon, the Precambrian, is compressed so that the figure fits on one page. The earlier portions of the Precambrian Supereon and the Hadean Eon are not shown in this figure.

These divisions are not arbitrary. Geologists and stratigraphers have spent decades decoding the boundaries between the eons, eras, periods, and epochs, in an ongoing process. Divisions of the geological time scale must have a definite start, a specific endpoint, and a physical presence in the geologic record that is clear, distinctive, and large enough to be recognized. To be considered, the formation must have a particular "type" (characteristic) of section of rock or outcrop, denoted by the placement of a Global Stratigraphic Section and Point (GSSP), also known as the "Golden Spike." Alternately, the rock unit may have a designated time boundary that best characterizes the particular limit in geologic time, as identified by the ICS, founded in 1961. The ICS agrees on a physical location that best represents the defining example of the physical boundary, with a latitude and longitude. The ICS physically marks the location with a Golden Spike and publishes the location. In 2003 the ICS ratified the base of the Turonian Epoch and in 2013, a new geologic boundary was marked with a Golden Spike in Colorado, near Pueblo, designating the boundary in the Mesozoic Era, Upper Cretaceous Period, marking the base of the Turonian Age was emplaced.[18]

Humankind has so influenced and changed the face of the Earth that some geologists and stratigraphers, specialists in designating geologic time divisions within the ICS, have been discussing whether to officially introduce a new epoch called the Anthropocene, for the "Age of Humans." Paul Crutzen (1933–), a Dutch Nobel Prize–winning atmospheric chemist, and Eugene Stoermer (1934–2012), professor of biology at the University of Michigan, introduced the Anthropocene as a possible time division in 2000.[19] This new epoch, if adopted, would end the Holocene and acknowledge the impact of humans. The debate mainly centers on when to designate its start date.

Selecting the beginning of the Anthropocene Epoch and an end to the Holocene Epoch depends, for stratigraphers and geologists, on the rock record. Some sediments and rocks document human activities. Nuclear weapons testing is one example of human impact enshrined into strata, making geologic history. During weapons testing, tritium released into the atmosphere settled out of the air into water and sediments. Sediments become rocks—lithified—over time, and they mark the start of the nuclear age. Another possible marker time is the early 1800s CE, the beginning of the Industrial Revolution and the wide-scale burning of coal in Europe. Plastic in the environment may be another clear indicator of the beginning of the Anthropocene. Several years ago, rocks made of an aggregate of plastic, volcanic ejecta, sand, and shells were found in Hawaii.[20] In 2019, the ICS had not settled the question of Anthropocene as a formal time division. However, the Anthropocene Working Group, part of the Subcommission on Quaternary Stratigraphy, a constituent body of the ICS, voted to support its designation and will be forwarding its formal proposal for adopting the Anthropocene to the ICS by 2021.[21]

As geologists gathered more data about rocks and outcrops in space, the need for standardized colors and symbols became apparent. The ICS standardizes the colors of the time scale of the Earth for Europe. American geologists have a distinct coloring scheme to indicate specific time divisions. The ICS designated Silurian and Devonian Age rocks in blue and tan, respectively. American geologists color these ages as purple for Silurian and blue for Devonian rocks. Geologic maps and the geologic time scale are integral to each other.

The historical ideal concerning the use of color on the time scale and later geologic maps was based on a spectrum from red-purple-violet-blue-green-yellow, with the oldest geologic units in red and the youngest in yellow. John Wesley Powell, as USGS director, followed this general pattern of assigning colors. The international system is better adapted for use in Mesozoic and Tertiary rocks but not as clearly able to be seen with the older rocks of the Paleozoic System, such as in the Appalachian Mountains.[22] How the two systems of color-coding on the geologic time scale and geologic maps came about is a story of its own, tied in with the history of the science of geology.

In 1876, on the heels of the World's Fair in Philadelphia, a group of geologists met in Buffalo, New York, to look into a committee to form an International Congress of Geology. The American Association for the Advancement of Science (AAAS) had one goal for the committee—to determine the "propriety" of such a group—while members of the committee went forward anyway in creating it. This action caused hard feelings among some members of the AAAS and geologists who saw the need for the congress.[23]

In 1878, the First International Geologic Congress (IGC) was held in Paris and included the American Committee.[24] Items on the agenda were the unification of names and symbols, decisions on boundaries between units, delineation of faults and linear features on maps, the best way to use fauna and flora to make decisions about boundaries, and how to utilize mineralogy, chemistry, and the texture of rocks to determine their origin and age.[25] The groundwork was laid to come up with guidelines during this first meeting, but decisions on specific resolutions were put off until the Second Geological Congress, in Bologna, in 1881.

Powell became the second director of the USGS in March 1881, prior to the initial IGC later that year. Not waiting until September for decisions by European geologists, one of Powell's first memos stated directives for producing U.S. geologic maps. Thus, the "American color system" was born.

During the Second Geological Congress in 1881, rock names and geologic age classifications were agreed upon and formalized, as well as a general color scheme based on the time divisions becoming the "International Color System." Rules for naming species were also set forth. In 1885, when the Third Geological Conference gathered in Berlin, refinements were made to the various systems, including the color on geologic maps.

The American Committee appointed to the IGC met seven times over its early years and sent a delegation to each congress. After the sixth meeting of the committee, Powell objected to a significant portion of the work of the IGC. Persifor Frazer, secretary of the committee, remarked:

> The question which most concerns American geologists is, what attitude they propose to assume toward the body of their own creation. A Congress of experts from every country on the globe assembled to clear away the confusion arising from misuse of terms, synonymy, and local prejudices, and thus to facilitate the progress of science, is a distinctively nineteenth century idea. Not that it originated with the gentlemen who called this geological Congress into being; for the experiment had been tried on numerous occasions before; but no science, needed such assistance more than geology, which is peculiarly liable to death by drowning at the hands of any one who is provided with a few terms, the physical power to walk over the country in his own neighborhood, and an ambition to be seen in print.[26]

Frazer goes on to say:

> The plain moral is that every true scientific geologist the world over should enter into hearty sympathy with the work of the Congress, knowing that by the cooperation which it secures, the advance of the science must be immeasurably greater than it could be from the divided energies of a handful of leaders, each using the resources of his government to enable him to show that all the rest are hopelessly wrong and himself phenomenally and wholly right.[27]

It was a stinging rebuke of Powell's actions. Two years later, in 1890, a small group within the AAAS discharged the American Committee to the IGC, much to the dismay of the members.[28] The ICG meeting was held in Washington, D.C., in 1891, but divisions lingered within American geology. Repercussions follow to this day in the way colors and symbols are used on geologic maps. Rock formations and their ages are represented differently between U.S. geologists and their colleagues throughout the rest of the world.

Mapping the Geologic Time Scale

Geologic maps carry the colors of the time divisions on the geologic time scale. These are maps of bedrock or surficial geologic units that show rocks and groups of similar rocks with identical colors and symbols to the geologic time scale, making them integral in determining the geologic history of an area and, also, visually striking. Geologic mapping is performed by geologists in the field and also by analyzing satellite and other methods of remote-sensing data. Maps of geology and the geologic time scale are closely related, not just through the identification of the rocks and rock units but also because the rocks on the map are shown in the format of the geologic time scale as part of the legend of the map. The more geologists find out about rocks and rock units in the field, the more accurately geologic time is defined and refined.

Much of the basis for the geologic time scale is derived from the identification of rock units and formations in nature—the pick-and-hammer work of field geologists. This evidence enhances and confirms the time scale. The first map of the geology of England, Wales, and southern Scotland, made in the modern era of geology, was created by Smith in 1815 (see chapter 1). Smith's map was large—6 feet 2 inches by 8 feet 9 inches—and was printed at the scale of 1 inch = 5 miles, necessitating fifteen copperplate engravings to create the map at the scale Smith desired (see chapter 1, color insert 1.2). Further, the map was watercolored by hand to represent the different rock types. Of the 380 original maps, approximately 100 still exist. One of the first editions of the document was misplaced in the archives of the British Geological Society and was not rediscovered until 2015—just in time for the map's 200th anniversary. Smith's map contains a legend by which geologic units are shown in age order from youngest at the top to oldest at the bottom, carrying through the concept of Werner's early geologic time scale and his principles.

In the United States, the first geologic map made during the modern age of geology was created in 1809 by Scottish geologist William Maclure, who settled in the United States[29] (color insert 4.2). An earlier French map showing marl and clay layers from Cape Breton Island to the Gulf of Mexico was drawn before Maclure's map, in 1752, but it is not clear that the map was ever field checked or that it resulted from direct field mapping efforts; a mapmaker in France may have created it from reports by French officers.[30] Maclure, however, performed diligent geologic mapping in the field before creating his map. However, not satisfied with how the mountains were shown on his map, he commissioned a second map in 1817.[31] Relying on Werner's classification of rocks, Primitive rocks are shown in "siena [*sic*] brown" (correlative with the Appalachian Mountains); Transitional formations are shown in "carmine" (a narrow band of rock formations that he found to be dipping at an angle—to the west of the Primitive rocks); Secondary formations, flat-lying rocks of the Allegheny Plateau, are shown in "light blue" (west of the Transitional formations); and Alluvial deposits are illustrated in "yellow" (east of the Primitive rocks and correlate to coastal plain sediments). The map designates mountainous areas by hachures, representing orientations of slope and relief. However, the stratigraphic column on the map is not in geological order from oldest on the base of the stratigraphic column to youngest on the top but instead runs from the oldest to the youngest, on the bottom of the column. Thus, the map cannot be correlated in the strictest sense to geologic time.

Early reconnaissance geologic maps of the western United States date back to the 1860s and 1870s. Maps covering the contiguous United States in its entirety have existed since 1855; after the Civil War, color lithography improved geologic maps.

Geologist George Stose and cartographer Olof Ljungstedt, both with the USGS, developed a detailed geologic map of the United States in color in 1932[32] at the 1:2,500,000 scale, the basis for further geologic work and analysis for the next forty years; the map was last reprinted in 1960. Later, King and Beikman's geologic map, produced in 1974[33] for the conterminous (lower forty-eight states) United States, in two plates at a scale of 1:2,500,000 (color insert 4.3), replaced the Stose-Ljungstedt work.

Along with geologic maps showing strata in a two-dimensional sense, from a bird's-eye (or plan) view, are cross sections that examine geology in a third dimension: vertically. Most geologists find it intuitive and easy to see the third dimension of the Earth, depth, and they despair when their students are unable to see the world in a cross-sectional sense. Fortunately, students and others can acquire and develop this skill with practice. Cuvier and Brongniart presented an early cross section of geology with depth in their Paris Basin report (color insert 4.4).[34]

Cross-sectional views allow geologists to unravel the history of the Earth, to see what has occurred to strata over long stretches of time, whether they are folded or faulted, and also to examine relationships between rock units. Most geologic maps are shown from both perspectives—plan view and with one or more sections in the third dimension. The geologic units are arranged in stratigraphic order, with the youngest rocks on top and the oldest at the bottom of the column. The cardinal direction of north is always toward the top of the sheet and includes a north arrow, magnetic declination, and a scale. The location of the cross section is indicated on the plan view of the map by a line with letters, typically A–A' for one, or the first, cross section, and appear shown below the main map with the same letters for reference points. Features that are not apparent on the surface may be visible when looking in the third dimension.

The Grand Canyon in Arizona is one example of a large vertical outcrop of exposed rocks. Another is 11,000 feet of sedimentary strata located on the Crow Reservation at the Bighorn Canyon National Recreation Area on the southern edge of Montana (figure 4.1).[35]
Natural exposures of this extent are rare. Data used to create cross sections usually are obtained by projecting known information from the surface into the subsurface, drilling wells, or remote sensing, as is done for drilling for oil. When using data from wells and boreholes, educated guesses must be made between the points to fill in gaps. Types of information yielded from building cross sections are the location of ground water and the height of the water table, mineral resources, the nature of faults and other hazards, and geologic history of the landscape.

Another way geologists work at depth is through block diagrams, or 3D geologic models, which illustrate the surficial geology on the top of the picture and depth on the vertical faces of the cube. These computer-generated models allow exploration of geologic structures, otherwise not easily accessed from the surface. Block diagrams often are used in teaching and for illustrative purposes. They can be drawn by hand or generated in 3D by computer models. Some of the representations are very complex, and geologists use them for economic resource evaluation, the location of ground water contaminant plumes, and fault properties.

Figure 4.1 Exposed cliffs, Bighorn Canyon National Recreation Area, Montana and Wyoming, 2004. (Brian W. Schaller, 2004)

Deep Time: A New Orientation to Geologic Time

Understanding geologic time is the keystone to comprehending all other geological concepts. Such a vast time scale allows geologists to organize and frame information we learn about rocks and structures in the field into a cogent biography of the Earth. Geologists perceive geologic time in billions of years, from the birth of the Earth to the present. Gould in his 1987 book *Time's Arrow, Time's Cycle* says that "We live embedded in the passage of time," and deep time is "geology's greatest contribution to human thought."[36] Gould lays out two historical ideas about time. The first is that of an arrow, traversing from events in the past to those in the present, unique and nonreplicable. The second is that of a cycle, in which, rather than a linear function, time has cyclical, repeating properties. These Gould views as two end members, a "great dichotomy" in thought, and he argues that an appreciation of deep time necessitates both. Deep time, to many geologists, is both a science and a philosophical anchor.

The next three chapters explore the additional foundational tenets of the Earth: plate tectonics (chapters 5 and 6) and changes in life forms through evolution (chapter 7). These chapters cover both the scientists and geologists who discovered and worked on these ideas and, through examples and stories, the precepts themselves.

5

Plate Tectonics

History of the Revolution in Earth Sciences

How the Crust of the Earth Moves: Overview of Plate Tectonics

The second major principle of geology (after the geologic time), one that explains and underscores how the planet works, is plate tectonics. One of the most important scientific developments of the twentieth century, its significance as a unifying theory cannot be overstated. The process of plate tectonics is responsible for the continental and oceanic crusts, the destruction of ocean basins and continents, the building of mountain belts, and the production of trenches on the ocean floor. Further, it accounts for most locations and incidents of earthquakes, volcanoes, and ore bodies.

Geologists now know that the movement of plates, large pieces of the lithosphere (color insert 5.1), is propelled by heat in the Earth's mantle and core. Volcanic activity at mid-ocean ridges and rifts creates new ocean floor, and old plates are destroyed and consumed, swallowed up as they dive under other plates. Plates jockeying for position on the curved surface of the Earth create large transform faults—lateral, strike-slip (where the movement is predominantly horizontal) faults occurring at the boundaries between two plates—which extend for thousands of kilometers as the Earth's crust accommodates to the massive shifts. On average, movement is about as much as a fingernail grows over a year, approximately 2 centimeters. But this adds up over the course of millions of years, causing continents to collide, oceans to shrink and close up, mountain ranges to rise, new oceans to expand, and supercontinents to form and disperse. The Earth has seven major lithospheric plates, but nine plates that are greater than 20 million square kilometers, ten minor plates ranging in area from 1 million to 20 million square kilometers, and many microplates less than 1 million square kilometers. There is evidence that the overall number of plates that make up the rigid pieces of the Earth's crust has varied in number, but today the seven major plates make up 78 percent of the surface of the Earth. Adding seventy-three smaller plates brings the total to more than 99 percent of the surface. Less than 1 percent of the Earth's surface consists of "microplates" defined by forty-seven small, rigid pieces of crust, which move independently from the other

Song of the Earth. Elisabeth Ervin-Blankenheim, Oxford University Press. © Oxford University Press 2021.
DOI: 10.1093/oso/9780197502464.003.0006

plates. Tectonic plates consist of both oceanic and continental crust, but none of the plates is made up of continental crust alone.

As noted, plate movement is driven by heat and convection cells within the mantle (color insert 5.2). The crust of the Earth is a relatively thin layer, up to 100 kilometers under continents and thinner under the oceans. The lithosphere includes the crust and upper solid mantle material, below which lies the asthenosphere. These two layers, the lithosphere and the asthenosphere, are critical pieces of plate tectonics, as the rigid lithospheric plates ride on the asthenosphere, which is heated from the Earth's core and flows in a plastic manner.

Before the plate tectonics theory was verified, the occurrence of earthquakes, volcanoes, and landslides did not relate to any known Earth processes. But after World War II, geologists plotted all known earthquakes in the world on top of what at the time were believed to be plate boundaries. The correlation was immediate and stunning: earthquakes defined the edges of the plates. Scientists have realized since the development of plate tectonics that many geologic hazards are directly related to the stresses at plate boundaries. Earthquakes are generated by crustal plate movements when faults rupture and transfer energy through the Earth to the surface. Landslides and the failure of soft sediments from shaking occur in association with earthquakes. Geologists have learned that volcanoes correlate to the edges of the large plates and form by a "melting" of the plate material, as the plate and entrapped ocean waters are drawn down in deep trenches at plate margins. Before this theory was confirmed, the location of economic ore deposits, mineable bodies of ore necessary for mineral extraction, seemed random. However, plate tectonics predicts that most ore deposits occur where pressures and temperatures are high along the boundaries where plates interact.

In other words, the plate tectonics theory produced a revolution in geology over the past sixty years. Many geologists now working can remember back to their university years when professors of geology did not even teach the theory of plate tectonics because it was so recent. And how the ideas moved from the hypothesis of continental drift to evidence-based proof of the unifying theory of plate tectonics is itself an instructive narrative.

Setting the Stage for Continental Drift and Plate Tectonics

Speculation on the shape of the Earth and the origins and permanence of the landmasses is rooted in the writings of the pre-Socratic philosophers.[1] Among other concepts, they addressed the "problem of change." These ideas seeded thoughts that led to continental drift and ultimately plate tectonics, millennia later when early explorers and scientists observed the shapes of the continents and how they mirrored each other as the age of discovery in the sixteenth century ensued, during the period of sea exploration and mapmaking. Flemish cartographer Abraham Ortelius[2] and other explorers, including Ferdinand Magellan, circumnavigated the globe and further recorded this relationship. Questions arose about the shapes of the continents and how they might be pieced together. Ortelius created the first modern world atlas, *Theatrum Orbis Terrarum*. Students of early maps of the Atlantic Ocean (figure 5.1), for example, noted that Africa seemed to fit conformably between North and South America.

French zoologist and paleontologist Jean-Baptiste Lamarck first deduced that there must have been continental movement, based on his research in fossils (see chapter 1). Lamarck documented his thoughts on the configuration of the landmasses and ocean basins in the early 1800s and speculated in his article *Hydrogéologie*, published in 1802, that the continents traveled the globe in a westerly direction spurred on by global currents. He suggested that marine fossil distributions made more sense if the continents moved slowly through time and that the oceans had repeatedly submerged lands that were now dry. However, he cited no evidence for his hypothesis of the movement of the continents and could not find a publisher for *Hydrogéologie,* the sole article during his long career concerning geology.[3] Not unlike a contemporary self-published author, he decided instead to print the paper himself, but few copies sold. It fell into obscurity until researchers delving into the history of science rediscovered it and recognized its importance.[4]

In the mid-1800s, further ideas about the puzzle-piece nature of the continents emerged in the scientific literature. In 1858, French geographer Antonio Snider-Pellegrini was the first to publish two maps, one of which illustrated a single continent that had existed sometime in the past consisting of North America and Africa "before the separation" and the other a map of the continents "after the separation" in their current locations.[5] To support his argument, he observed that plant species of the Carboniferous Period were the same on either side of the Atlantic. However, he, too, proposed no viable mechanism for how the continents could have drifted apart. He fell back to the catastrophic argument of a Noachian flood to create the opening of the Atlantic, an argument that was out of favor even at that time.

Another set of theories in the nineteenth century concerned global continental processes and focused on ideas about mountain building. These came in several varieties. In one line of reasoning, scientists ascribed the formation of mountains to the expansion of the Earth, like a balloon being blown up, with uplifted areas and ranges made as the surface stretched. Another held that the entire Earth began as molten rock and gradually cooled, and as it shrank, ranges were created much like the skin on an apple wrinkles when it becomes overripe and withers. This was known as the "contraction" theory. Continents were thought to rise as ocean basins sank. Two camps of thought developed, one in the

United States and the other in Europe, advocating two opposing versions of the contraction theory of mountain building.

American geologist James Wright Dana (see chapter 3) was the chief proponent of the contraction theory. His version held that minerals cooling at different rates was what formed the landmasses and ocean basins. Elements in the heavier minerals—iron and magnesium—formed the deeper oceans, whereas elements in lighter minerals—silica, sodium, calcium, and potassium—created the continents riding higher in the crust.[6] Dana postulated that once created, the continents and basins did not move and were fixed in their locations. This U.S. view on contraction theory was also called the "permanence" theory.

Eduard Suess and His Tectonics

Meanwhile, Austrian Eduard Suess, the leading proponent of the European school on contraction theories, was an expert in the geography of the Alps and applied the theory of contraction to that range as a causative factor of its formation in *Ueber den Aufbau der mitteleuropäischen Hochgebirge* (*On the Structure of the Middle European High Mountains*), published in 1873. Dana's work in the United States profoundly influenced Suess, and he was the first to use the word "tectonics" (from *tecto*, Latin for "to build"), the idea that explains how large blocks of crust move, deform, and change shape over time relative to one another. This concept was a vital piece of what would ultimately become the plate tectonics theory. He described how the blocks jockey for position, causing shortening on the front sides of the blocks when smashed together and extension when they are pulled apart. This groundbreaking view of contraction theory was in direct contrast to the slow and steady processes of uniformitarianism discussed by Lyell.

The raw natural materials for mountain ranges had to come from somewhere. Accordingly, the "geosynclinal" theory championed by American geologist James Hall linked Dana's permanence theory with ideas on sedimentary basins. Hall suggested that large, downwarped basins full of sediments ultimately became lifted by contraction, or sliding of blocks, to create ranges. These basins were assigned awkward names: "geosyncline," "miogeosyncline," and "eugeosyncline." Nevertheless, like hard-to-weather stones, some of these words have persisted in the geologic literature.

Several problems and inconsistencies were endemic in contraction theory. If it was valid, then all the mountains would have formed at the same time and have been of the same age, which geologists were not able to verify. Additionally, temperatures within the Earth should have been cooling, but data supported increased heat with depth—known as the geothermal gradient—produced by

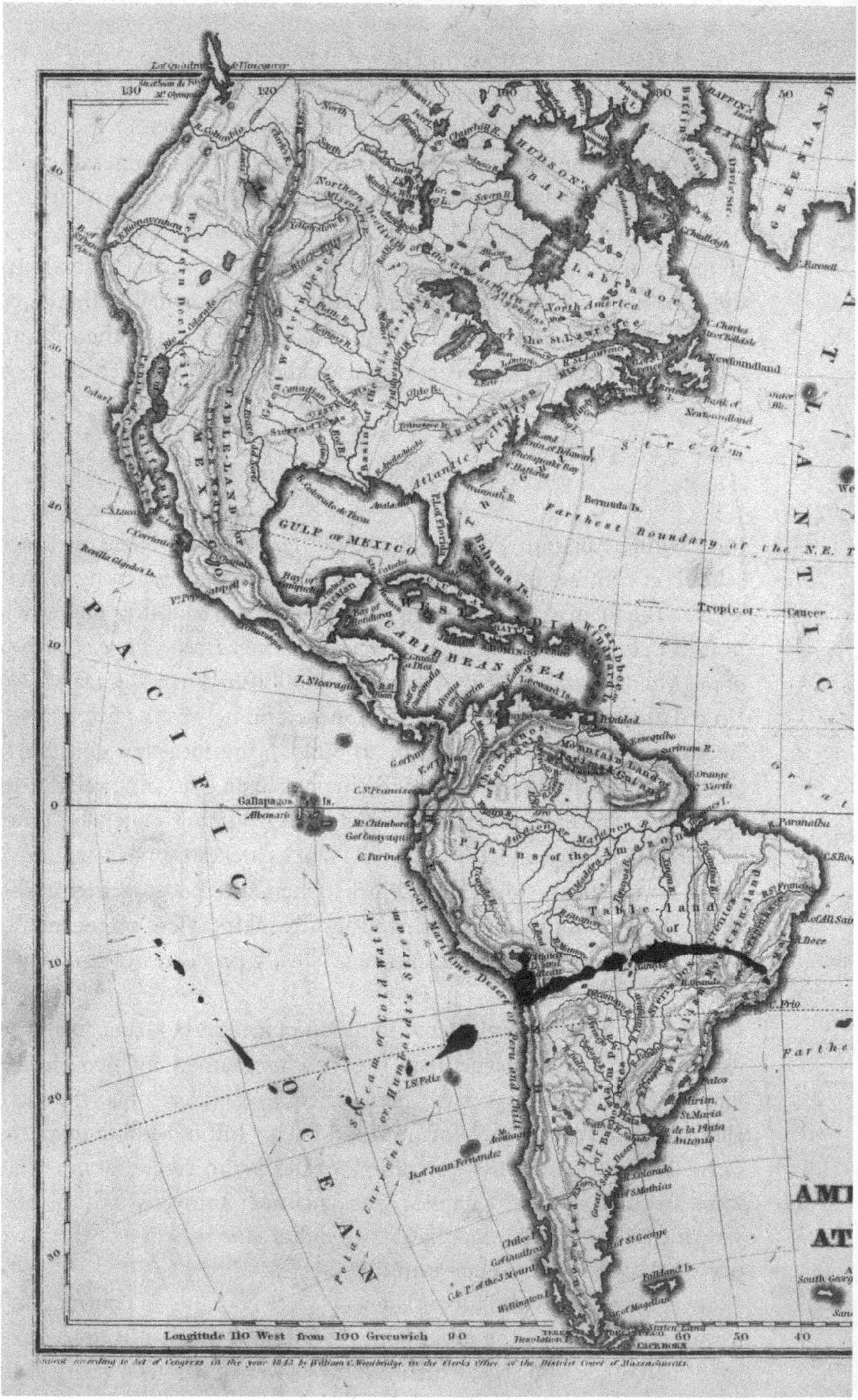

Figure 5.1 Physical map of America and Africa and the Atlantic Ocean, 1849, 24 by 28 inches (no scale shown). (Woodbridge, 1838, courtesy of David Rumsey Map Collection, David Rumsey Map Center, Stanford Libraries, https://purl.stanford.edu/zh174hj7174)

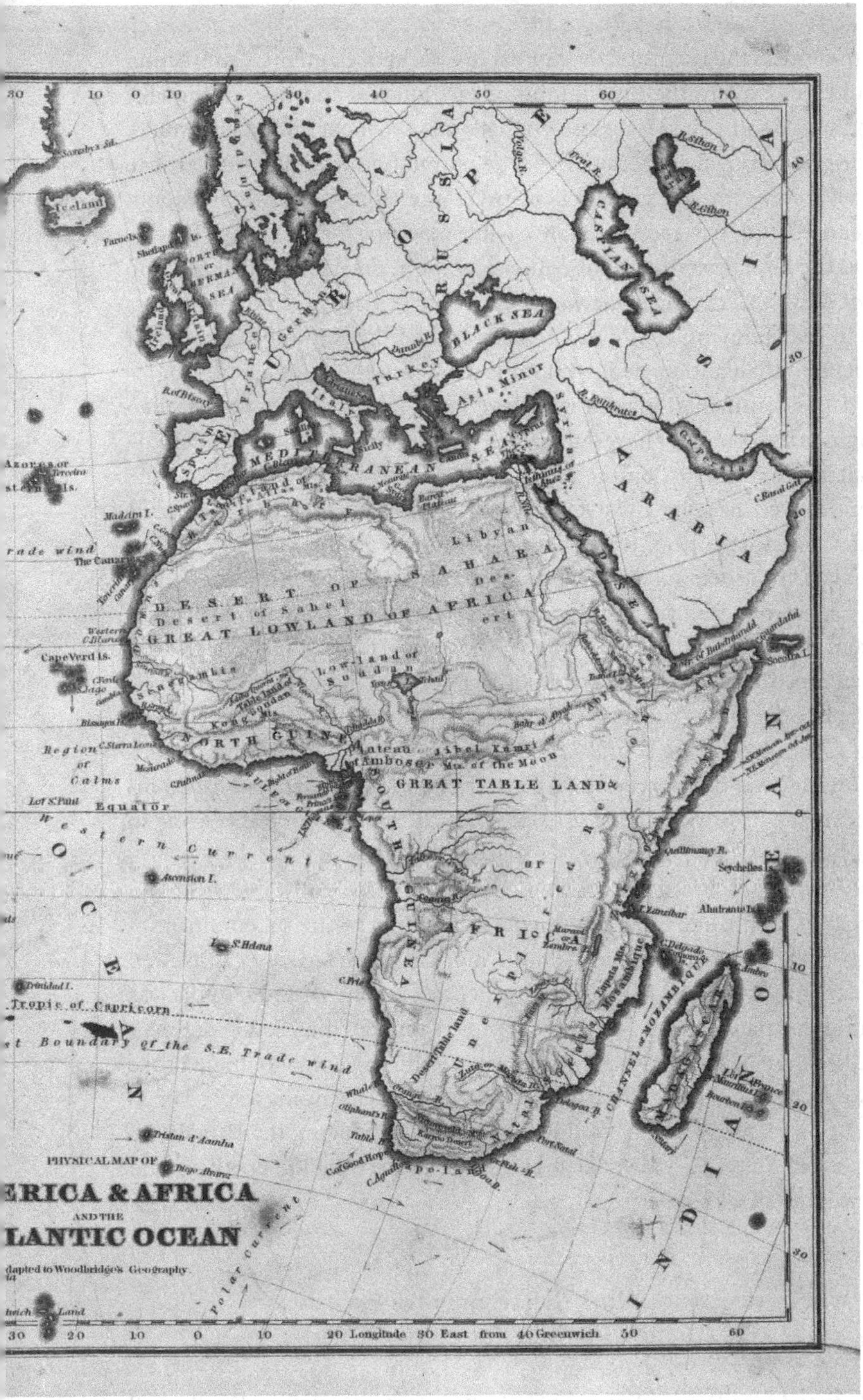
PHYSICAL MAP OF
ERICA & AFRICA
AND THE
LANTIC OCEAN
dapted to Woodbridge's Geography
Iceland
RUSSIA
EUROPE
ASIA
BLACK SEA
CASPIAN SEA
Turkey
Asia Minor
Italy
Sicily
MEDITERRANEAN SEA
ARABIA
Libyan
DESERT OF SAHARA
Desert of Sahel
GREAT LOWLAND OF AFRICA
Low-land of Soudan
Senegambia
NORTH GUINEA
Mts. of the Moon
GREAT TABLE LAND
SOUTH GUINEA
AFRICA
Madeira I.
The Canaries
Cape Verd Is.
Region of Calms
Equator
Western Current
Ascension I.
St. Helena
Tropic of Capricorn
Boundary of the S.E. Trade wind
Tristan d'Acunha
Polar Current
Seychelles Is.
Zanzibar
Mozambique
CHANNEL of MOZAMBIQUE
Port Natal
C. of Good Hope
Socotra I.
INDIAN OCEAN
OCEAN
20 Longitude 30 East from 40 Greenwich

heat from the core of the Earth and the radioactive decay of elements. Contraction theory could not explain the distribution of earthquakes and non-earthquake zones. Besides, scientists had long known that a large mass of mountains would, due to their gravitational attraction, deflect a plumb bob from its true vertical. Indeed, Welsh surveyor and geographer Sir George Everest, for whom Mount Everest is named, did such reconnaissance while mapping for the British Empire in India and the Himalaya Mountains. His calculations determined that a plumb bob did not deflect as much as one would expect based on the amount of mass contained in his eponymous mountain. He reasoned that Mount Everest must have less density than anticipated, with roots deep into the crust supporting its mass. But if the continental masses were less dense than the ocean crust, then the mountain crust could not have sunk to become ocean crust, and vice versa—another nail in the coffin for contraction theory.

Suess, moreover, became most famous for proposing that the continents we see today must have been joined in one continent in the distant past. His "Gondwanaland" was named for the Gondwana strata of Upper Carboniferous to Lower Cretaceous sediments in India. The continuous sequence of rocks includes coal deposits and a plant of note—*Glossopteris*, the genus designation of an ancient seed fern—which turned out to figure prominently in the development of plate tectonics theory. *Glossopteridales*, the extinct order of plants containing *Glossopteris*, are found in coal swamps from the Carboniferous Period through the end of the Permian Period (figure 5.2). Suess published his findings in his major work starting in 1883, with the final volume appearing in 1909,[7] which is also when the English translation was released.[8]

The distribution of *Glossopteris* was not only in the Gondwana rocks but spread across at least three other continents in addition to the subcontinent of India—Australia, South Africa, and South America—as noted by E. A. Newell Arber, a professor at Cambridge specializing in paleobotany, in 1905.[9] India was part of Gondwanaland in the late Paleozoic and separated as a subcontinent during the breakup of Pangea starting about 130 million years ago. India broke off from Gondwanaland and rafted northeast toward Asia, gaining a speed of 20 centimeters per year, and finally collided with Asia, producing the Himalayas. Arber referenced Suess's Gondwanaland in his map, showing the northern and southern varieties of *Glossopteris*.

Antarctic Exploration, the Terra Nova Expedition, and Plate Tectonics

Notably, Antarctic explorers played a role in the story of determining the full range of *Glossopteris*, confirming the supercontinent of Pangea and, ultimately,

Figure 5.2 *Glossopteris* fossil seed fern leaves from the Permian of Australia (field of view 12.4 centimeters across). (James St. John, 2014)

"continental drift," and then plate tectonics. Pangea (also spelled Pangaea, Greek for "all land") was the supercontinent at the end of the Permian Period. The discovery occurred on Robert Falcon Scott's ill-fated "Terra Nova" expedition. As Scott and his party traveled back from the South Pole in 1912, Edward Wilson (known by the team as Uncle Bill and referred to by Scott as Bill), Scott's chief scientist and geologist, found *Glossopteris* fossils, of the southern type, near the Beardmore Glacier in coal seams interlayered with the Beacon Sandstone (figure 5.3).

Although their dogs and ponies had long since expired, and the team was on foot, snowshoeing and skiing while dragging "sledges," Wilson and his colleagues collected 17 kilograms (approximately 35 pounds) of fossils and samples from the formations (figure 5.4). Scott noted in his diary:

> I decided to camp and spend the rest of the day geologising. It has been extremely interesting. We found ourselves under perpendicular cliffs of Beacon sandstone, weathering rapidly and carrying veritable coal seams. From the last Wilson, with his sharp eyes, has picked several plant impressions, the last a piece of coal with beautifully traced leaves in layers, also some excellently preserved impressions of thick stems, showing cellular structure. In one place we

saw the cast of small waves on the sand. To-night Bill has got a specimen of limestone with archeo-cyathus—the trouble is one cannot imagine where the stone comes from; it is evidently rare, as few specimens occur in the moraine. There is a good deal of pure white quartz.[10]

Figure 5.3 Map of Robert F. Scott's British Antarctic expedition, Track Chart of Main Southern Journey, mapmaker unknown, circa 1930, National Library of Scotland (original map from Scott, 1913, p. 418)

Figure 5.4 Members of the Terra Nova expedition at the South Pole, on skis and towing a sled of provisions, January 1912. (Library of Congress, 1912; photo by Herbert G. Ponting)

The Beacon Sandstone, interlayered with coal and shale, is a relatively flat-bedded deposit formed under at times dry, semiarid conditions, and at other times in wet, swampy conditions. Sand grains in the Beacon Formation are well rounded and have ripple marks indicating wind-blown dune deposits. Weathering, cited in Scott's quote, is a phenomenon that occurs either physically with water or by chemical reactions causing the rock to start to disintegrate.

Scott and his team members perished not long after the discovery, in March 1912, after becoming exhausted and running out of food. A party from base camp at Terra Nova was sent out in a rescue attempt but had to turn back in late March because of severe weather. The searchers, no longer leading a rescue but now in a recovery effort, did not reach the camp until November 12, 1912, when they found the bodies of Scott and his colleagues, their diaries, and the weighty geological samples that Wilson had insisted they carry for the last fifty days of their journey. Of course, those specimens may have slowed the explorers down as they left the Beardmore Glacier. The searchers brought back to England the diaries, accounts, some gear, and the geologic specimens.

Among the geological samples at Scott's last camp, researchers found many pieces of fossil *Glossopteris*.[11] These particular specimens provided fundamental data for continental drift and were pivotal in the understanding of Gondwanaland. Climate and conditions of deposition of this Beacon Sandstone unit were vastly different from current Antarctic conditions, suggesting the movement of the continent. Indeed, the fossils confirmed that the seed ferns were found on the Indian subcontinent and the continents of Australia, South Africa, and South America but now also in Antarctica. This significant discovery, at the cost, arguably, of the explorers' lives, gave further strong support to the idea that the continents had indeed once been joined. Today, the expedition's materials can still be viewed at the Scott Institute of the British Museum.

In 1914, British paleobotanist Albert Charles Seward (1863–1941) published a geological report on the *Glossopteris* fossils, their discovery, and their importance.[12] He realized the implications of finding the ancient plants—the work of Suess on Gondwanaland was already widely known. Previously, there had been only possible impressions of fossil plants in Antarctica. British geologist Hartley T. Ferrar, for example, had collected some specimens in South Victoria Land on the first Scott expedition in 1901–1904, and in 1908, members of Sir Ernest Shackleton's expedition found traces of poorly preserved wood. However, none of the samples was identifiable by species or type. The Australian Antarctic Expedition of 1911–1914, led by Sir Douglas Mawson, collected specimens of carbonaceous shale (a rock high in carbon) with plant "markings" from near the Beardmore Glacier, but these samples, too, were indistinct and unidentifiable. Seward noted in the conclusion of his report:

> It is by a comparative study, in the light of recently acquired data, of the southern extremities of South America, South Africa, and Australia, and what remains of the continent within the Antarctic circle, that progress may be expected towards a better understanding of the geographical and geological problems of Antarctica. The heroic efforts of the Polar Party were not in vain. They have laid a solid foundation: their success raises hope for the future, and will stimulate their successors to provide material for the superstructure.[13]

Alfred Wegener, for one, responded to that call.

Alfred Wegener and Continental Drift

Wegener was a German meteorologist, climatologist, and oceanographer, born November 1, 1880. He resurrected the idea of continental drift, which he termed "continental displacement." He held that all the continents appeared to have fit together at one time and later drifted apart into their current configurations. Wegener's research brought scientific rigor to the examination of these ideas; he gave the hypothesis flesh on what had been bare bones.

Wegener received his PhD in astronomy in 1905 and also had a deep interest in meteorology and studies of climatic zones. He noticed that the mountains on the east and west sides of the Atlantic Ocean were close to the edges of the continents and seemed similar in origin and rock types to each other. He was inspired by reading the early research and mapping by Alexander du Toit, a South African geologist, and the writings of H. Keidel, director of the Geological Survey of Argentina. Wegener examined the shapes of the continents and the ocean and ultimately concluded that only continental displacement could explain the occurrence of the similar mountain belts on opposite sides of the ocean. His theory revealed how the continents mirrored each other in a geologic fit.

On January 6, 1912, Wegener presented a lecture at the general meeting of the Geological Society (Geologishe Vereinigung) of Frankfurt, outlining his initial ideas about continental displacement (later called continental drift) and some of his preliminary evidence, including geophysical data, information on Permian Period glaciation patterns, and polar wandering. His ideas were met with skepticism and dismissed by most scientists. Yet as Wegener was giving his talk in Frankfurt, Scott and his team of British Antarctic expeditioners were approaching the South Pole.

Wegener served in the German army during World War I, during which he was twice wounded and finally discharged after being diagnosed with a heart ailment.[14] In 1914, while recuperating in a military hospital from his injuries, he began refining his theory of continental displacement. Why had geologists found his theory objectionable? Of course, Wegener was primarily a meteorologist and an outsider to the field of geology. Accordingly, he determined to develop geologic evidence to support his theory.

In the first version of his theory, Wegener focused mainly on a geophysical argument for continental displacement. Although he had briefly mentioned fossil evidence in lectures, he mentioned no specific species.[15] Once back on his feet, he sought out his colleague, Hans Cloos, a German structural geologist, who guided him through geologic publications and taught him how geologists look at the world. Then, in 1915, he published his work, *The Origin of Continents and Oceans (Die Entstehung der Kontinente und Ozeane)*.

But geologists again criticized continental drift on the basis that Wegener did not provide enough convincing evidence of a geological nature to support his tenets. Though he cited du Toit's geological argument for drift, based on the similarity in ore deposits in South Africa and South America as his principal source, the geology community at large remained unconvinced. The theory of continental drift, in general, and Wegener, in particular, faced stiff opposition, and he was derided especially by American geologists.[16] "True" geologists viewed him not only as an outsider but also as an interloper. The fact that he was German, in a time of deep anti-German sentiment, may also have been a factor. In addition, his works were not translated into English for some time.

Still, the fact of the matter was that Wegener could not provide a robust, scientific explanation for how continental drift could occur. Other early-twentieth-century scientists bandying about proposals for how the continents could move proved no more satisfying:[17] one such hypothesis, called isostasy, concerned the vertical movement of continental blocks rising or sinking, such as occurs when the underside of an ice sheet melts due to its own weight and the land it covered rebounds in elevation. But if continents could move vertically, perhaps they could move horizontally as well, as Wegener stated, like icebergs floating in ocean water.

Wegener racked his fertile mind for data to prove his continental displacement theory. He used geophysical methods, along with his meteorological skills, to try to find evidence of the movements of the continents over geologic time. To do this, he employed myriad techniques, including those of geodesy (using astronomical data or, more recently, global positioning satellite, or GPS, data) to determine precise locations on the Earth's surface.[18] Wegener used the principles of geodesy to map the movement of plates, particularly in Greenland.

Besides being a meteorologist and climatologist, Wegener was an explorer during the great age of research and mapping of the Arctic and Antarctic regions. He had been on several expeditions to Greenland (figure 5.5), where more opportunities for exploration existed. British and Australian explorers claimed the lands in Antarctica in the race to the South Pole. Wegener believed Greenland was drifting westward based on his calculation of the position of the Earth's wandering magnetic poles through time, which track the movements of the landmasses. In addition, Wegener tracked the movement of Greenland based on the geographic positions of stations he set up, whose locations were determined from lunar measurements and the transits of various stars—geodetic data. The station locations were recorded first by chronometers and then over time with increasing accuracy with telegraphic-time and radio signals.[19] One problem was that the stations from which he made these readings were inaccurately located and sometimes kilometers off from their actual positions. In Greenland, Wegener and fellow scientists also made many measurements of ice thickness using seismometers and generated seismic waves to measure the extent of the ice.

Today we know that among Wegener's writings and thoughts were some seeds that would bloom into the theory of plate tectonics. Yet during his lifetime, critics, for the most part, seized on his more impractical ideas. For example, one

Figure 5.5 Alfred Wegener (left) and Rasmus Villumsen at Station Eismitte. This is their last photo; both died around November 16 on the way back to the coast. (Fritz Loewe, Johannes Georgi, Ernst Sorge, Alfred Lothar Wegener, Archive of Alfred Wegener Institute, 1930)

untenable hypothesis held by Wegener was that the tides caused by the attraction of the sun and the moon drove the movement of the continents. Another was that the centrifugal force from the rotation of the Earth caused the landmasses to move. In 1925, the American Association of Petroleum Geologists went so far as to organize a symposium devoted to criticisms of his work. By 1929, most other scientists had dismissed Wegener's theory of continental drift and his ideas. Wegener was viewed as an outsider whose concepts were unverifiable.

Wegener tragically died on an expedition in Greenland in 1930 at the age of fifty, before his work in the theory of continental drift was verified. He had traveled to explore Greenland and to study the Greenland ice cap by calculating its rate of drift. Later explorers found his grave marked by a pair of skis in the ice and determined that he, a heavy smoker, must have died of a heart attack while searching for food for his fellow researchers. His bereaved wife decided to leave his body buried under the ice of his beloved Greenland, but replace the skis with a cross to mark his final resting place.

If he had lived longer, Wegener probably would have been able to support his theory further as new evidence arose. But as it was, his theories would not be revisited, let alone proven, until the 1960s. Although recognition from his peers eluded him in life, his *The Origins of Continents and Oceans* (fourth edition published in 1929) nevertheless demonstrated his scrupulous adherence to a broad scientific method:

> Scientists still do not appear to understand sufficiently that all earth sciences must contribute evidence toward unveiling the state of our planet in earlier times, and that the truth of the matter can only be reached by combing all this evidence. . . . It is only by combing the information furnished by all the earth sciences that we can hope to determine "truth" here, that is to say, to find the picture that sets out all the known facts in the best arrangement and that therefore has the highest degree of probability. Further, we have to be prepared always for the possibility that each discovery, no matter what science furnishes it, may modify the conclusions we draw.[20]

Data for Continental Drift

Wegener was certainly right about one thing: the advancement of geology is inextricably linked to natural scientists of many different disciplines. Five lines of evidence bolster the theory of continental drift: the same vertebrate fossils found on far-flung continents; the remains of ancient seed trees, *Glossopteridales*, distributed on different landmasses; glacial grooves and the patterns of ice-sheet distributions; rocks and deposits of similar climatic zones that lined up across different landmasses; and mountain ranges on either side of the Atlantic Ocean that mirrored each other.

The first set of information supporting the drift of the continents came in the form of large swimming and terrestrial animals, which had evolved in the late Paleozoic and early Mesozoic Eras. Some of these creatures split off, through evolution, from their reptilian forebears to fill new ecological niches and roles. The ancestors of dinosaurs and mammals were among those branching out. *Lystrosaurus*, *Cynognathus*, and *Mesosaurus* were land or shallow freshwater-dwelling animals found in Antarctica, South America, Africa, India, and Australia, providing critical pieces of data substantiating the theory of continental drift.

The second line of data was from the distribution of *Glossopteridales*, identified widely across all of the southern continents. Fossils from the four species—three vertebrates and one plant—were found in rocks of the same age on different continents, bolstering continental drift theory.

Lystrosaurus was mostly a land-dwelling herbivore with several aquatic species that lived during the late Permian to early Triassic Periods (figure 5.6). American paleontologist and comparative anatomist Edward D. Cope (1840–1897) named this genus *Lystrosaurus* in 1870.[21] Though *Lystrosaurus* walked and carried its body off the ground, it had sprawling limbs, more like those of reptiles.[22] The largest of the species was 2.5 meters (8 feet), but most were slightly smaller than 1 meter (3.2 feet). The upper jaw of *Lystrosaurus* had tusks adapted for digging roots, and it boasted a horny beak designed for cutting vegetation. Geologists have uncovered *Lystrosaurus* fossils in Antarctica, India, and South Africa and some in China.

Cynognathus, named by British paleontologist Harry G. Seeley in 1895,[23] lived from the late Permian–Triassic Periods and was a land-dwelling early carnivore approximately 1 meter (3.2 feet) long, with a short, squat body and strong jaws for tearing meat (figure 5.7).[24] *Cynognathus* had a more upright manner of movement, with limbs more under its body, as evidenced by the bone structure of its legs. Paleontologists have found bones of *Cynognathus* in Antarctica, Argentina, Africa, and China.

Mesosaurus, first named by French paleontologist and entomologist François Louis Paul Gervais around 1864, were initially considered para-reptiles and later classified as reptiles. They lived in the early Permian Period in freshwater environments and were highly specialized, with diminutive, needle-like teeth, presumably eating small fish and invertebrates. Mesosaurs grew to be approximately 1 meter (3.2 feet) in length and were similar in shape to a giant lizard (figure 5.8). Their tails were designed for swimming, and they may have developed webbed feet. Fossil discoveries reported in 2012 in Brazil and Uruguay of adults with embryos found within suggest they may be the first-known animals to give live birth and possibly nurture their young.[25]

None of these vertebrates—*Lystrosaurus*, *Cynognathus*, or *Mesosaurus*—could have traversed the long distances in the open ocean between the continents if the landmasses were as far apart as they are today. Moreover, the

(a)

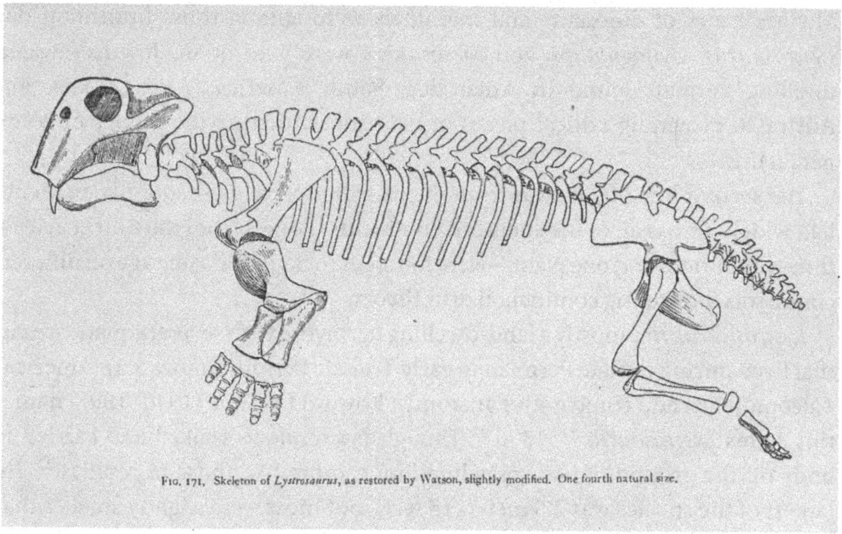
Fig. 171. Skeleton of *Lystrosaurus*, as restored by Watson, slightly modified. One fourth natural size.

(b)

Figure 5.6 (a) *Lystrosaurus georgi* skeleton. (Williston, 1925) (b) *Lystrosaurus* life drawing. (Nobu Tamura, 2016)

fossil finds line up in zones when the continents are placed back together into one supercontinent.

The three vertebrate species and *Glossopteris* often are illustrated together in the literature of plate tectonics and continental drift. The original figure was not strictly accurate, because the distributions of the three vertebrates has changed as additional fossils were found. Updated here (color insert 5.3), the stylized map shows distributions of the various fossils in zigzag patterns covering the subcontinent of India and the major southern continents: South Africa, Australia,

(a)

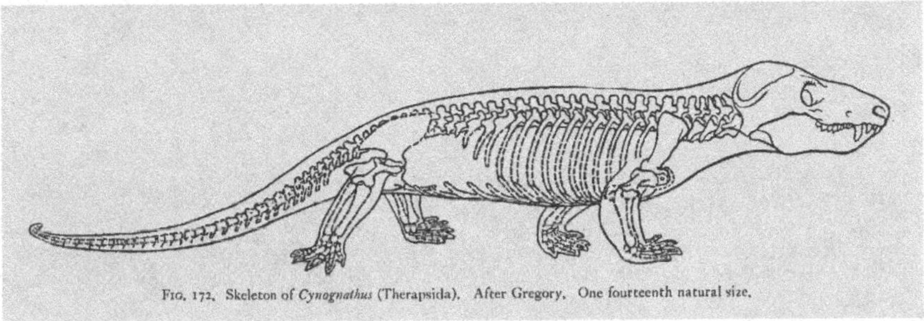
Fig. 172. Skeleton of Cynognathus (Therapsida). After Gregory. One fourteenth natural size.

(b)

Figure 5.7 (a) *Cynognathus* skeleton. (Williston, 1925) (b) *Cynognathus* life drawing. (Nobu Tamura, 2016)

South America, and Antarctica, based on work by Colbert published in 1973.[26] The vertebrate species lived from the Permian to the Triassic Periods and did not overlap in the patterns for the original ranges of the species, as illustrated. Still, these vertebrates and *Glossopteris* are critical pieces of evidence in support of continental drift.

A third type of data adding to the argument for continental drift consisted of information from ancient glaciers. Patterns produced by these glaciers supported the theory that landmasses amalgamated into the supercontinent of Gondwanaland. Wegener noted glacial striations in the southern continents of Gondwanaland. The glacial grooves in India, for example, suggest that ice sheets flowed away from the current coast, inland toward the north; striations in South America show that the glaciers traveled westward; rock grooves in South Africa outline glacial movement in all the compass points; and in Australia, grooves reveal that the ice sheets trended eastward. Once reconstructed as a supercontinent,

(a)

(b)

Figure 5.8 (a) *Mesosaurus tenuidens (Mesosaurus brasiliensis)* skeleton. (Karl Volkman, 2009) (b) *Mesosaurus* life drawing. (Nobu Tamura, 2007)

the striations fit together and make a coherent pattern, with a massive ice sheet centered on what now is Antarctica. Ice sheets and glaciers flowed out from the high mountains, at that time in Antarctica, down to lower elevations.

As a fourth line of reasoning for drift, Wegener noted that the paleoclimate, as indicated by rocks, showed swaths of dry, arid conditions and regions that were wet and swampy. And of course, each produced particular rock types. Deserts arose in the arid inland areas, leading to mega-dunes, ultimately captured in the rock strata, along with salt deposits that resulted when ancient seas evaporated as the climate became more arid. These salts turned into deposits of halite or gypsum over time. Rocks produced in the wetter areas created coal deposits interbedded with nearshore sandstones and clays. As a climatologist interested in ancient conditions, Wegener realized that the various environments aligned only when the continents were placed back together.

Fifth, Wegener examined the composition of the mountains on either side of the Atlantic Ocean, as geologists had done before. Du Toit and others had noted rock similarities in the mountain ranges of South America and western Africa—similarities that suggested to Wegener that these ranges were once contiguous. Thus, although five lines of evidence supported Wegener's theory of continental drift, the theory languished after Wegener's death, as few scientists were willing to pursue it.

Mechanisms for Continental Drift: Early Ideas

Arthur Holmes (1890–1965), a British geologist who studied geochronology and the origins of rocks, was one of those early few to support Wegener's theory of continental drift. In the 1920s, his studies of the Earth's outer layers led him to speculate that there may be slow-moving convection cells caused by the decay of radioactive elements in the mantle and that these could drive continental drift, creating new ocean floor and consuming old material. He wrote about his theory in his famous textbook published in 1944, *Principles of Physical Geology*.[27] The last chapter was devoted entirely to the topic of convection cells and continental drift. Holmes stated that his idea was speculative; he did not have any data to support it at that time, but some respected him as a thinker with broad views. Like those of Wegener, his ideas were ignored for more than thirty years, though he taught them to his students at the University of Edinburgh. In the end, Holmes's mechanism was found to be very close to the final version of how plates actually move.

Another proponent of continental drift was du Toit, whose previous work had a profound influence on Wegener's thoughts. In 1937, du Toit published an exposition of geology on either side of the Atlantic Ocean in *Our Wandering*

Continents.[28] A proponent of Wegener and his ideas, du Toit relied upon correlation maps of the geology of South America and western Africa to illustrate that these continents had once been joined and then drifted far apart, like pieces of a newspaper torn but from the same page. Still, this information had little impact on a still-resistant geology community, and European and American geologists generally ignored the South African du Toit's findings.

Indeed, the evidence for continental drift would not quite be revealed until the advent of technological developments following World War II.

Post–World War II Exploration of the Ocean Floor: A Unified Theory of Plate Tectonics

Before World War II, the geography of the ocean floor was essentially unknown. Scientists theorized that it was mainly a flat, featureless plain. Geologists did not give much thought to how old it was or how it formed. Early surveyors had documented intimations of oceanic ridge structures in the mid-1800s when mapping the Atlantic for laying the transatlantic telegraph cables. They collected bottom samples via an instrument called the Brooks sounding machine, accompanied with depth soundings (dropping lines down as far as possible in the ocean to determine depth). The first measurements at this time of what later became known as a mid-oceanic ridge were made in 1853 by U.S. Navy Lieutenant Otway Berryman, who located a ridge fragment north of the Azores when mapping the ocean floor.[29] Some oceanographers interpreted the bathymetric data to include a "telegraphic plateau," an elevated flat area, located in the mid-Atlantic Ocean, but these early measurements recording depth had inaccuracies, and shortly after oceanographers collected the data, disputes abounded about the nature of the plateau and even whether it existed. Initial theories held that the ocean floors rose to create continents and the continents sank to form oceans, through contraction theory, but none of these was borne out with the new information on the ridge and rift that became available.

Early in the war, the German navy gained the upper hand on the seas, resulting from its development of a perfected, efficient killing machine: the U-boat. Because of German submarine warfare, researchers developed sonar (echo) sounding to locate submarines and magnetometers to study the ocean basins. Sonar operates by sending down waves from a ship to echo-locate the features in the water column of the ocean (such as subs) or, of course, the bottom of the ocean. When scientists first applied sonar to examine the ocean floor, it was found to have structures that included mountain chains with peaks taller than Mount Everest, deep-sea trenches with depths as great as 11,000 meters (36,070 feet), and, indeed, some flat areas called abyssal plains. Oceanographers measure relief in the ocean by bathymetry, similar to how geographers and geologists record topography on the land surface. Oceanographers started to make bathymetric maps for all ocean floors and found that vast mountain chains ran down the middle of the Atlantic Ocean between North and South America and Europe and Africa, in addition to other areas.

Several scientists, including Harry Hammond Hess (1906–1969), a petrology professor at Princeton University, and William Maurice Ewing, a geophysicist at Lehigh and later Columbia University, applied remote sensing to figure out the structure and nature of the ocean floor. Hess had enlisted in the U.S. Navy before World War II and used echo sounding in the Pacific to identify flat-topped mountain structures under the ocean which he named "guyots" for one of his professors. He theorized that these were ancient volcanoes that sank below the ocean's surface and were flattened on top by waves. He later went on to bring together the many lines of evidence in support of seafloor spreading (the way in which new ocean floor is created along mid-oceanic ridges, causing the continents to move farther away from each other), but first, extensive seafloor data had to be collected; moreover, in some cases, new instruments had to be developed to record the information.

Prior to World War II, Ewing and his students were tasked by the U.S. Coastal and Geodetic Survey and Princeton University to develop not only the new field of geophysics on land, using dynamite for seismic refraction studies, but also geophysics applied to the continental shelf and ocean basins. He was well prepared to do so, having researched remote sensing applications for oil and coal exploration. But because no equipment existed to undertake such endeavors, Ewing designed and made most of the instruments himself. In 1935, using his own remote sensing techniques, Ewing figured out that the continental shelf, including the continental slope and rise, was made up of sediments 4,000 meters (13,123 feet) thick. Additionally, the continental basement underlies the sands, muds, and gravels of the shelf and slopes evenly down to the ocean floor.

Ewing was the first director of the Lamont Geological Observatory,[30] now the Lamont-Doherty Earth Observatory, which was founded in 1949. The Lamont family estate, Torrey Cliff in Palisades north of New York City, hosts the observatory. It currently houses the most extensive collection of deep-sea cores and ocean sediments in the world, along with ocean bathymetric profiles, magnetic surveys and profiles, and heat-flow measurements. Ewing earned his PhD from Columbia University in physics and brought physics to bear on challenges faced in geology, notably how to measure and record data from subsurface features or structures well beneath the land surface where the eye cannot see.

Lamont, at the time, was the most recently founded of the U.S. oceanographic research facilities. The Scripps Institution of Oceanography was chartered in La Jolla, California, in 1903 as the Marine Biological Association of San Diego. In

1912, it became part of the University of California with a name change in recognition of the benefactors who provided much of the funding. Woods Hole Oceanographic Institute was founded in 1930 as a summer endeavor but became year-round when research and defense needs increased in relation to impending war efforts.

In the early 1950s, several researchers, assistants, and graduate students of Ewing at Lamont provided critical information on the structure that ultimately would be named the mid-oceanic ridge. One was geologist and cartographer Marie Tharp (1920–2006), hired as a research assistant to Ewing. Tharp's supervisors consigned her to the lab, as, predictably, they did not allow women to participate in the work on the research vessels collecting data. She mapped six long east-west traverses of the North Atlantic based on the seismic and sonar data provided by the scientists on those ships. Tharp realized that a rift valley, a steep-sided valley produced as extensional forces pull apart crust, was located in the center of a massive north-south trending mountain range in the middle of the North Atlantic.

Arguments and doubt followed. Tharp replotted all of the data from the soundings and arrived at the same results. American geologist Bruce C. Heezen started mapping the epicenters of earthquakes in the North Atlantic shortly after and found that the earthquake locations lined up with Tharp's structure in the middle of the ocean, centered on the rift. This development meant that the fissure was active because mid-ocean ridge earthquakes accompany the creation of new seafloor as magma rises. In the 1950s, Heezen and Ewing were spurred by the linkage of the underwater temblors to the ridge in the North Atlantic to examine more accurate ways to plot earthquake data and also to extend the study to other ocean basins. Along with some of his research assistants, notably Heezen, Ewing published a paper in 1956 positing the existence of a worldwide mid-oceanic ridge and its associated earthquakes. Heretofore, the ridge had been found only in the North Atlantic. Other scientists immediately were skeptical. They tried to prove the theory wrong, but the presence of the ridge was verified where Heezen predicted it would be. The ridge and its rift valley running down its middle led ultimately to the theory of seafloor spreading.

Through his sonar soundings over the ocean basins of the world, Ewing realized that mid-ocean ridges are not separate systems but a nearly continuous underwater range and actually make up a global ridge of more than 65,000 kilometers (40,390 miles) in length, by far the most extensive mountain range on the Earth. The ridge is mostly underwater in the ocean but emerges in places like Iceland and East Africa (color insert 5.4).[31] It wraps around the globe like seams on a baseball and is a vital structure for understanding plate tectonics.

Ewing also discovered near the ridges that little sediment had accumulated, indicating new ocean floor being produced there. These facts helped scientists start to think about the age of the ocean basins and processes along the ridges.

By the early 1950s, geologists and other scientists had discovered the majority of the pieces leading to the theory of plate tectonics. However, bits of research were scattered in different places, published by a variety of authors in disparate subfields of geology and oceanography and at various institutions. Scientists did not know how they fit together or that they would be critical in developing the new theory of plate tectonics. One area of such research was heat flow from the continents and ocean basins.

Geologists had long recognized that the continents consisted of lighter, silica-rich rocks, like granites and rhyolites, and that those rocks had more associated radioactive elements that could decay, producing higher heat-flow measurements. They realized, too, that compared to the continents, the ocean basins were made up of a darker, denser, and silica-poor rock: basalt. Accordingly, heat-flow calculations should differ between masses of different compositions—what Ewing would refer to as a "brutal" fact, which he defined as one that appears contradictory but needs to be addressed to shape the progress of science.

Much of the heat-flow research was conducted at Scripps by Roger Revelle (1909–1991), along with Sir Edward Crisp Bullard (1907–1980), a geophysicist from Cambridge University, and their student Art Maxwell. Before their study of heat-flow measurements in the oceans, Bullard had researched continental heat flow. It was a great surprise, then, when, as described in a paper published in 1956, they found that heat-flow measurements in the ocean basins were of the same magnitude as those over landmasses.[32] Geophysicists theorized that the continents should have higher heat flow based on the types of rocks present, the light-grained igneous rocks, so they were surprised at the findings. The much-higher-than-expected heat reading over oceanic crust indicates the upwelling of the hot asthenosphere as it produced magma at the mid-oceanic ridges.

Another of Revelle's students, Richard P. Von Herzen (1930–2106), subsequently with UNESCO and Woods Hole Oceanographic Institution, compared ocean-floor heat-flow studies from the mid-oceanic ridge at the East Pacific Rise and plotted it in 1959 on a map along with Bullard's data, reinforcing the findings that ocean-floor and continental heat flows were virtually identical.[33] Further, Von Herzen's later work in 1963, recording heat flow from the mid-oceanic ridge at the East Pacific Rise at levels five times higher than those of the surrounding ocean floor, provided a vital clue to the nature of the mid-oceanic ridges, as the asthenosphere rose up under the ridge.[34]

The first author to put together these disparate facts and propose seafloor spreading was Hess, who in 1962 published "History of Ocean Basins," in which he laid out the arguments for spreading of the ocean floor from the creation of new ocean floor at the mid-oceanic ridges.[35] He attributed the idea of spreading in the seafloor to the earlier work of Holmes, calling it an essay on "geopoetry," a nod to what he thought was an outlandish idea and to capture his audience's imagination.

Hess proposed that convection currents in the upper mantle are driven by heat from radioactive decay in the Earth's mantle and core generated by hot material from deep in the Earth. As the magma surfaces and cools, it sinks, much like the action of a lava lamp. He thought these convection cells could be accountable for spreading of the ocean floor from the mid-oceanic ridges in a 200- to 300-million-year cycle with the continental portions of the continents riding passively along. Hess further discussed the relatively young age of the ocean basins compared to the age of the continents. He also noted that there is little sediment in parts of the ocean as related to seafloor spreading because of the destruction of ocean crust, through what came to be known as tectonic processes. He considered mid-oceanic ridges to be ephemeral features, just like the ocean floor, because they, too, can get recycled and destroyed over geologic time. Hess also noted "fault scarps" and displacement along the faults on the ocean floor.

Robert S. Dietz (1914–1995), an oceanographer and geophysicist with the Scripps Institution, often is mentioned in the same breath with Hess when it comes to codifying the theory of seafloor spreading. However, these two never published together, and Dietz's 1961 note in *Nature*[36] was released after Hess's concept of the seafloor and its origins appeared as a preprint in 1960. Hess was troubled by the similarity of Dietz's note to his idea, as well as the ordering of the topics following the structure of his paper.[37] In a 1968 article, Dietz conceded that Hess was the first to come up with the idea and should be considered the primary contributor to the theory of seafloor spreading.[38]

Dietz's views on the fracture zones perpendicular to the mid-oceanic ridges, however, did vary from Hess's in that he thought they were the result of uneven convection, whereas Hess did not think they related to the ridges. In time, neither was proved correct, but that is a story to come. Dietz suggested that the lithosphere, not just the crust as proposed by Hess, was riding on and being moved by the plastic asthenosphere. In the end, Hess's theory of seafloor spreading was supported and further defined by the research of Dietz.

One of Dietz's colleagues at Scripps was Henry William Menard (1920–1986). Together they published five papers on the Pacific Ocean floor and ocean processes, including information on submarine escarpments and the impacts of turbidity currents. In 1955, Menard wrote an article on large fracture zones that were found perpendicular to the mid-oceanic ridges in the Pacific and were associated with them.[39] He identified four of the fracture bands spanning up to 1,000 meters (3,300 miles) in length and parallel to one another. These results also laid the groundwork for one of the linchpins of plate tectonics later in the 1960s.

The theory of spreading of the seafloor was now in place, and the overarching theory of plate tectonics was well on its way to being formulated. The former has been called the most critical development in geology since Hutton's uniformitarianism emerged as a geologic principle and Darwin published the *On the Origin of Species*. Now what was needed was further confirmation. Not too many years were to follow before such confirmation came, in an unexpected way, in the form of bar strips, like zebra stripes, found on the ocean floor emanating from and parallel to the mid-oceanic ridges.

In 1963, Fred Vine and his professor Drummond H. Matthews, British geologists at Cambridge, published in *Nature* the results of a magnetic survey made the year before in the Indian Ocean along the Carlsberg Mid-oceanic Ridge.[40] The study showed an area of low magnetic values parallel to the ridge and in the center, corresponding to the deepest part of the rift valley. As magma cools into rock, iron-rich elements within the material align to the Earth's current magnetic polarity. Parallel to the ridge, the results of the magnetic survey displayed alternating areas of normal magnetism and reverse magnetism leading away, symmetrically, on either side of the ridge axis (figure 5.9). The pattern develops over time (a in the figure) as the magma emerges from the mid-oceanic ridge, creating new ocean floor as it cools (c in the figure). Vine and Matthews postulated that there must have been changes or reversals in the magnetic field of the Earth to account for what they saw in their data.

Another geologist researcher, Lawrence Morley (1920–2013) of the Geological Survey of Canada, independently proposed that the zebra-striped pattern, as

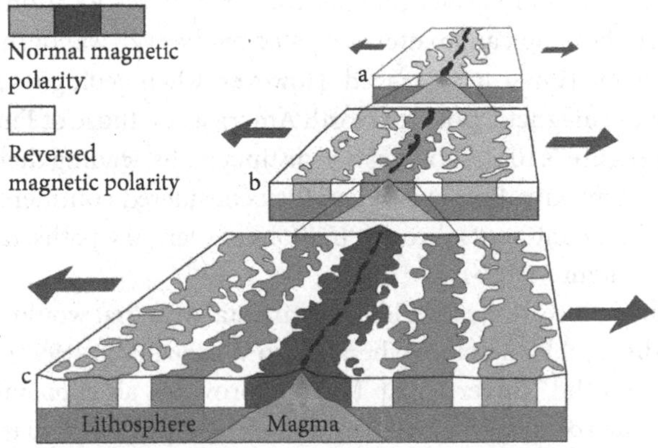

Figure 5.9 Graphic model of magnetic stripes on the ocean floor at the mid-oceanic Ridge. (USGS, 1999)

shown by magnetic reversals on the seafloor, was related to the spreading and widening of the ocean basins at the mid-oceanic ridge. He based his work on U.S. Navy research done by two Scripps oceanographers, Ronald Mason and Arthur Raff, working in the Pacific Ocean along the U.S.-Canadian border in what is known as the Juan de Fuca Ridge. Their work generated "barcode"-striped patterns spreading out from the ridge.

Mason was unsure what the patterns represented at the time, but Morley examined the data and submitted to *Nature* in February 1961 an article on the theory of magnetic imprinting and seafloor spreading based on a qualitative interpretation of the data. He was unable to include the magnetic map from Mason's earlier article from 1958,[41] later refined by Mason and Raff in 1961, because the data and the chart, at that time, were classified information. Accordingly, the journal rejected the paper. In addition, Morley did do a quantitative analysis, and without the map, the reviewers did not support its publication. He submitted it to another journal with the same discouraging results. Still, Morley's work was an early confirmation of the seafloor spreading theory, sometimes referred to as the Vine-Matthews-Morley hypothesis, the first scientific test of seafloor spreading.

Although sets of zebra stripes showing changes in the magnetic signatures of the seafloor rocks are evidence in support of spreading at mid-ocean ridges and continental drift, geologists use other types of information as well to confirm the theory of plate tectonics.

For example, those who have used a compass may know that there's a difference between geographic North—the latitude of 90 degrees, the axis upon which the Earth rotates—and magnetic North, the location where the Earth's magnetic field points directly downward. As a result of the spin of the nickel-iron core in the Earth's interior, the Earth's magnetic field changes over time; there's always been variation in the location of the planet's magnetic poles. And, just as with the zebra-striped ocean-floor basalts, igneous rocks record the location of the magnetic field when they cool, and patterns created by the ever-wandering magnetic poles over geologic time can be traced. However, when geologists first plotted the changes in the magnetic poles of North America and those of Eurasia,[42] they did not match (figure 5.10a), showing two distinct paths, leading them to scratch their heads in perplexity. But later, when they considered continental drift and the locations of the continents through millions of years, the paths, rather miraculously, aligned (figure 5.10b).

In 1965, a final piece of the puzzle of continental drift that would lead to plate tectonics as a theory fell into place when J. Tuzo Wilson (1908–1993), a Canadian geophysicist from the University of Toronto, provided an explanation for the large seafloor faults discussed by Menard, which exist perpendicular to the mid-oceanic ridges and which he named transform faults.[43] These faults offset the mid-oceanic ridges and also offset the zebra-striped patterns on the ocean floor,

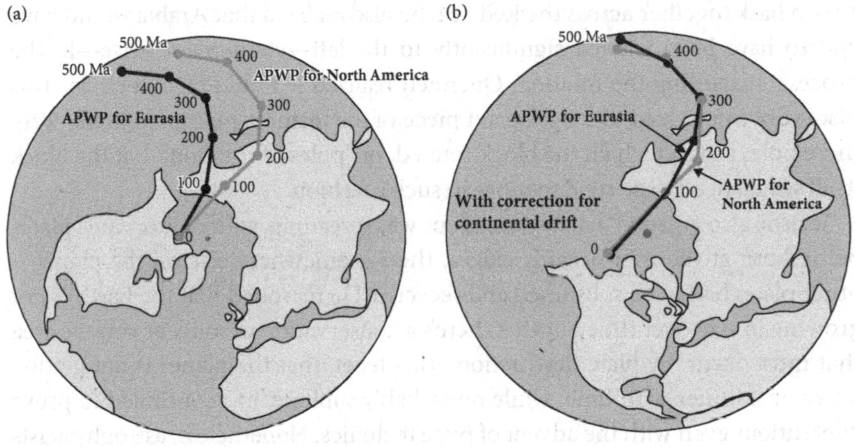

Figure 5.10 Polar wandering over 500 million years and continental drift, (a) before accounting for movement of continents, (b) accounting for continental drift. (APWP stands for Apparent Polar Wander Path). (Modified by the author based on Runcorn, 1959)

so Wilson theorized they must have formed after or concurrently with the ridge. Before 1961, Wilson had been skeptical of continental drift and held a view of the rigid Earth. His thoughts evolved, however, as the data and evidence mounted, and he became a leading proponent of and contributor to the theory of plate tectonics. Even after 1961, as he accepted the idea of drift, he was not convinced that it occurred before the Mesozoic. Wegener, for one, had held that continental drift only arose in Cenozoic time.

Wilson is also known as the first to use the term "plates" and also to realize that volcanic islands like the Hawaii-Emperor Seamount chain formed when the oceanic plate on which they rode moved over a "hotspot" rising from the mantle. Two years before his paper on transform faults, he published research on hotspots and the origin of the Hawaiian Islands. He based his work examining the linear volcanic chain on the geology of the ocean floor, geophysics, magnetism, and heat-flow measurements. A number of journals rejected his paper as too controversial, but it ultimately was published in 1963 in the *Canadian Journal of Physics* and became a seminal paper in the field of plate tectonics.[44]

Though Wilson named transform faults, other geologists had described the structures before. The "father" of these faults is acknowledged to be New Zealand geologist Albert M. Quennell (1906–1985). In 1956, Quennell gave a paper at the International Geological Conference in Mexico and published the results two years later.[45] In the paper, Quennell described the opening of the Red Sea through the Dead Sea fault—a large, land-based strike-slip fault extending to the north of the Red Sea—which was later conformed. When piecing Arabia and

Africa back together across the Red Sea, he also realized that Arabia would have had to have been rotated significantly to the left—counterclockwise—in the process. Extending the rotation, Quennell realized it would form a circle. This discovery would lead to a significant piece of the tectonic puzzle: not only were there poles around which the block rotated, or "poles of rotation," but the block itself would need to be rigid to move in such a fashion.

Wilson also grasped that if the seafloor was spreading, with new oceanic plates being born at the mid-oceanic ridges, then—somewhere else on the planet—other plates had to be subsumed and recycled. He reasoned that the Earth is not growing in size over time but that there's a conservation of mass or surface area that must occur, by plate destruction. This tenet, that the planet is not getting fatter or skinnier with time, while remarkably sublime, has continued to prove contentious even with the advent of plate tectonics. Nonetheless, astrophysicists have shown that the diameter and radius of the Earth have not changed in any statistically significant way since measurements have been made by NASA.

Wilson published again in 1968, detailing his thoughts on how repeated episodes of plate tectonics could occur and lead to the formation of different features and rock types relative to which phase a region is.[46] This process of the repeated opening and closing of ocean basins through the geologic record was named the Wilson cycle after its founder. The sequence operates on an approximately 500-million-year cycle.

In 1965, Cambridge's Bullard, along with two of his graduate students, James E. Everett and Alan G. Smith, circled back to the problem of how the continents fit together, where Wegener had begun, but with a new paradigm and novel methods, and wrote their famous paper "The Fit of the Continents around the Atlantic."[47] These authors relied on early computer modeling to match the edges of the reconstructed world—one in which all the continents were fitted back together—as in the last time the Earth witnessed Pangea. They argued that the coastlines of the continents did not create the best fit because a small rise or fall in sea level dramatically impacts the location of the coastlines. Instead, the authors found that the ideal accord between the continents exists at a depth along the continental shelf of 500 fathoms (900 meters, or 3000 feet). The continental shelf is the "real edge of the continent," and other factors, such as river erosion, have much less influence below sea level. Indeed, Wegener had used the coastlines in his analysis and noted that the continental shelf might provide a better match in some areas, but he did not pursue that line of thought.

A critical aspect of the Bullard paper is the use of Euler's theorem, a mathematical formula describing the displacement of a sphere or the approximation of where the sphere would be if it were rotated based on a starting point and a slope (a rising or falling surface). Bullard realized from Quennell's work that because the continental blocks described circles as they moved, this theorem could apply to the problem at hand—that is, matching the continents back together—in a brilliant leap. The theorem is applicable because the Earth is a sphere, not a flat surface. In sum, he stated rotation around a sphere's axis describes how a rigid plate moves on the sphere's surface."[48] Every rotation has an axis. Bullard and his colleagues measured the fit produced by using Euler's theorem, because while the southern continents came together reasonably effortlessly, the northern ones were more challenging to piece together as a result of complex plate movements in the north. They employed a least-squares fit analysis run on early computers by coauthors Everett and Smith. These first machines were essential for carrying out the analysis. Several areas defied a best-fit solution, including the area off of southwestern Ireland and mid-ocean ridge segments along with Iceland and the Faroe Islands.

The 1970s witnessed the amalgamation of the data and information from the prior decades into a sound theory. Still, doubters of plate tectonics thought the proposed mechanism was not feasible; they were called "fixists." The "mobilists" were in the other camp as proponents of continental drift. As with other times and other theories as they emerged, divisions in thought led to considerable discussion, debate, and sometimes acrimony. Before 1970, it was considered a travesty to be a mobilist; after that, as the data and reports poured forth, it was thought backward-looking to be a fixist. The standing of departments in colleges and universities rose and fell depending on which side of the argument their faculty supported.[49] The reputations of academics rose or fell as well, depending on which side of the argument they adopted. Some geologists were concerned that accepting plate tectonics would mean the "end" of the field of geology because it threatened the status quo. The opposite has turned out to be true, of course, as the theory has opened up new vistas of research, exploration, and study. Ewing, the great explorer and researcher of the ocean floors, himself held to his fixist views into the 1970s. He finally came to accept the emergence and development of plate tectonics—of which his research was a linchpin.

The research by Bullard and his colleagues provided the bridge from continental drift to a theory of plate tectonics and later was extended to the planetary scale. Investigations to more fully understand the finer details of the theory continue to this day. In the 1990s, work by geophysicist, Seiya Uyeda professor at Tokai University in Japan in 1929 shed light on the role of deep-sea trenches that had been thought to be passive structures, simply where two plates came together without active dynamics of the trench itself. Other Japanese researchers, in particular Akiho Miyashiro (1920–2008), a metamorphic petrologist at SUNY Albany, sorted out how various types of metamorphic rocks form in different tectonic settings related to subduction zones (places where one plate dives under another, recycling and destroying some of the plate material and often resulting in ocean basins closing) and island arcs (long chains of volcanic islands caused by magma rising up from convergent plate margins). In 2000, Richard G. Gordon (1953–), a geophysicist at Rice University, described diffuse plate boundary zones.[50] These zones differ from the more traditional plate boundaries in width, amount of strain, and movement per year. An example of a diffuse plate zone is the Basin and Range Province of the western United States.

Geologists verified the Wilson cycle in the Precambrian rocks of the Grenville mountain-building period, proving that, indeed, plate tectonics had occurred before the Mesozoic Era. This cycle represents a culmination of global plate tectonics and illustrates how the different plate boundaries fit together, how continents collide, how ocean basins open, and how material subducts down trenches.

The Wilson cycle process follows two stages: opening and closing (declining) of an ocean basin. During the opening part of the cycle, a hotspot may emerge in the middle of a continent where the crust is stretched and starts to split the landmass apart, creating a rift, such as the East African Rift Valley. Wilson called this the embryonic stage. This phase is followed by basins opening up, as with the Red Sea. Next is the full development of the ocean in Wilson's mature stage, as seen with the Atlantic Ocean basin. At some point in time, the forces reverse as they can no longer undergo extension given the fixed size of the Earth, and the declining phase begins. A subduction zone forms at any location in the ocean basin, but the most straightforward example would be along one of the continental edges. Ocean plates get subducted into a trench, and the basin shrinks in size, as with the Pacific Ocean. The ocean continues to close in the terminal phase of the Wilson cycle, with compression and uplift, as is found in the Mediterranean Sea. During the terminal stage of the cycle, mountains can form at the beginning of collision, with the subduction zone acting as a ramp for materials to slide up the edge of the other plate. Volcanoes appear as the subducted material of the downgoing plate melts. Metamorphic rocks are commonly found in association with this stage as well. The final stage is a relict scar from the continents colliding, such as is seen along the Indus Line of the Himalayan Mountains formed when the subcontinent of India smashed into Asia starting 50 million years ago.

Research in the 1970s by Donald Forsyth of Lamont-Doherty Labs and Uyeda illustrated the finer details of plate tectonics in terms of two new ideas: ridge push and slab pull.[51] Ridge push is generated at the mid-oceanic ridges as a force that creates an outward push of the newly formed oceanic crust from the ridge. For many years, deep-sea trenches were theorized to be passive structures merely absorbing and recycling the crust that went down them, but their research pointed to a more active role in the subduction zones, with pulling on the oceanic crust from subduction zones, termed a slab pull. Both ridge push and slab pull appear to be critical drivers of seafloor spreading through the forces of gravity, in addition to crust creation at the mid-ocean ridges.

Direct evidence for the movement of the tectonic plates had to wait until the 1990s and the advent of GPS data, which coincidentally, and perhaps ironically, are dependent on geodesy to work—the type of confirmation Wegener had been seeking. By the close of the twentieth century, Wegener's theories had been proven: the continents, indeed, do drift upon a mechanism of convection currents deep in the asthenosphere. If the Earth is a song, plate tectonics is its most elegant melody.

6
Plate Tectonics
Oceans, Continents, Plates, and How They Interact

Plate tectonics, the second of the foundational principles in geology, is the grand unifying theory of the Earth that explains multiple events and processes on the planet, including the location of earthquakes and volcanoes, why particular rocks exist where they do, the age of the ocean floor, the positioning of the continents and ocean basins, the location of critical economic minerals and ore bodies, the placement of mountain chains, the distribution of fossil remains, and why mountain ranges and different types of faults occur in specific locations. It has revolutionized the field of geology and with it the understanding of how the Earth works.

As set forth by the science of plate tectonics, earthquakes occur in specific areas where plate margins interact or where new plates are formed. Unlike mere decades ago, geologists today can predict the placement of volcanoes and mountains because of their positions related to plates. Also, before the plate tectonics theory had been developed, the location of mineral deposits made little sense. Now, based on this new theoretical understanding through plate tectonics of pressures and temperatures at various geological boundaries, geologists can more fully explain the rock cycle, which documents how rocks transform from one type to another through processes such as burial, melting, uplift, weathering, and heat and pressures (igneous to sedimentary, sedimentary to metamorphic, metamorphic to igneous). Among the many practical applications of such science, for example, are improved methods to better predict where in the world valuable economic ore deposits are located.

There are several ways in which scientists determine the sizes of quakes in order to compare them. The gauge most people are familiar with is the Richter scale, named for seismologist and physicist Charles Richter (1900–1985) of the California Institute of Technology, who developed his scale in 1935 as a quantitative way to measure earthquakes in Southern California.[1] The Richter scale is based on the height (amplitude) of the wave generated by a particular earthquake as captured by a seismometer (an instrument used to record earthquake waves). The height of the wave is correlated with the distance of the seismometer from the epicenter to calculate the magnitude of the earthquake. The scale is logarithmic; a change in magnitude from 4.0 to 5.0 on the Richter scale, for instance,

Song of the Earth. Elisabeth Ervin-Blankenheim, Oxford University Press. © Oxford University Press 2021.
DOI: 10.1093/oso/9780197502464.003.0007

is approximately 10 times more intense, whereas an earthquake with a magnitude of 6.0 is 1,000 times more intense than a 4.0 temblor.

As the number of seismometers increased over the world, scientists realized that the Richter scale was not fully describing all of the earthquakes recorded, in particular underestimating the magnitude of subduction-zone temblors. Geologists and seismologists developed other scales to measure earthquakes in the years following Richter's work, such as those to measure particular types of earthquake waves through their amplitude, but these methods had their limitations, too. The most comprehensive scale is the moment magnitude scale, developed by USGS seismologist Thomas Hanks and Hiroo Kanamori, professor of seismology at the California Institute of Technology, in 1979.[2] The moment magnitude scale is a measure of the energy released during an earthquake and accounts for parameters not considered in earlier scales, including the length of the fault rupture and the amount of displacement (slip) on the fault. The scale is logarithmic, just as for the other measures.[3]

News reports often, incorrectly, will state that an earthquake was of a certain "Richter" magnitude, but in reality, they are talking about moment magnitude, as conveyed by the USGS's National Earthquake Information Center.[4] Reporters frequently confuse changes in amplitude in earthquake waves with changes in magnitude. For instance, for every increase in the magnitude of an earthquake, the amplitude of the wave increases 10-fold. However, the energy released is based on a log function, so an increase from an earthquake magnitude 4 to magnitude 5 represents a 32-fold increase in energy released, and a magnitude 4 to magnitude 6 is approximately a 1,000-fold increase in energy released. Without a correct understanding of the difference between amplitude and magnitude, the public may underestimate the perceived impact of an earthquake.

Plate Tectonics Theory

The big lithospheric plates of the Earth made up of oceanic crust, or oceanic and continental crust, interact at their edges where they meet one another. Depending on the tectonic mechanism that exists at a particular location, either new plates are created, old plates are destroyed, or two plates slide past each other.

One primary type of plate boundary occurs when new plate material is produced on ocean floors at mid-oceanic ridges (see chapter 5, color insert 5.1), also called divergent boundaries or spreading centers. Many of these are in the oceans, but sometimes they emerge through the continental crust, gradually splitting the land apart, as in the case of the East African Rift Valley. Iceland sits on top of the mid-Atlantic ridge, as well as on top of a mantle plume—a localized column of magma.

At another major type of plate boundary, old plate material is consumed in deep-sea trenches (see chapter 5, color insert 5.1), when one tectonic plate, made up of oceanic materials, subducts (dives down) under a lighter-density plate, made up of some oceanic and mostly continental material, at what is called a convergent boundary. Examples of this process include the Aleutian Trench near Alaska and the Cascadian Subduction Zone in the Pacific Northwest. Alternatively, mountains are built up when two plates with continents collide, such as the Himalayas.

An additional major type of boundary occurs where plates slide past each other (see chapter 5, color insert 5.1), such as on the San Andreas Fault in Southern California, forces are conservative, and rock is neither created nor destroyed. These boundaries are called transform or strike-slip boundaries;

Earth processes confine these three types of boundaries to extensive but relatively narrow reaches of the planet's surface. Contrasting the edge-dominated zones of plate interaction, diffuse plate boundaries (see chapter 5) can be hundreds to more than a thousand kilometers wide and occur in either continental or oceanic plates.

Scientists once thought that the ocean basins were as old as the continents, but tectonic research and study of the ocean floor have led to the understanding that ocean basins are much younger than the continents. Through numerical age-dating using isotopic analysis of the igneous rock basalt, geologists have determined that the oldest ocean floor is in the Mediterranean Sea, at merely 340 million years. Indeed, most ocean-basin rocks are younger than 200 million years old, as all ocean-basement rocks older than the Cretaceous Period have been subsumed, and essentially recycled, at convergent margins of plates.

Compared to the human experience of time, ocean basins are old but not as ancient as the rocks that make up the continents which can be dated to more than 4 billion years in the continental shield areas of the crystalline bedrock of Canada and Greenland, as previously mentioned. But because the ocean floors are either subsumed into deep-sea trenches or smashed into the edges of continents, they do not persist like continental rocks. Accordingly, new ocean-floor rocks created along rifts are the Earth's youngest rocks, shown as red in color insert 6.1. The colors trend to green, blue, and violet (the oldest oceanic rocks), demonstrating increasing ages as ocean layers proceed away from ridge axes.

Associated with the predominant plate boundaries are faults—breaks in the crust of the Earth across which there is movement (figure 6.1). Faults are critical features in the history and current state of the Earth. The three major types are normal faults, which occur from extensional forces, reverse faults from compressive forces, and strike-slip faults from lateral, sliding forces. When faults rupture because of the buildup of strain (deformations in the rock caused by the stress of various forces), enormous energy is released to create earthquakes.

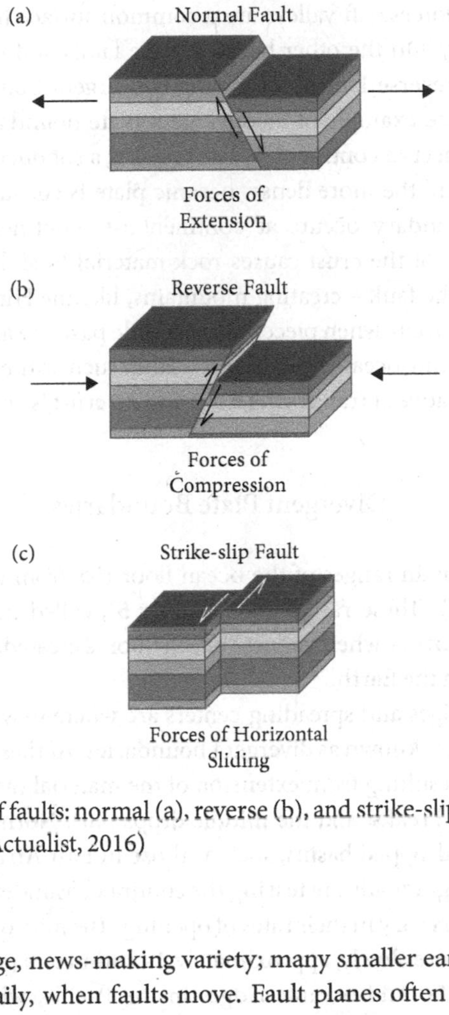

Figure 6.1 Types of faults: normal (a), reverse (b), and strike-slip (c). (Modified from diagrams by Actualist, 2016)

Not all are the large, news-making variety; many smaller earthquakes occur all over the world, daily, when faults move. Fault planes often are rough surfaces and get stuck as the plates jostle, building up strain from plate movement until the pressures overcome the strength of the rock, causing a sudden rupture and resulting earthquake. Earthquakes also can occur when magma moves around underground below volcanoes. Human activities can induce earthquakes as well, such as in the pumping of waste (produced) water back into the ground in the process of extracting natural gas and lubricating faults. Seismographs have recorded frequent earthquakes of this type in Oklahoma and Colorado.

Particular faults are associated with specific types of tectonic plate boundaries (color insert 6.2).[5] Normal faults are found at mid-oceanic ridges, spreading centers, and divergent boundaries where new plates are created, and extensional forces dominate. Such pulling apart of the crust causes one side of rock material to move down relative to the other side of the fault. When continents are split apart by tectonic forces, rift valleys are a common surface feature, like the East African Rift Valley. On the other hand, reverse faults and thrust faults (a type of shallow-angle reverse fault) accompany convergent boundaries, with compressive forces. One example of a convergent plate boundary would be where an oceanic plate meets a continental plate creating a subduction zone and deep-sea trenches. Where the more dense, oceanic plate is consumed. Another type of convergent boundary occurs at continental-to-continental crust margins smashing together of the crust causes rock material to slide up relative to the other side across the fault—creating mountains, like the Himalayas. Transform (strike-slip) faults occur when pieces of crust slide past one another, such as perpendicular to the mid-oceanic ridges or on land, such as the San Andreas Fault. Forces are conservative at transform faults, and no crust is created or destroyed.

Divergent Plate Boundaries

Some of the mountain ranges of the ocean floor rise from their bases to 2,500 meters (8,200 feet). These ranges (see chapter 5), called mid-ocean ridges or spreading centers, occur wherever new ocean floor is created, the most extensive mountain range on the Earth.

Mid-oceanic ridges and spreading centers are where new ocean plates made and pushed apart are known as divergent boundaries. At these locations, normal faults are created resulting from extension of the material (as if pulling a cookie apart, which often breaks, and the middle drops out). Normal faults create rift valleys and down-dropped basins, such as those in East Africa where the faults and associated rift gradually are tearing the continent asunder.

Spreading centers vary in their rates of opening. The mid-oceanic ridges in the Atlantic expand more slowly, approximately 4 centimeters per year (1.6 inches) on average, than the mid-oceanic ridges in the Pacific, which move about 9 centimeters per year (3.5 inches) on average. In other words, the Pacific is shrinking at more than twice the average rate at which the Atlantic is expanding.

Not only are the rates of spreading centers different, but the shapes of the ridges contrast as well. Fast-spreading ridges have smooth sides on their flanks, while slow-spreading ridges have been shown to have rough ridge flanks. Differences exist because of how hot waters circulate within them and also depending on whether there are persistent magma chambers percolating beneath them.

Among the more fascinating and easily observable mid-oceanic ridge points is located in Iceland, which bisects the mid-Atlantic ridge. On one side, the North Atlantic plate moves westward, and on the other, the Eurasian plate pulls eastward. The surface expression of the mid-Atlantic ridge is a rift valley that runs down its center with a split. Iceland hosts volcanoes that form along the mid-Atlantic ridge, including the notorious Eyjafjallajökull volcano that disrupted flights in the northern hemisphere when it sent up clouds of ash in 2010. Ridge volcanoes are directly related to the formation of new plate material. They are beautiful, as well as dangerous, and also provide geothermal heat resources and hot springs. Iceland is unusual not only because it rides atop the mid-Atlantic ridge but also because it is above a mantle plume, also known as a hotspot. Volcanic activity in Iceland is undoubtedly higher because of the conjunction of the hotspot and the ridge.

Evidence for Iceland's mantle plume comes from changes observed in remote sensing techniques using seismic waves. Seismic values under Iceland are low, indicating a difference in the materials at depth compared to most areas on mid-oceanic ridges. Geophysicists speculate that the plume originates 2,880 kilometers (1,790 miles) deep, at the boundary of the outer core and the mantle. Recent data show that the mantle plume under Iceland rises and spreads out, with fingers, or petals, radiating from a central core as the changes in heat and viscosity (stickiness) of the material interact with the cooler upper asthenosphere. These fingers of hot rock extend out to Norway and Scotland and might be the reason for uplift in these regions. Scientists continue their research on mantle plumes to answer questions about whether they are relatively fixed or move over time.

Other geologic hazards occur as a result of the presence of the mid-Atlantic ridge in Iceland. Seismometers record scores of earthquakes each week. Most of these temblors are less than 3.0 in magnitude, but the propensity for more massive earthquakes exists in the South Iceland Seismic Zone. Some of these quakes are related to magma moving about under the volcanoes, while the remainder of the temblors are a result of the two tectonic plates moving away from each other. In fact, Iceland is gradually splitting apart at a rate of 2.5 centimeters (1 inch) per year.

Another area dominated by rifting and extension is the East African Rift Valley. In this region, the mid-oceanic ridge does not emerge on the land, but forces are pulling, or rifting, the Nubian (African) plate from what will be the new Somalian plate when it breaks off. Valleys and ridges created from normal faulting characterize the surrounding region. The down-dropped parts of the faults form basins where water can pool, creating lakes that have come and gone, some of which have been very deep. The weather and surrounding land have been prone to rapid changes of dry and wet periods, which impact the presence of lakes.

Adjacent to some coastlines, such as North America's east coast, broad, sloping continental shelves made up of sediments extend under the ocean to an average of 65 kilometers (40 miles) and at depths of 60 meters (200 feet). Buried beneath the shelf sediments are normal faults, formed from the extensional forces and rifting that made the adjacent ocean basins. These margins are considered passive and are on the trailing edge of the continent, having been pushed apart by the new crust formed at the mid-oceanic ridges. Geophysicists use remote sensing techniques to see these faults.

The Basin and Range region, extending through most of the western United States and into northwestern Mexico, is an additional area of normal faulting related to tectonics. Here, and including the active volcanic hotspot currently under Yellowstone National Park, the crust of the Earth is thin and undergoing extension. The down-dropped blocks form valleys known as grabens, and the uplifted ridges are called horsts. They trend north-south perpendicular to the extensional forces.

Convergent Plate Boundaries

The second type of plate boundary is the convergent variety. Reverse faulting is associated with these because these boundaries have compressive forces where two plates push, dive, or collide against each other. Depending on the types of plates, various things can happen. When two pieces of oceanic crust grind together, island arcs and trenches form, with the islands aligned in an arc because of the spherical nature of the Earth.[6] The subducting plate creates an incision; the smaller the angle of the subducting plate, the greater the curvature of the feature at the plate margin. The island chain of Indonesia is an example of this type of boundary. Where an oceanic plate meets a plate with continental material, the oceanic crust dives under, because of its greater density, and creates a deep trench, such as the one that lies between the Pacific plate and the North American plate known as the Cascadian Subduction Zone in the Pacific Northwest. Two pieces of continental crust can merge as well, causing mountain ranges.

Convergent forces dominate in the Pacific Ocean area, creating the "Ring of Fire," a horseshoe-shaped structure 40,000 kilometers (25,000 miles) long consisting of almost continuous oceanic trenches as a result of tectonics. Volcanoes, earthquakes, and tsunamis are prevalent in this area. The Pacific plate is being eaten up, so to speak, driven down under other plates as the entire ocean basin shrinks. Surrounding the Ring of Fire and directly caused by the subducting plates are 452 volcanoes making up 75 percent of all the active volcanoes in the world. More than 90 percent of all earthquakes on the Earth are in this area. The ring may extend farther to the south, to Antarctica, where, until recently, "extinct" volcanoes have been buried 2 kilometers (1.2 miles) under the ice sheets. With changes in climate and warming related to increases in carbon dioxide, the massive ice sheets in the Antarctic are melting. Accordingly, geologists are detecting the vast nature of these volcanoes for the first time and have identified ninety-one new volcanoes, in one of the most significant volcano fields of its type.[7]

Studies of the Pacific and other oceans revealed another astounding feature: deep-sea trenches plunging from 7,000 meters (22,960 feet) to the lowest spot known on Earth, in the Mariana (or Marianas) Trench, called Challenger Deep, 10,916 meters (35,814 feet) below sea level. For comparison, Mount Everest rises 8,848 meters (29,029 feet). Magma from the melting of material of the subducting plate "bubbles up" to form the island arc to the southeast of Guam. The Mariana Trench has a length of 2,500 kilometers (1,500 miles), five times that of the Grand Canyon, and it is 69 kilometers (43 miles) wide. It is a protected reserve of the United States, situated in U.S. waters.

Speculation about what is down in the trench has spurred stories in movies and books, as well as real-life adventures—such as filmmaker James Cameron's successful mini-submarine voyage to its bottom in 2012. It was not known until submersibles descended that, incredibly, there are life forms deep in the Mariana Trench. No legendary "Krakens" have emerged, but scientists have found other types of life in the form of amphipods, crustaceans similar to shrimp but much larger. These creatures are 17–30 centimeters (7–12 inches) long, almost colorless because of the lack of light, and can withstand immense pressures. Also seen and collected by oceanographers are varieties of sea cucumbers, jellyfish, foraminifera, and bacteria. The pressures are too high for hard limestone shells to exist, but some of the foraminifers use sand grains made of quartz (SiO_2) to build structures and protect themselves. Other trenches have life forms, and some even more abundantly, because food sources fall into the trench, especially when close to land, such as in the New Britain Trench south of Papua New Guinea in the Solomon Sea. Not only are the varieties and species of the trench dwellers more abundant, but the walls of the trench are lined with white anemones.

Despite their vast depths, these environments are not untouched by human activities. Recent studies of amphipods show that they contain high concentrations of plasticizers and other chemicals, some of which governments have banned, including PCBs.[8] These chemicals are not found in nature; they are certainly products of industrial and other anthropogenic waste. Radioactive compounds also have been found in these settings. As scientists study trenches and their role in ocean chemistry, through such efforts as the Hadal Ecosystems Studies (HADES) project funded by the National Science Foundation and the European Union Research Council, they're asking critical questions about how much of the material on the bottom of the seabed and in ocean trenches gets recycled by microbial action.

The Peru-Chile Trench is located on the western side of South America in the circum-Pacific Ring of Fire and has produced the highest volcano in the world, the Nevados Ojos del Salado, at 6,879 meters (22,500 feet), located along the

Chile-Argentina border. As the Nazca plate subducts under the South American plate, it carries with it cold ocean water. At depth, around 660 kilometers (410 miles), the plate starts to melt because of mantle heat and produces magma that bubbles back up to the surface and erupts in the form of volcanoes. Earthquakes go along with the reverse and thrust faulting generated at these convergent boundaries. Indeed, geologists have recorded some of the most massive and destructive earthquakes in the world along this particular boundary. In 1960, the Valdivia earthquake in Chile was pegged at a magnitude of 9.5 by the USGS—the strongest earthquake ever measured. This Great Chilean Quake, as it is known, produced a series of tsunami waves that traveled across the Pacific Ocean at between 200 and 500 miles an hour and created destruction as far away as Japan and the Philippines. The Chilean government estimated that as many as six thousand people were killed, and 2 million people were left homeless as a result of tsunami waves.

Transform Plate Boundaries

The last type of plate boundary is the transform type. These occur with faulting that is vertical, or nearly vertical, and the plates move horizontally past each other in what is called strike-slip motion. New plate material is neither formed nor destroyed at these faults, and they represent a conservative boundary, unlike divergent and convergent boundaries. Transform faults occur between plates and are a mechanism by which plate motion is transferred, for instance, from one type of margin to another as a type of pressure accommodation.

Although no new plates are created or destroyed at transform faults, this does not mean that significant earthquakes do not occur there. One well-known location where a strike-slip fault cuts through the continental crust, for example, is the San Andreas Fault and faults related to it. To the west of this fault system is the Pacific plate, and to the east is the North American plate, with the Pacific plate moving to the north relative to it. The sense of motion on the San Andreas Fault Zone is termed right-lateral strike-slip. If an observer is standing on one side of the fault zone, the side opposite appears to have moved to the observer's right. This sense of motion always is relative to the observer. This zone of faults is 1,827 kilometers (800 miles) long, 16 kilometers (10 miles) deep, and made up of interrelated faults with branches and offshoots. The San Andreas Fault Zone traverses a semiarid region and can be observed on the surface. Along with the surface expression of the fault zone are linear valleys, bays, lakes, ridges that have been cut off by faulting, stream drainage patterns that line up with the fault and are linear, and the presence of springs where the fault interrupts ground-water patterns.

The Great San Francisco Earthquake of 1906 is one of the most famous and most destructive quakes ever emanating from the San Andreas Fault. The temblor is estimated to have been 7.8 magnitude, with seven hundred deaths, but this number most likely is low. The quake fired blazes that burned for three days in San Francisco as gas lines were breached, destroying 500 blocks in the city. Movement on the fault caused a rupture in the Earth 477 kilometers (296 miles) from northwest of San Juan Bautista to Cape Mendocino, the location of a triple junction (where three plates meet). The amount of rupture on the fault and the degree of shaking from the earthquake astounded geologists of the time. Besides extensive damage in San Francisco, the earthquake offset fences and other structures as the plates moved past one another and caused one hundred and eighty nine additional deaths.

Transform faults exist in several places on continental plates besides Southern California. The Enriquillo-Plantain Garden Fault is a sizable left-lateral (opposite side of the fault moving to the observer's left) strike-slip fault in the Caribbean, located between the North American and Caribbean plates. The countries of Jamaica, Haiti, and the Dominican Republic sit astride this fault. In Haiti, the surface expression of the fault is a linear east-west valley seen from the air and satellite photos south of Port-au-Prince. The Caribbean moves to the east, and the North American plate is moving west.

In January 2010, a magnitude 7.0 earthquake occurred on the Enriquillo-Plantain Garden Fault, killing an estimated 316,000 people, with about the same number injured. Many people died as the concrete-block buildings collapsed, even those that were reinforced, particularly in the urban areas. More than 300,000 buildings collapsed, with 3 million people becoming homeless. Unlike wooden structures that flex with ground shaking, these buildings disintegrated. The temblor also destroyed much of the country's infrastructure. Haiti is one of the poorest countries in the western hemisphere and was unprepared for such a disaster. Some sources cite this earthquake as having rendered the most significant human impact of any recorded in modern times.[9]

Transform faults occur more typically in ocean basins than on the continental plates and cut the ocean floor perpendicular to the mid-oceanic ridges. These features can be up to several thousands of kilometers long, cut valleys in the ocean floor, and offset the ridge axes. The active part of the fault is between the ridges where the sides of the fault are moving in opposite directions, and the ridgelines get split by the fault. The apparent motion is opposite that which defines most strike-slip faulting because the ridge is expanding at the same time as the transform fault is created. The inactive part of the fault (inactive fracture zone) gets carried farther away from the mid-oceanic ridge because of the formation of new ocean floor, where both sides move together, though the trace of the fault still can be seen and actually may produce a deep canyon. The farther away from the ridge axis, the more the plate material cools and sinks.

Geologists and physicists posit that transform faults provide accommodations in the crust to relieve tension related to the sphericity of the Earth. Along mid-oceanic ridges, there is not only the movement along the transform fault but also spreading of the ridge as well, at the same time. Thus, the apparent movement on the transform fault in between the ridges is the opposite of how it appears. The ridge will appear to be offset to the left, but the active fault segment between the ridge axes actually will have moved in a right-lateral direction.

Hotspots and Diffuse Boundary Zones

Other types of features besides the three major boundaries exist in theory of plate tectonics—hotspots, diffuse boundaries, and unknown or speculative boundaries. Hotspots occur where magma from deep in the mantle wells up in a mantle plume, creating volcanoes and other features. A prime example is the Emperor Seamount Chain, which includes the islands of Hawaii and the Hawaiian Ridge. The Pacific plate is moving northwest at approximately 7 centimeters (2.7 inches) per year, and the hotspot has been relatively stable for the past 47 million years. Volcanoes are the surface expression of the hotspot; the older ones are to the northwest, recording the track of the plate over the mantle plume. With time, they sink under the ocean surface and become cut off from the hotspot, and the tops are sheared by waves, creating seamounts and guyots. The younger islands, such as Hawaii, the "Big Island," are to the southeast, with a new lobe of volcanic land currently being added to this main island of Hawaii. A bend of 60 degrees occurs in the Emperor Chain of islands and guyots, related to both the hotspot changing location 47 million years ago and the Pacific plate trajectory moving.[10] The Hawaiian Islands, from the older islands with extinct volcanoes in the northwest to the volcanically active island of Hawaii in the southeast, follow the trend of the Pacific plate over the varying hotspot locations.

Another prominent inter-plate hotspot is the Yellowstone hotspot, located on the North American plate. This feature is expressed in the series of calderas (volcanoes collapsed back down in sunken craters) starting to the southwest of, and merging with, the Snake River Plain, ending with one of the most famous supervolcanoes in the world: the caldera of Yellowstone. The path of the Yellowstone hotspot is recorded by the 800-kilometer (497-mile) track, with the oldest, the McDermitt caldera in the southwest, age-dated at 16.5 million years. The Yellowstone caldera is the most recent manifestation of the hotspot, and geologists age-date the structure to 2.1 million years. The North American plate is moving southwest over a mantle plume or hotspot 1–2 centimeters (0.4–0.8 inch) per year. Geologists report that sediments and younger lava flows obscure the older calderas, but the track of the hotspot is visible and is one of the best examples of inter-plate volcanism that progresses with age. Scientists can detect the mantle plume and the Yellowstone caldera, in particular, by changes in the gravitational field of the Earth. The hotspot has produced a topographic high made up of a 500-meter (1,640-foot) by 400-kilometer (249 mile) broad swell, the Yellowstone Plateau, above the surrounding area, with thirty times the heat flow compared to other parts of the continent. Geologists also record geochemical changes in the geothermal springs and plumbing system of Yellowstone—the most extensive hydrothermal system in the world.

The topography in the area is distorted, making it difficult to see the Yellowstone caldera and the nearby calderas from the ground because the volcanoes expelled their contents in a series of massive eruptions and collapsed back into calderas. The Yellowstone caldera covers an expansive area of 3,884 square kilometers (1,500 square miles). Situated under the crater are two magma chambers—the upper more enriched in silica and smaller in volume, measuring 10,200 cubic kilometers (2,447 cubic miles). Connected to this chamber is the larger one below, measuring, based on recent seismic data, 46,000 cubic kilometers (11,036 cubic miles).[11] The contents of the second chamber consist of denser and darker minerals—iron and magnesium—and geophysicists think the structure is situated at the base of the Earth's lower crust on top of the upper mantle. Within the magma chambers, the rock exists in a spongy, semisolid state. Only a portion is liquid magma. Below these two magma compartments is the hotspot reaching down as far as 660 kilometers (410 miles) to the mantle-core transition zone.

Supervolcanoes are those that erupt with a volume of deposits exceeding 1,000 cubic kilometers (240 cubic miles). The Island Park caldera, adjacent to Yellowstone, spewed out ash, debris, and rock 2.1 million years ago in a supereruption covering some 2,450 cubic kilometers (600 cubic miles), 2,500 times the amount of material produced from the 1980 Mount St. Helens eruption in Washington State, considered the most destructive volcanic event in recorded U.S. history. The Island Park supereruption resulted in the Huckleberry Ridge ash fall that cooled, solidified, and compressed into a volcanic rock called tuff, covering most of the western mountain states. Another supereruption, the Lava Creek ash fall, was deposited 1.3 million years ago in a massive eruption of 1,000 cubic kilometers (240 cubic miles), covering most of the western interior and mountain states to the present-day Mississippi River. And a third gigantic eruption 1.3 million years ago from Yellowstone formed the Mesa Falls ash fall with a volume of 280 cubic kilometers (67 cubic miles). Geologists can figure out how much material ejects from a particular volcano by examining the thickness and extent of volcanic ash fall and other deposits. These shrink down over time and become volcanic rock.

The Yellowstone hotspot and its calderas are integral to and have had a profound effect on the geology and topography of a significant portion of the western United States, not only from the ash-fall deposits and voluminous lava flows but also from the hotspot's impacts on other tectonic structures and the surface expression of those stresses. From the Snake River Plain and Yellowstone Dome, the hotspot is associated with the extensional forces creating the Basin and Range area of Nevada. For example, the flood basalts of the Columbia River plateau may have the same source as the Yellowstone caldera—the mantle plume. Modeling suggests the original location of the hotspot was coincident with the Juan de Fuca oceanic plate, a subducting feature currently located off the Pacific Northwest coast.[12] Flow in the upper mantle travels eastward, counter to the migration of the North American plate. The plume formed and rose behind the Juan de Fuca plate, where the plate shielded it from flow in the upper mantle. Twelve million years ago, conditions changed when the subducting plate no longer protected the plume and became entrained in the eastward flow of the upper mantle. Along the way, it was carried under the continent, where it rests today in a relatively stable location, geologically speaking, and where it has produced some of the most majestic vistas in the world, along with some potentially most dangerous hazards.

In addition to Yellowstone, there are two other supervolcanoes in the United States: the Long Valley caldera in California and the Valles caldera in New Mexico. Although the last supereruption was 640,000 years in the past, Yellowstone has been far from quiescent. Lava flows from this caldera have occurred as recently as 70,000 years ago on the Pitchstone Plateau, on the southwestern corner of Yellowstone National Park. Another series of nonexplosive lava releases poured out within the caldera about 160,000 years ago. Smaller, less explosive eruptions transpired, including one 174,000 years before the present that created the West Thumb of Yellowstone Lake.

Along with lava flows and smaller eruptions, the area undergoes 1,000–3,000 earthquakes a year, mostly of a relatively low magnitude, less than 4, still strong enough to be felt by most people. One massive earthquake in 1959, recorded as a 7.5 magnitude west of the park near Hebgen Lake, killed twenty-eight people, mostly caught in an earthquake-induced landslide, and caused $11 million of damage. Earthquakes at Yellowstone are primarily a result of magma moving around deep in the Earth and some adjustments of strain along related faults.

If the Yellowstone caldera were to generate a supereruption, it would release immense clouds of hot ash, gas, molten rock, and other material. The ash alone would cover most of the continental United States, varying in thickness from more than a meter (3.3 feet) in the surrounding 50 square miles to estimates of

3–40 millimeters (0.1–1.6 inches) beyond an 800-kilometer (497-mile) radius, reaching New York, Dallas, Toronto, Los Angeles, and Chicago.[13] One millimeter of ash can impact airports and flights, causing them to shut down, produce damage to vehicles and houses, destroy crops, and contaminate water supplies. It would be a devastating event. A "kill zone" around Yellowstone extends hundreds of miles and would be dependent on whether an umbrella cloud forms, carrying the material high into the upper atmosphere, as has been indicated in other supereruptions.

Umbrella clouds are less impacted by wind and weather patterns, creating more of a bull's-eye pattern. Smaller eruptions produce fanlike distributions of ash and particles, are lower in the air column, and are more dependent on prevailing winds. Sulfur dioxide and particles from the supereruption would cool the weather by blocking out sunlight, resulting in a volcanic winter. Models show that ocean water temperatures would drop by 3 degrees C, which would have negative impacts on ocean life and global currents. Effects could last for decades.

Geologists think such a massive supereruption has a low chance of occurrence and is unlikely to happen in the next ten thousand years or so. But there is too little data on which to calculate a recurrence interval for an eruption based on past eruptions, and information from other supervolcanoes is rare. The problem is that baseline conditions are mostly unknown. No supervolcanoes have erupted since Lake Taupo on New Zealand's North Island 26,500 years ago. Additionally, the magma chambers at Yellowstone are relatively crystalline, and the rock within is only partially melted, leading to a lower chance of a massive eruption.

Smaller eruptions are more likely to occur and would include explosive hydrothermal releases and the expulsion of lava. Volcanologists monitor Yellowstone and the surrounding area through studies undertaken with the Yellowstone Volcano Observatory, a consortium of institutions led by the USGS, including the National Park Service; the University of Utah; the state geological surveys of Wyoming, Montana, and Idaho; and UNAVCO (for remote sensing needs) funded by the National Science Foundation. For thirty years now, scientists on this project have measured earthquakes, changes in the elevation of the land surface, and hydrochemistry using a series of stations and sensors. And while Yellowstone has been active throughout its 140 years of monitoring, it is nevertheless relatively stable.

Diffuse Boundaries

Diffuse boundaries are just as they sound, seismically active areas spread over extensive distances, hundreds to thousands of kilometers wide, but relatively far from tectonic plate margins. They can occur in either the oceanic or continental crust and account for about 15 percent of the Earth's surface. They move and change more slowly than the traditional plate boundaries, a maximum of 1.5 centimeters (0.6 inch) per year. The strain across these regions is 100 times less than that of their more narrow plate boundary cousins but 10,000 times more than the stable parts of plates (the craton).[14]

As noted, one example of a diffuse boundary is the Basin and Range region, which covers much of the western portion of the North American continent, including Nevada; parts of Idaho, Oregon, and Utah; southern Arizona; and south-central New Mexico. It extends to Texas and beyond the United States to Mexico. It makes up a separate physiographic province (areas grouped by similar surface features) and is bounded on the southeast by the Colorado Plateau, the northeast by the Rocky Mountains, the north by the Columbia Plateau, and the west by the Cascade and Sierra Nevada Mountains. The crust here is stretched over a large area, thousands of kilometers wide, so tectonic forces are pulling away from one another, even though no specific plate boundaries are associated with the region. Recent research suggests a pool of liquid carbon deep in the mantle, causing thinning of the crust there.

Extensional forces dominate in this seismically active zone, causing north-south trending basins (grabens) and ridges (horsts). Normal faulting creates down-dropped basins of flat valleys bounded by ranges rising on either side as much as 3,048 meters (10,000 feet). In fact, this region has been stretched to the extent that its original width has doubled. The crust is thinner than other parts of the continent—approximately 30–35 kilometers (18.6–21.7 miles) thick on the eastern side of the province and 45 kilometers (28 miles) proceeding to the west. If the inexorable rifting forces continue, the area will be split off from the rest of the North American plate far in the future.

The Basin and Range Province consists of the northern region, called the Great Basin, and a southern region made up of the Sonoran Desert, the Salton Trough, the Mexican Highlands, and the Sacramento area. Elevations in the southern Basin and Range are lower than in the northern region. Of particular interest is the Great Basin. Here, in this approximately 517,998-square-kilometer (200,000-square-mile) expanse, all precipitation drains internally into the basin (known as the Hydrographic Great Basin) with not a drop of water reaching the ocean. Though arid and seemingly barren, this land supported Native American tribes for thousands of years. It includes Death Valley, the lowest elevation on the North American continent, -86 meters (-282 feet) in Badwater Basin.

Unknown or Speculative Boundaries

Earthquakes, as noted previously, mark the boundaries and assist geologists in determining where plate edges exist. However, sometimes earthquakes occur away from apparent boundaries, or plate boundary zones, in the middle of landmasses and continents. These temblors are known as intra-plate earthquakes. One of these localities is the New Madrid area in southeastern Missouri.

In 1811 and 1812, three large earthquakes, estimated to be magnitude 7.3 to 7.5, followed by hundreds of aftershocks, rocked the surrounding area, sending tremors up the spine of the Appalachian Mountains, which are made of relatively soft sedimentary rocks. The power of these atypical Missouri quakes even rang bells in Boston. People reported that the Mississippi River "ran backward," and though the river did not actually reverse its course, waves of water sloshing back and forth (seiches) made it appear so, as a result of the shaking and subsidence of surrounding land. Sand boils appeared in five surrounding states, and the earthquakes so disrupted some of the ground that farmers could not work the topsoil. The quake destroyed homes in the two small towns around the epicenter and would have brought down more, but the population in the area at the time was low. Buildings as far away as St. Louis were damaged.

The New Madrid Fault and associated earthquakes may be the result of a failed rift, called the Reelfoot Rift, but it is far from any plate boundaries. Other speculations center on the area being a hinge line between the continent and the soft sediments of the Gulf Coastal Plain and the deposits of the Mississippi River. A second seismic zone to the northeast of New Madrid is the Wabash Valley Seismic Zone associated with the Cottage Grove Fault.

Earthquakes are just one of the hazards associated with plate boundaries and tectonic forces. These processes occur regardless of the human footprint on the planet, but the impact is far more significant because of population distribution. Socio-economics plays into the extent of an earthquake catastrophe, such as in the case of Haiti, which had been considered to be the most impoverished country in the western hemisphere when the massive earthquake occurred. Another factor in the magnitude of the disaster was the lack of any earthquake awareness or preparedness. No building codes existed for such exigencies, and thus many did not survive the quake. More than a third of the population of the country was directly and immediately impacted by the temblor and continues to be today.

Consequences of Plate Tectonics

The human and ecological impact of tectonic forces that result in massive earthquakes, volcanoes, landslides, and tsunamis should not be underestimated.

One of the deadliest volcanoes in recorded history is Mount Vesuvius in Italy. It has erupted numerous times, most notably 79 CE, when it spewed ash and dust in a column 32 kilometers (20 miles) high. The towns of Pompeii and Herculaneum were devastated. Tephra and ash 3 meters (10 feet) thick covered Pompeii, and a devastating pyroclastic flow (a dense mass of ash and volcanic debris) overrode Herculaneum, resulting in 23 meters (75 feet) of volcanic material. Residents felt earthquakes, lasting nineteen hours in a two-phase event, before the eruption, with 4 cubic kilometers (1 cubic mile) of ejecta from Vesuvius. It was also the first eruption in history with a written eyewitness account.

Gaius Plinius Caecilius Secundus, later known as Pliny the Younger (61–113 CE), was raised by his uncle, Pliny the Elder, after his father died. During the eruption of Mount Vesuvius, when Pliny the Younger was eighteen, the elder Pliny, commander of the Roman fleet and stationed at Misenum attempted to rescue friends and other victims. The uncle set sail with five of the fleet's warships across the Bay of Naples, from Misenum 35 kilometers (22 miles) to the northwest. The sailors had the wind in their favor, but as they approached Vesuvius, ash and cinder rained down. Whether they were not aware that the volcano was erupting or unable to return because of the winds, they stayed the night in nearby Stabiae. People at the time were accustomed to earthquakes related to Vesuvius but did not realize it was a volcano, because it had been quiescent for many generations. The elder Pliny urged calm. Unfortunately, the situation became dire, and he died on the beach two days after he had set sail, either from the volcanic gases and ash or possibly from a heart attack or stroke.

Pliny the Younger wrote of the event in two letters. The first was to the historian Tacitus, describing what his uncle had seen and related to him:

> he immediately arose and went out upon a rising ground from whence he might get a better sight of this very uncommon appearance. A cloud, from which mountain was uncertain, at this distance (but it was found afterwards to come from Mount Vesuvius), was ascending, the appearance of which I cannot give you a more exact description of than by likening it to that of a pine tree, for it shot up to a great height in the form of a very tall trunk, which spread itself out at the top into a sort of branches; occasioned, I imagine, either by a sudden gust of air that impelled it, the force of which decreased as it advanced upwards, or the cloud itself being pressed back again by its own weight, expanded in the manner I have mentioned.[15]

Described precisely by Pliny in this account is the great height to which the pyroclastic column erupted and then collapsed back down upon itself as a result of gravity and the weight of the expelled material. The pyroclastic flow of hot ash and gases up to 1,000 degrees C rushed down the side of Vesuvius at as much as 50 meters (164 feet) per second, equal to 180 kilometers (112 miles) per hour.

These high-altitude eruptive columns are known as the Plinian type, after Pliny's depiction.

Centuries later, archaeologists identified two thousand casts of bodies, but the number killed by the falling ash and pyroclastic flow may have been much higher. Between ten thousand and twenty thousand people lived in the area and abandoned both cities after the catastrophe. Pompeii lay buried and forgotten for centuries until the late 1500s, when locals digging ditches rediscovered the city. But the site was pillaged and not studied in a systematic way by archaeology until the mid-1700s.

Vesuvius is a stratovolcano, one of the most dangerous types of volcanoes because, instead of creating only lava that flows out in quieter eruptions, like the Hawaiian shield volcanoes, these erupt either explosively because of higher silica content in the magma or at other times more quietly with lava. When volcanoes form on a continent, they often melt some of the silica-rich material of the landmass. Vesuvius lies within another caldera structure of Mount Somma—on the map (color insert 6.3), the crescent-shaped outline can be seen to the north, hugging the vent of Vesuvius. Mount Somma last erupted 17,000 years ago, creating the purple-colored beds of lava and welded scoria. Green areas surrounding Vesuvius on the map are ash fall and pyroclastic deposits from the 79 CE episode. Lava flows in pink and red overlie the deposits that destroyed the two cities.

Not only did earthquakes precede the eruption at of Vesuvius, but a tsunami also may have occurred. Pliny reports "the sudden retreat of the sea" as his uncle's ships approached the shoreline near Stabiae. In his second letter, Pliny noted:

> The sea seemed to roll back upon itself, and to be driven from its banks by the convulsive motion of the earth; it is certain at least the shore was considerably enlarged, and several sea animals were left upon it.[16]

Tsunami is Japanese for "harbor wave." Tsunamis have no resemblance to wind-driven waves. In a tsunami, the entire column of ocean water moves, not just the surface. They are generated from movement on a fault, often at a convergent boundary displacing the seafloor, to create a shock wave that transfers through the surrounding ocean water. Before a tsunami hits the shore, the ocean water will draw far back and expose the seafloor.

Since 79 CE, Vesuvius has erupted approximately thirty-six times, with the last active period lasting from 1913 to 1944. During this span of thirty-one years, 3,500 people were killed, mainly by ash fall. Since 1944, it has been relatively quiet, though the tectonic forces continue as the African plate pushes under at a complex convergent boundary with the Eurasian plate. The African plate is moving at 2–3 centimeters (1 inch) per year, causing the Mediterranean Sea, which rides upon it, to slowly collapse and close upon itself.

In more recent times, tsunami hazards related to moving plate boundaries make international news, in particular, the Indonesian tsunami of December 26, 2004—the Great Indian Ocean Tsunami. It was caused by the third-largest recorded earthquake in the world since 1900—the Sumatra-Andaman Earthquake—magnitude 9.1 on a megathrust fault between the Burma microplate and the India plate off the west coast of northern Sumatra. The temblor was the longest ever recorded at some seismic stations, lasting between from eight to ten minutes. The actual rupture on the fault was three to four minutes and started slowly. It triggered other faults to move, with resulting earthquakes extending to Alaska. Its power was so immense that the USGS estimated the earthquake released energy equivalent to 23,000 Nagasaki atomic bombs. The quake offset the ocean floor by 15 meters (49 feet), with the majority of the vertical rupture concentrated within 400 kilometers (248 miles) of the overall 1,300-kilometer (808-mile) rupture length. The break in the fault down-dropped the ocean floor and propagated multiple tsunamis.

The impact of the tsunamis was felt from east Africa to Thailand, across the Indian Ocean, and killed more than 250,000 people in fourteen countries, mainly in Indonesia, Sri Lanka, India, and Thailand. The tsunami was measured at 30 meters (98 feet) high and was so devastating because of the significant numbers of people living in vulnerable coastal areas. Scientists recorded the impact of the tsunami around the world by the wave amplitude (height) chart by NOAA, as far as the east coast of the United States.

Many had no warning of the impending disaster—in 2004, there were only thirteen seismometers and four coastal sea-level stations in the entire Indian Ocean. After the tragedy, UNESCO, through a consortium of countries, developed the Indian Ocean Tsunami Warning System. As of 2015, there are now more than 140 seismometers, 100 core coastal sea-level stations, and nine deep-ocean tsunameter networks deployed throughout the Indian Ocean, all producing and sending real-time data for any earthquake registering greater than 6.5 magnitude.[17] Earthquakes of 7.5 magnitudes or higher are necessary to create tsunamis. Today twenty-four countries have tsunami warning centers, most centered on the Pacific Rim.

Past and Present Tectonics

Plate tectonics, from recent research, appear to have been very different in the deep past of the Hadean and early Archean Eons compared to tectonic processes from 2,500 million years ago to the present, and more like tectonic conditions on other planets. The Earth at that point was much hotter, creating a higher geothermal gradient—the increase in temperatures with depth through the layers of the planet—two to three times more than now. A particular kind of convection dominates in the mantle under these conditions, creating a layer on top called the "stagnant lid." The heat from the stagnant lid in the early Archean was radiated to the surface by conduction, much the way the handle of a pot gets hot from the heat on a stove.[18] This process led to the formation of a shell of basalt—a single plate—covering the Earth. Scientists think there were no subduction zones and no moving plates on the early Earth. Stagnant lid conditions are thought to be present on the terrestrial planets in the Earth's solar system, including Mercury, Venus, and Mars, as well as on the moon.

Continents initially formed on the Earth in the Archean Eon 3,500 to 2,700 million years ago. Geologists think these first landmasses were created by mantle processes 100 kilometers (62 miles) deep, but geochemical computer models have disproved this scenario. Researchers in 2012 showed instead that the early continents originated from the crust at depths of 30–40 kilometers (19–25 miles), much shallower than supposed.[19] The study was conducted in West Greenland on ancient basalts, early Archean in age.

Over time, through repeated heating events, the dark basalt separated into lighter materials that migrated toward the surface, with the heavier minerals settling at depth through differentiation. Geologists have found evidence for this process in western Australia in the rocks of the 3,500-million-year-old East Pilbara Terrain.[20] Both lighter granites and basalts are present in this area, and researchers have found trace elements in the granites indicating that the basalts are their parent rock. Data show that there are grandparent and great-grandparent rocks, suggesting that the development of the granites, which became the continental masses, took an extended period, unlike the way in which ocean and continental plates form through tectonic processes on the Earth today.

Tying It All Together

In the deep past, supercontinents, when all the landmasses aggregated together, formed and dispersed in multiple episodes. Two of the supercontinents existed in Precambrian times during the Proterozoic Eon—Rodinia and the short-lived Pannotia—and the supercontinent Pangea formed at the end of the Paleozoic Era and lasted 100 million years.

Geologists noticed that mountain belts frequently are close to and have a similar orientation to that of the edges of the continents. Once plate tectonics developed as a theory, the reason for this placement became apparent, from repeated collisions of landmasses through time impacting the borders of the plates. The eastern coast of the United States is now a passive margin in tectonic terms, meaning that the energies along the coast are lower than those of an active margin. As an effect, for example, the continental shelf along the Atlantic is more extensive and contains more sediment, and barrier islands can form because of the quieter tectonic setting. Nevertheless, the Appalachian Mountains and other east coast mountains are parallel to the coast and not far inland. The reason is that at one time, the east coast was an active margin. Active margins are high-energy environments where mountain building and uplift occur along with subduction zones forming. The west coast of the United States currently is an active margin. Repeated cycles of the opening and closing of oceans caused the formation of the Appalachian Mountains—a complete, if complicated, Wilson cycle.[21]

The edges of the supercontinents and continents were and are subject to stresses resulting from plate tectonics forming mountain belts, collision zones, and welding on of material by accretion. Some of the blocks propelled onto the continents are called exotic (or suspect) terranes (pieces of a continent broken off from another landmass and sutured onto another through tectonic forces). Exotic terranes are groups of rocks formed in one area and moved to another. They consist of material scraped off oceanic plates or island arcs emplaced on the edges of the active margins of the continent and are very unlike other rocks of the surrounding area. Ocean-floor sediments, ophiolites (figure 6.2), made of nodules of manganese, pillow basalts, ultramafic mantle rocks, and volcanic and sedimentary deposits from island arcs, are examples of exotic terranes. Faults

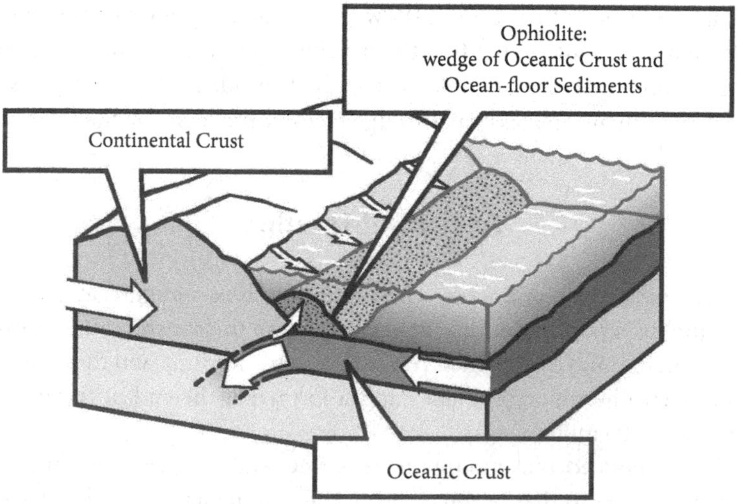

Figure 6.2 Active tectonic margin and exotic terranes. (Jim Houghton/ Paleontological Research Institution, 2000)

separate these pieces of crust at their base from rocks below. In California, the Coastal Ranges are made up of exotic terranes of this nature, as are the Alps. Indeed, finding materials derived from the ocean floor high in the mountains puzzled geologists for a long time until the knowledge of tectonics emerged. Some exotic terranes contain precious minerals and are of industrial importance.

The interior of continents—the cratons—generally are undeformed because they are far away from rifting or compressive forces. The North American craton covers a significant portion of the center of the continent, but little of it is exposed at the surface. The part of the craton exposed at the land surface is the "shield," made up of crystalline basement rocks. These areas are of low relief, with few mountains, and contain the oldest rocks on the planet, billions of years old, because they have not been subject to recycling or destruction. The North American craton includes the Canadian shield covering most of Canada, the northern tier of the United States, and Greenland. Surrounding the shields are platform areas where sediments made into rocks of Paleozoic and Mesozoic age bury the craton.

The Appalachian Mountains and the mountain ranges of central Europe, the Urals, and Iberia formed contiguously and underwent three significant mountain-building times during the Paleozoic Era. The story of these peaks begins even earlier, in the Precambrian Supereon, when the Grenville orogeny occurred and Rodinia—a supercontinent—was created 1,200 million years ago in the middle Proterozoic. All was well on the supercontinent until the forces of tectonics intervened, and incipient rifting started to occur around 750 million years ago. As tectonic forces tore the supercontinent apart, an irregular coastline formed, with promontories made up of the areas of Alabama, New York, and the St. Lawrence region, and embayments composed of Tennessee, Pennsylvania, Quebec, and Newfoundland, forming a scalloped pattern.

Rodinia broke up into three continents of the late Proterozoic: North Rodinia, South Rodinia, and Congo. Only 150 million years later, in an event called the Pan-African collision, a new supercontinent formed—Pannotia (Greater Gondwana)—when the continents rotated and collided over a span from 650 million to 560 million years ago. But almost as soon as Pannotia came together as a supercontinent, it broke up into the four continents of the Paleozoic Era—Laurentia (North America), Baltica (Europe), Siberia, and Gondwana. The Pan-African collision creating the brief tenure of Pannotia made mountains, and the climate was colder, which dropped the sea levels. Although ice sheets were present at both poles, it is thought that the ocean was free of ice in the region of the equator; these are known as Ice House conditions.[22]

By 565 million years ago, the basin east of the North American continent was riven and filled with a new ocean. Sediments from the continent washed down and began to create the continental shelf and an extensive carbonate platform.

Volcanic deposits associated with the split were farther away from the shoreline. As the new oceanic crust pushed away from the mid-oceanic ridge, it cooled and started to subside. But then, 420 million years ago, tectonic forces changed from extension to compression, and the plates containing North America (Laurentia) and Europe and Africa (Gondwana) started back toward each other. The next 50 million years witnessed two periods when island arcs were squeezed in between the compressive powers of the approaching landmasses. Material from these arcs was glued onto the continents—the Taconic mountain-building event, followed by the Acadian episode. The third phase of mountain building occurred with the continental collision of the two giants—Laurentia and Gondwana—270 million years ago in the Alleghenian orogeny that built mountains higher than the contemporary Himalayan Range. These mountains, remarkably, are now eroded and buried. The collision was of such intensity that it generated a megathrust fault resulting in the Blue Ridge Piedmont and drove folding to the west of that range, still evident today in the relatively soft rocks of the Ridge and Valley Province of the Appalachian Mountains. Weather-resistant sandstone makes up these ridges, while limestone, worn away over time with rainfall and underlain by faulting, creating the valleys. This last collision of giant continents produced the supercontinent Pangea. The just-created mountains were located at the interior of Pangea until it started to split apart, 180 million years ago, in the Mesozoic Era. Accordingly, the entire process of creating the Appalachians and associated structures took 490 million years from the breakup of Pannotia to the formation of Pangea.

7
Life on the Earth
Evolution, Extinctions, and Biodiversity

The third unifying principle of geology, along with geologic time and plate tectonics, is evolution—how life has changed throughout the biography of the Earth. Not only has evolution shaped the forms and varieties of life, but extinction events also have had a dramatic impact. Both forces result in the patterns of biodiversity through geologic time. Additionally, the Earth and the life it supports have shaped each other throughout the long span of deep time. This chapter examines the history of how scientists came to understand natural selection and the additional mechanisms of evolution, and genetics, leading ultimately, to the unification of the lines of evidence into the "modern synthesis of evolution. Following this discussion is an example of a well-known species in the fossil record—the horse and its ancestors—and, after that, an illustration of human alterations to the environment resulting in natural selection in the peppered moth. The chapter wraps up with sections on mass extinctions and biodiversity through the geologic record.

Development and Change of Life on the Earth: Evolution and Natural Selection

Life present on the Earth today makes up only 0.1 percent of all life that ever existed.[1] In other words, of the life forms that have lived on this Earth, only 1 in 1,000 species are alive today; at least 99.9 percent are extinct. Organisms, in their many forms, have varied tremendously through geologic time, as shown in the fossils of the rock record. The placement of those fossils helped geologists deduce the sequence of geologic time, as discussed previously. Early geologists and stratigraphers, notably Charles Lyell and William Smith (see chapter 1), perceived that fossils embedded in rock layers progressed from more simple forms to those that were more complex. As we have seen, these observations led to the principle of fossil and faunal succession—recording how organisms change through time—a fancy phrase for the idea that particular fossils occur in particular layers of rock and are significantly different in rocks located stratigraphically higher or lower. This principle of faunal succession formed the basis of the

Song of the Earth. Elisabeth Ervin-Blankenheim, Oxford University Press. © Oxford University Press 2021.
DOI: 10.1093/oso/9780197502464.003.0008

geological time scale before there was knowledge of radiometric dating of rocks that provided numerical age dates, as discussed in chapter 3. Faunal succession allows geologists to correlate strata through time and space. However, this does not mean there is a direct line from more simple life forms to more intricate ones. A better analogy is a large tree, or even a vine structure, with many limbs, branches, and leaves. Some branches of the tree, or vine, of life do not persist, having become extinct.

Geologists have found signs of early life, albeit indirect, from northeastern Canada, near Quebec, in rocks dated 3,950 million years old. There, deposits of graphite, a mineral made of pure carbon found in metamorphic rocks, had an isotopic signature that indicates the carbon was made by organisms.[2] Other initial forms of life, "extremophiles," some of which were single-celled organisms, may have been the first life forms on the planet, emerging initially from deep within the Earth at the mid-oceanic ridges. These organisms do not appear to have needed oxygen and were able to adapt to inhospitable environments.[3]

The first substantial evidence of life, which geologists find as fossils, appeared approximately 3,700 million to 3,400 million years ago in the form of single-celled organisms, similar to modern cyanobacteria. Paleontologists think these bacteria may have existed individually at first and later in their evolution formed colonies, creating mounded forms called stromatolites (figure 7.1a). Stromatolites still exist on the Earth today, in Mexico, Chile, Brazil, the west coast of Australia, and offshore Florida near the Bahamas, and are composed of calcareous (calcium-carbonate-rich) material and trapped mud. A critical feature of these bacteria is that they were photosynthesizing, harnessing the energy of the sun to make food for themselves. Significantly, therefore, as they breathed, or respirated, they took up carbon dioxide in their tissues and released oxygen. This process had a fundamental impact on the chemistry of the late Archean to early Proterozoic oceans and, eventually, the Earth's atmosphere. This "Great Oxygenation Event" (discussed further in chapter 8) occurred over nearly a billion years. The oxygenation of the waters profoundly impacted the future of the planet, causing the iron in the oceans to "rust" and the dissolved iron to precipitate and form extensive banded-iron formations. Later, as the oceans became saturated and could not absorb any more oxygen, it filled the air, ultimately creating the oxygen in the modern atmosphere.

Geologists have discovered the oldest stromatolites, the first organic-sourced rocks, from Greenland in metamorphic rocks that still had their sedimentary structures, dated 3,700 million years ago. Often found with stromatolite formations are oolitic limestones (figure 7.1b). These particular limestones are made of carbonate particles called ooids, which consist of concentric layers built up around a spherical particle. The ooids become rounded by being swirled to

(a)

(b)

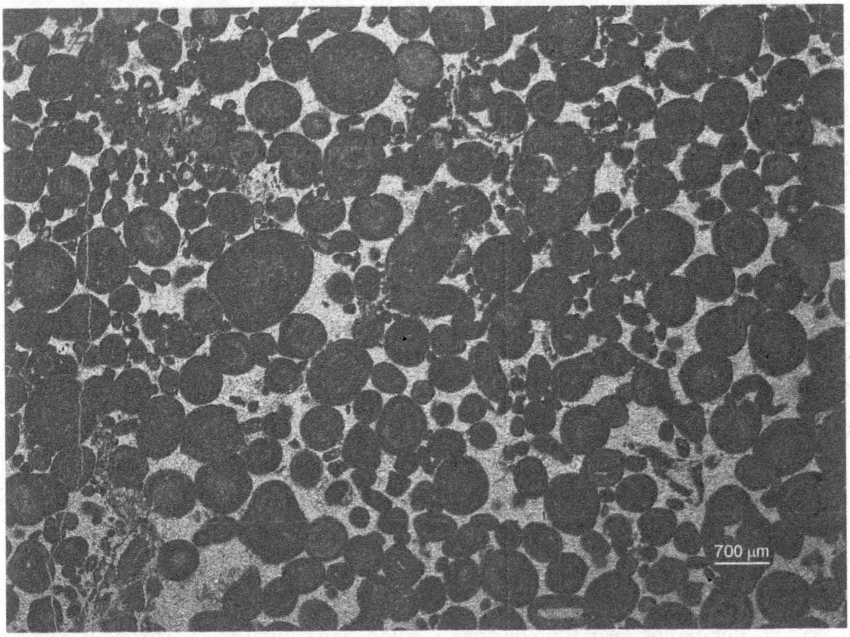

Figure 7.1 (a) Fossil stromatolites: detail of a fossil stromatolite in the Old Heeseberg Quarry, part of the Heeseberg Nature Reserve in Lower Saxony, Germany. (Ginganz-in, 2014) (b) Oolites, a constituent in limestone. (Ralvarezlara, 2013)

and fro in turbulent, shallow waters in environments like the "swash zone"—an area of a beach where the ocean water rushes up and back.

As early geologists and scientists started to understand the rock record, how geologic processes occur, and the vast age of the Earth, the ideas led to questions about organisms. Once they realized that fossils represent ancient life forms and that some of those life forms are not around now, another thought emerged: that species must not be static and unchanging but are shaped through time. This work became the domain of greats such as Charles Darwin, Alfred Russel Wallace (1823–1913), and others to follow. The question they asked was, how did life diversify and branch out from the bacteria and archaea to life on the Earth today?

Charles Darwin, the Journey of HMS *Beagle*, and Natural Selection

Scientists, famously Darwin and others, developed the ideas around evolution and natural selection through the observation of life, species differentiation resulting from changes in the environment, and variations in fossils in the rock record. So much has been written and discussed concerning Darwin and his theory of natural selection that it is difficult to know where to start with an account of his contribution to science in general and geology specifically. Few other scientists have had such a significant impact on the understanding of life on the Earth. However, Darwin was not the first to think about how species change over time and adapt to their environment. For one, French biologist Jean-Baptiste Lamarck proposed a common ancestor for all species.[4]

Darwin was born in 1809 in England to a wealthy family. Geologist Lyell's work greatly influenced the young scientist. Indeed, Volume 1 of the Lyell's *Principles of Geology* was in the HMS *Beagle* library on its legendary voyage in 1831. Lyell corresponded with Darwin and was his friend and mentor. In a letter sent from the *Beagle* in early August 1835, Darwin wrote to his second cousin William Darwin Fox, a clergyman and naturalist, about being a "zealous disciple" of geology:

> I am glad to hear you have some thoughts of beginning geology.—I hope you will, there is so much larger a field for thought, than in the other branches of Nat: History.—I am become a zealous disciple of Mr Lyells views, as known in his admirable book.—Geologizing in S. America, I am tempted to carry parts to a greater extent, even than he does. Geology is a capital science to begin, as it requires nothing but a little reading, thinking & hammering.[5]

Many consider Darwin a geologist first and foremost.[6] In 1859, the prestigious Geological Society in London awarded Darwin the Wollaston Medal, the highest honor in the Society, mainly for his contributions to the study of coral reefs, barnacle fossils, and South American geology. Lyell's influence on Darwin and Wallace is well acknowledged, and to them, "Lyell gave the gift of time. Without that gift, there would have been no *Origin of Species*."[7] Lyell is renowned for his work on geologic time and showing evidence of the great age of the Earth, through gradual geologic processes, or uniformitarianism, as discussed in chapter 2. These small changes accumulated over the span of the geologic record, adding up to significant variations on the Earth. Darwin applied Lyell's principle to species.

Biologists define a species as a group of individuals making up a gene pool, capable of breeding and passing on their genes to offspring. However, there is a long history, back to classical times, of what does and does not constitute a species.[8] The term usually applies to those organisms that require two sexes to reproduce; that definition would not include bacteria, as one example where the meaning is lacking.

Darwin, along with British ornithologist Wallace, studying in the South Pacific, pioneered the work on natural selection, a key mechanism of evolution. Natural selection occurs when one trait or set of genes is favored over another by the environment for survival of organisms because it provides an adaptive advantage or coping strategy. It operates on the species, not the individual, level.

In 1859, Darwin published his theory of natural selection based, in part, on his research into variations in the finches of the Galapagos Islands, which he had studied as the naturalist on the HMS *Beagle*'s journey in 1831. During the *Beagle* voyage, Darwin spent five years researching geology, plants, and animals from Cape Verde, a chain of volcanic islands, north along the coast of South America, to the Galapagos Islands and beyond. During that time, he corresponded with family and his Cambridge University mentor, John Stevens Henslow, a biologist and geologist, sending his observations and specimens back from far-flung British ports and establishing himself as an eminent naturalist. His questions about the range of species he saw provided the grounds for his future theory. In the Galapagos, Darwin collected examples of the birds and animals on each of the islands. Initially, he was more interested in the tortoises and mockingbirds in the Galapagos; he did not recognize the patterns of variation in finches until the specimens he collected were examined by British ornithologist and artist John Gould, who identified them as finches.[9]

Back in England, once Gould had identified the specimens, Darwin began thinking about the patterns of speciation in the finches and the fact that some species lived only on one island, which led him to ask how that distribution could occur. In 1845, when Darwin included his comments in Gould's publication, he did not identify the particular islands from which the various finch specimens were collected. However, in the years after the voyage, Darwin pondered the many types of finches, their differences, and their geographical locations. The finches had specialized beaks, adapted for microenvironments that differed depending on the food available on the particular islands. The finches that survived and thrived were those that could eat the available food source and pass their genes on to subsequent generations. Though varying in appearance, these finches had a common ancestor, leading Darwin to develop his theory of natural selection. Darwin gathered additional support for his idea until he was ready, some twenty years later, to reveal his theory of natural selection to the world. He alluded to his theories in several articles before completing his famous book.

While Darwin was completing his research and analysis, Wallace was working on a similar hypothesis. Then, in 1858, as Darwin was preparing his masterwork for publication, Wallace corresponded with him, sending Darwin a copy of his paper that Darwin was shocked to see mirrored his own theory. Through the agency of Lyell and botanist Joseph Dalton Hooker, the Linnaean Society invited Darwin and Wallace to present their findings jointly at a session later that year.

Darwin went on to publish his theory in 1859 in *On the Origin of Species*[10] in England and a year later in the United States,[11] while Wallace focused his studies on biogeography—examining the geographic distribution of plants and animals. The only diagram included in Darwin's book illustrates the divergence of life in a treelike form for a particular species going back to a common ancestor (figure 7.2). Letters A–L show variations and varieties of species; the horizontal lines represent as many as tens of thousands of generations.

The diagram is sometimes referred to as the "Tree of Life," but this is thought to be a misnomer.[12] Darwin was interested mainly in the mechanism of evolution, which he termed "natural selection," as compared to the patterns of evolution. The Tree of Life concept and use of the phrase goes back to the Bible. Darwin referenced the simile of a branching tree,[13] but he used it as an analogy for how a tree grows, with individual branches becoming more prominent (successful) relative to other branches that wither and die out. He applied the analogy of a tree to competition between species. Darwin himself never referred to this diagram as the Tree of Life.

Philosophers in ancient cultures and medieval times talked about the "Great Chain of Being," or *scala naturae*, the scale of nature. The Great Chain of Being promoted the idea that the Earth and inanimate forms are situated at the bottom

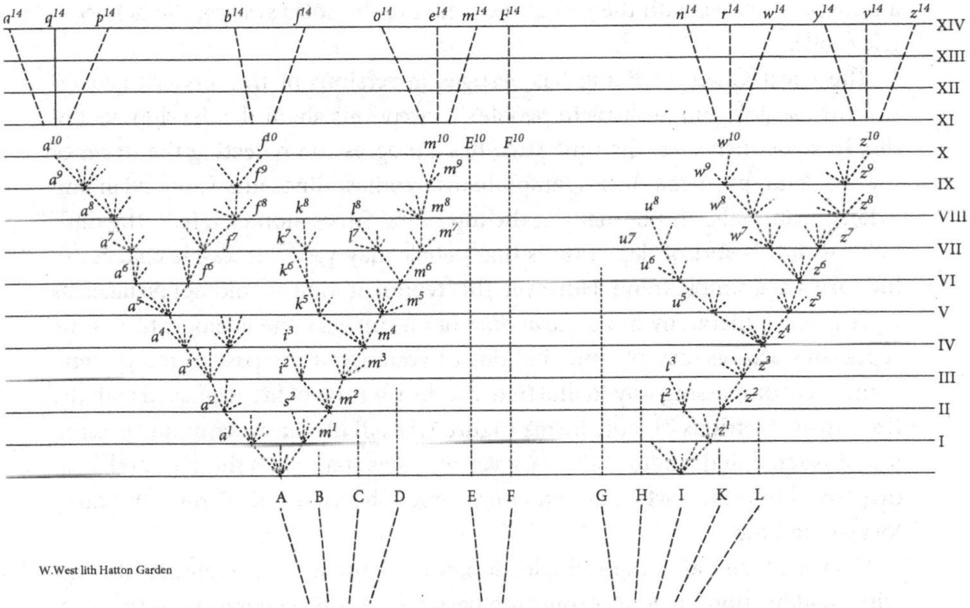

Figure 7.2 Diagram of branching lines of life. (Darwin, 1860, p. 108–109 foldout)

of the ladder structure, followed by plants, animals, humans, and, ultimately, a divine being. The forms higher up on the ladder have authority over forms beneath them. The structure has had a significant impact on moral, political, and scientific ideas and mores through time; for instance, in a monarchy, the king, as ruler of his subjects, had authority and ownership over all his kingdom.[14] Early concepts related to the Great Chain of Being were around for thousands of years before the Greek philosophers codified and defined it. It relates even to the development of the geocentric model of the universe, attributed to Ptolemy, who speculated that the Earth was the center of the universe around which all other planets and the sun rotated.[15] This view was held until the work of Copernicus in the sixteenth century, bringing the solar system and the Earth's place in it into a new perspective, with the sun at the center of the solar system, the heliocentric model.

The Great Chain of Being has vestiges in versions of the modern geological time scale[16] and perhaps in people's conceptions about the development of life. In some instances, the time scale has a progression reflecting the stages of life: bacteria, protozoa, fishes, amphibians, reptiles, dinosaurs (non-avian and avian), mammals, and humans at the top of the Succession of Life.[17] The difficulty with this sort of depiction is that people may view successive changes in life forms as a simple trend. However, this representation should not be taken as a progressive hierarchy in the *formation* of life. Instead, the geologic time scale represents a trajectory of time, billions of years from the past to the present. A more encompassing way to illustrate the distribution of life is charted with the time span of various organic forms (figure 7.3). All life has a common ancestor, not illustrated in the figure. The dark vertical lines start when the ancestral form diversified into the particular taxonomic order, shown in the picture, and carry forward in time.

Darwin's focus, after natural selection, was the theory of descent, or "descent with modification," whereby traits are passed from one generation to the next, resulting in species changing over time and the origination of new species from

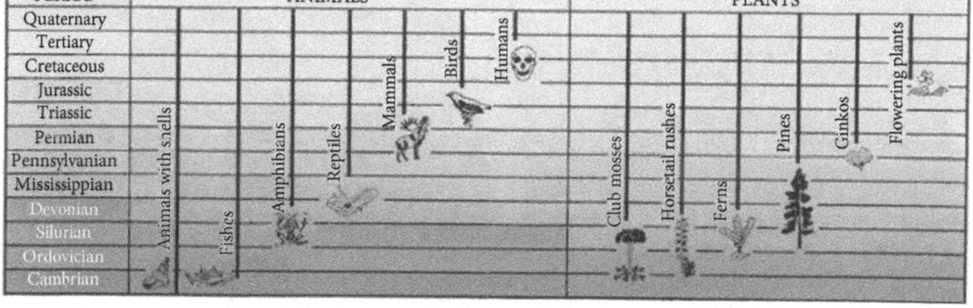

Figure 7.3 Geologic time scale with fossil succession. (USGS, 1997)

old species based on natural selection. Through both natural selection and descent with modification, Darwin proposed a common ancestor for all beings. He wrote that natural selection was the most critical, but not the only, way by which evolution occurs. Darwin took the development of evolutionary theory further than Lamarck and others had, tying in the mechanism of natural selection. Darwin's ideas were revolutionary and shook the foundations of species as static, placing humans "in the stream of animal evolution."[18]

Fossils and Preservation of Ancient Life

Critics of Darwin cited the apparent incompleteness of the fossil record, which he admitted readily. The problem is that fossilization is a rare process and has a relatively low chance of occurring[19] and also that geologists must locate and excavate the fossils.

Most organisms do not become fossilized upon their demise. Once an animal or plant dies, it usually undergoes decomposition, if there is enough oxygen. Rarely, however, conditions exist for the preservation of biological material in the fossil record. These unusual circumstances include rapid burial by sediments, such as through a flood event, along with a low-oxygen environment and calm conditions, so the body might not be fragmented up and washed away. Sometimes, though, if the bones are separated, paleontologists can reassemble them. The chance of an organism becoming a fossil is, hypothetically speaking, one in a million, or a minuscule probability. A specialized group of paleontologists called taphonomists—scientists who study a fossil specimen from its birth, life, death, burial, diagenesis (physical and chemical processes that occur when sediments are made into rocks), and, finally, recovery as a fossil—specialize in the transition of the living organism to a fossil being entombed in rocks. For example, one of the best locations for dinosaurs or other animals to be preserved is an arid environment where shifting rivers would cover the body, such as on a floodplain. Humid or tropical settings, by contrast, do not favor fossilization. Another location is underwater, in the ocean or lakes, where rapid burial by sediments and an anaerobic (oxygen-poor) environment would occur. Preservation favors the durable parts of creatures—the calcium-rich bits, such as bones, teeth, and shells—more easily than the soft parts. Geologists or paleontologists, mainly, look for sedimentary layers exposed on the Earth's surface after uplift from the depths and erosion, but many fossils are yet to be discovered. Thus, the chance of an animal, plant, or bacteria being preserved and then found is relatively low. How, then, could geologists reckon and fill out the missing fossil record in search of evidence for evolution?

The Historical "Bone Wars" of Marsh and Cope: Filling Out the Dinosaur Fossil Record

Enter two famous (and warring) American paleontologists, Othniel Charles Marsh (1831–1899) and Edward Drinker Cope in the mid- to late 1800s. Marsh was the head of vertebrate paleontology at Yale's Peabody Museum of Natural History and the museum's director. Cope worked mainly in Philadelphia with the Academy of Natural Sciences. Both came from families of means. Marsh was considered an "armchair" paleontologist by some, as he was not known for extensive fieldwork. He held a prominent position at Yale, with the department of paleontology endowed through his family.

Cope was from a Quaker family and was sent by his father to study in Europe around the time of the Civil War, where he met Marsh. Before that time, he studied fish, reptiles, and fossils. In 1858, Cope joined the Academy of Natural Sciences in Philadelphia and studied under the famous paleontologist Dr. Joseph Leidy (1823–1891), considered the founder of vertebrate paleontology in America. Leidy was self-taught as a paleontologist and was a physician by training; his research included cancer and parasitology. He authored a treatise on the *Cretaceous Reptiles of the United States* in 1865[20] which included a description of the original *Hadrosaurus foulkii*, located in 1858 by W. Parker Foulke, another member of the Academy, in the marl pits near Haddonfield, New Jersey. The dinosaur was named for Foulke and was given its generic name in honor of Haddonfield.

Most of southern New Jersey consists of coastal plain sediments, including marls, deposited in shallow waters or a near-shore environment—the perfect place for dinosaurs. Marl is a general term for calcium-carbonate-rich muds that farmers and others excavated for use as fertilizer. Workers digging in the marl pits often found bones and teeth of animals, which are now are known as dinosaurs from the Cretaceous Period. The marl deposits of Haddonfield are dark blue-black, limey clay with mica, exposed by a stream-cut bank, which made them more accessible for Foulke to find his dinosaur.

Foulke sent the *Hadrosaurus foulkii* bones to the Academy in Philadelphia, and in 1868, Leidy directed the bones to be assembled by British artist Benjamin Waterhouse Hawkins. It was the first lifelike dinosaur mount ever created and displayed to the public. Paleontologists everywhere heard of the marvel, and it sparked the interest of Marsh back at Yale.

Cope invited Marsh to visit the marl pits in Haddonfield, where he had discovered numerous bones while working with Leidy. Cope and Marsh explored the area and found two partial skeletons of different dinosaur remains in 1866. The "Bone Wars," a vivid time of disagreement and quarrels between the two, began shortly thereafter when Marsh persuaded Cope's workers to send the fossil finds to him rather than give them to their employer. When Cope found out, he was livid.

Relations between the two devolved from there. Marine saurian bones were sent to Philadelphia from a fossil location in western Kansas for restoration. The bones were found by Dr. Theophilus H. Turner, an Army surgeon near Fort Hayes, in 1867.[21] At first, Cope thought the bones were from two individuals but then realized it was just one creature as he reconstructed it. Soon after, Leidy noticed that vertebrae connecting the neck to the head were in the wrong place on the reconstruction and rebuked Cope for placing the head of the *Plesiosaur* on its tail rather than its neck.[22] Even worse, Cope published his initial findings, complete with diagrams, as a preprint on the swimming reptile *Elasmosaurus* (a type of *Plesiosaur*).[23] He tried to buy up all the copies of the paper but was unsuccessful, and though he asked the journal to issue a correction, it had no date on it, giving the impression (and probably rightly so) that he was attempting to cover up his mistake. Like posting a screenshot of a deleted tweet, Marsh refused to relinquish his copy of the original article and periodically brought up the embarrassing incident. Cope was forever known as having been responsible for the "backward *Plesiosaur*."

The Bone Wars heated up, and the dispute continued for thirty years with dirty dealings. Both men headed to the western United States, where dinosaur fossils were easier to find among the treeless plains, prairies, and foothills of the mountains. One area was Como Bluff, east of the Medicine Bow Range in Wyoming Territory. In 1872, Marsh began to search for dinosaur bones in a place Cope considered his "turf."[24] Five years later, railway workers found fossil bones in a dipping outcrop called Como Bluff. They wrote to Marsh, who sent a colleague, Samuel Williston, to investigate. Williston reported back that the bones extended seven miles and probably weighed in the tons. It was a treasure trove of Jurassic Period dinosaur bones, including *Allosaurus*, *Diplodocus*, *Stegosaurus*, *Camptosaurus*, and others. Marsh tried to keep the site secret, but Cope got wind of the find and sent scouts to infiltrate the area. He even appeared at Marsh's site himself and tried to charm Marsh's employees. Each tried to steal workers from the other, and they spied on each other. Marsh, according to lore, snuck into a site and planted an extraneous dinosaur skull where Cope was excavating, which researchers did not realize until years later. The rivals had their employees dynamite sites after reports of significant finds, in order that no one else (then and in the future) could access them. They fought in the press, through journal articles, and in meetings.

When Cope and Marsh started collecting dinosaur fossils, only 18 species from partial specimens, mainly teeth and vertebrae, were known from North America. However, as the Bone Wars came to an end, more than 130 new species had been described, greatly benefiting paleontology, albeit through poor practices in the field, acrimony, and reputational assassination, in an altogether dishonorable process. Many of the finds were later reidentified and assigned to the proper species; so rapid were searching and naming by Cope and Marsh that they often had mix-ups and gave different scientific names to different specimens now known to belong to the same animal.

Marsh was an adherent of Darwinism, and through his effort, he was able to support Darwin's theory by providing specimens to fill in gaps in the fossil record. He met Darwin in England in 1862 and 1865. Marsh wrote up his data on odontornithes, or birds with teeth in 1875;[25] two years later, in 1877, he proposed a theory that birds descended from dinosaurs, and he published his paper on the theory in 1883.[26] Cope, on the other hand, was a neo-Lamarckist who recognized evolution as a linear pattern and considered acquired characteristics as traits passed on to descendants.

Genes, DNA, and Quantitative Biology

Gregor Mendel (1822–1884), the father of genetics, studied changes in organisms (peas) resulting from characteristics related to independent units of inheritance. Mendel described the transmission of genetic traits by studying units of inheritance that later became known as genes. He studied hereditary characteristics and developed his Laws of Inheritance.[27] Darwin developed and wrote about natural selection without knowing about genes. Danish botanist Wilhelm Johannsen defined and used the word "gene" in 1909.[28] However, forty years prior, the first scientist to identify DNA was Swiss physiological chemist Friedrich Miescher (1844–1895) in 1869.[29] Miescher studied white blood cells and isolated a new substance that was neither a protein nor a lipid (a class of organic compounds made up of fatty acids), which he called nuclein and was later named deoxyribonucleic acid (DNA). Miescher had no idea that it had anything at all to do with inheritance. The fact that DNA was the building block of genes and chromosomes was not established until 1944, by Canadian-American molecular biologist Oswald Avery.[30] American biologist James Watson and English physicist Francis Crick became famous in 1953 for the discovery of the structure of DNA, the double helix, which suggests how genes replicate.[31] They famously said, "It has not escaped our notice that the specific pairing we have postulated immediately suggests a possible copying mechanism for the genetic material."[32] DNA contains very long chains of nucleic acids composed of four building blocks. These nucleotides—A, adenine; C, cytosine; T, thymine; and G, guanine—govern the making of proteins though a multitude of matches and possible combinations.

DNA provides additional evidence on evolutionary processes. Species whose ancestors diverged from one another further back in time have more variations in their DNA through mutations, along with genetic recombination, the "shuffling" of alleles of different genes during meiosis (cell division), than do species that are genetically closer on the evolutionary tree. Scientists speak of a "molecular clock"—first discussed by American biochemist Linus Pauling and French biologist Emile Zuckerkandl in 1962—a way to measure by gene mutations how far apart species are in an evolutionary sense. The rates for the molecular clock vary, depending on the species.

In 1959, one hundred years after Darwin's groundbreaking work, the theory of evolution was further bolstered by discoveries in molecular biology. Researchers found that the proteins of hemoglobin and myoglobin are similar in all organisms and can be sequenced, giving weight to the idea of a common ancestor. Based on these findings, biologists developed a family-tree pattern of variations in these two proteins, built on information from particular species showing the relationship of life forms to each other.[33] This work confirmed the same organization of life, consisting of the three domains of life—archaea, bacteria, and eukarya—which the study of fossils and anatomy had already determined. Scientists today use other molecules, such as cytochrome c, for the same purpose.

In 1975, a team led by British biochemist Frederick Sanger was the first to sequence DNA. He and his colleagues were looking for a particular virus that attacked bacteria. Sanger was awarded the Nobel Prize in Chemistry in 1980 for his work. Scientists realized that sequencing DNA, which involves determining the patterns of the molecule and all its variations, was an immense task. The Human Genome Project, an international effort, began in 1990, led by Watson of DNA-structure fame, with the goal of mapping human DNA.[34] Researchers completed the mapping in 2003. It has been called a history book of the hominin species, pointing back in time, and maybe a roadmap forward to unlock the power of gene therapy for diseases, as it sets forth where specific genes lie in the map and what they do.

So what do genetics and DNA have to do with the fossil record? For one thing, the fossil evidence and DNA findings from living organisms initially appeared to be at odds. Some species and genera are not well preserved in rocks because of their body types or where they lived, died, or were buried. Back-calculating species abundance based on their DNA indicates that certain species should have diversified in greater numbers over time, not accounting for any extinction events or loss. Nevertheless, the account of life does not support this conclusion—most types of organisms have undergone multiple extinctions and have boom and bust times.

It was not until 2011 that scientists found they could correlate the fossil record and modern DNA sequencing from cetacean (marine mammals, such as whales, dolphins, and porpoises) ancestors through the application of a slight, but critical, change to mathematical techniques previously employed.[35] Currently, there are 89 species of cetaceans. The fossil record for these creatures is well known because of their excellent preservation on the seafloor. Whales and their cetacean cousins arose about 50 million years ago, and by 10 million years ago, the genus included 150 species, more than twice its current number of 65 species. Knowing this, and by changing the modeled rate of diversity from a fixed value to a more realistic one that allows fluctuation and negative numbers, researchers indeed were able to match the data from the fossil record.

DNA preserved in the rock record, 700,000 years old or younger, and in amino acid sequences, is part of a newly recognized group of fossils called molecular fossils that reveal organic molecules of past life. Along with DNA, these rare finds include proteins, lipids, and carbohydrates. Lipids are the most resistant to decay. The most common environments for preservation are kerogen deposits (sedimentary rocks high in organic matter, which over time produce oil or coal), made up of organic-rich muds and shales, shells, and bones. Most molecular fossils, 80 percent, are located in these types of deposits. Kerogens form mainly from plankton in the oceans, dying and settling to the seafloor, where they become buried. Over time and with burial pressure, they undergo a process that severs the chemical bonds, allowing the material to flow and become a hydrocarbon—for example, in some cases, petroleum. The study of molecular fossils may provide clues in the identification of the earliest forms of life in the Great Oxygenation Event 2,400 million years ago by examining the evolution of metabolic pathways.[36] Researchers are trying to separate the history and development of the three domains of life based on the molecular information and how the primary domain—the eukaryotes—emerged from their ancestors.

Toward a Unified Synthetic Theory of Evolution

Not all change in a population is adaptive or based on natural selection. Biologists, geneticists, and paleontologists have expanded upon the theory of evolution and enlarged it to include additional mechanisms since Darwin's pioneering work. These mechanisms include genetic drift, gene mutations, and migration in populations as causes that promote evolutionary changes over time. In the late 1920s, American biologist Sewall Wright (1889–1988) and colleagues in England, geneticist Sir Ronald Fisher (1890–1962) and biologist J. B. S. Haldane (1892–1964), forged the study of evolutionary genetics, including how small mutations spread through a population. They formulated the idea of genetic drift. These scientists performed traditional breeding experiments after the research of Mendel. Critically, Fisher and Haldane applied mathematics, including models and statistics, to the field, which provided quantitative data on how the mechanisms function, founding the field of population genetics. Genetic drift occurs when only specific individuals pass on a portion of the genes of a population through reproduction. Other members of the species may have met with early death by chance accidents or may not be able to reproduce; thus, their genes are removed from the pool of genetic material. One resulting impact is that genes that are the most adapted to an environment or specific need may not survive to exist in different environments in subsequent generations. In the 1930s, geneticist Theodosius Dobzhansky (1900–1975) and his colleagues proposed that mutations accumulate in a species, leading to formations of new species, and developed their observations into what they called the synthetic theory.[37] The term "synthetic theory" was first used by British evolutionary biologist and eugenicist Julian Huxley in 1942.[38] In addition to genetic drift and mutations, another process causing evolutionary change is the movement and migration of individuals from one geographic area to another, bringing along a subset of genes, gene flow, or drift to a new location, thus causing a winnowing effect and different organisms.

Recognition of these new mechanisms led to the realization that both Darwin's and Mendel's theories could explain different parts of evolution. Darwin wrote about the process of evolution—natural selection—and Mendel wrote about the facts of evolution. This new, synthesized version of evolutionary theory integrated the tenets of natural selection, genetic variations, and reproductive and geographical isolation. It addressed population research done by Mendel, including that on gene pools and gene frequency.

The modern synthesis, in addition to the work of Darwin and Mendel, included paleontology, thanks to the influence of vertebrate paleontologist George Gaylord Simpson (1902–1984), professor of zoology at Columbia University and curator at the American Museum of Natural History from 1945 to 1959. Simpson wrote *Tempo and Mode in Evolution* in 1944[39] and was one of the foremost paleontologists of his time. He is credited with the modern version of the synthetic theory of evolution,[40] which states, based on the fossil record, that evolution is variable. It sometimes occurs in radical steps, which he termed "quantum evolution," and sometimes in small ones. As research into DNA and evolution proceeded, a new theory has emerged, the "extended synthesis."[41] Extended synthesis addresses the role of epigenetics (variations in organisms caused not by a change in DNA but by modification of gene expression) and other advances in genetics. This theory ultimately led invertebrate paleontologist, evolutionary biologist, and Harvard professor Stephen Jay Gould (1941–2002) and Niles Eldredge (1943–) from the American Museum of Natural History to publish the idea of "punctuated equilibrium" in 1972.[42] The theory of punctuated equilibrium states that a species will remain stable for millions of years and then suddenly undergo a rapid burst of evolutionary change rather than undergoing gradual change (phyletic gradualism). The fossil record, according to Gould and Eldredge, usually will not capture the rapid burst, because speciation occurs on the fringes of the species, where population numbers are smaller and species members are less likely to be preserved as fossils.[43] Punctuated equilibrium was debated and discussed for decades. However, the theory matured, and most biologists and paleontologists began to accept it as a complement to gradualism, twenty years after it was first published.[44]

Gould was one of the most influential biologists and invertebrate paleontologists in the latter part of the twentieth century, many say since Darwin. He wrote popular books on paleontology for the public, including *Wonderful Life: The Burgess Shale and the Nature of History*. Gould was a sometimes controversial figure, interlacing politics, Marxist leanings, and science;[45] he died of cancer at the age of sixty.[46]

Horse Fossils, Their Geologic History, and Change in Environmental Conditions

Beyond dinosaurs, many other species are exemplars of evolutionary change in response to shifting environments and changing habitats. One of these is the family of the beloved (to many) horses, the Equidae. The evolutionary history of the modern horse is among the best known in the long biography of life.[47] The account of horses is given here to illustrate the principles of evolution over the geologic record. Georges Cuvier, the French naturalist, whose work was discussed in chapter 1, found horselike animal fossils in the Eocene rocks of the Paris Basin in 1804 and named them *Palaeotherium*.[48] Sir Richard Owen identified *Hyracotherium*, the earliest ancestor of the modern horse, from the Eocene-aged deposits of the London Clay in 1841.[49] He classified two teeth found to be the same genus discovered by Darwin while journeying on the *Beagle*.

The first scientist to describe odd bits and pieces of ancient teeth and bones, found near Natchez, Mississippi, as the first undisputed horse ancestor in America, was Leidy when he wrote on *Equus americanus*.[50] This publication, Leidy's first on vertebrate paleontology, clearly demonstrated that horses had been present in North America during the geological past. It was preceded by physician and polymath Samuel Mitchill's report of potential horse fossils in the collection of the Lyceum of Natural History of New York (later the New York Academy of Sciences) in 1826.[51] However, it is not clear that the horse fossils reported by Mitchill were fossils or, if they were fossils, that they belonged to horses present in North America before their reintroduction by the Spanish.

During the Neogene Period of the Cenozoic Era, sea levels were lower, exposing land bridges between the continents, particularly Asia and North America. Horses and other mammals utilized land bridges to disperse between the two continents. The Eocene Epoch recorded the spread of horse species throughout North America, in parallel with *Palaeotherium* in Europe. After this epoch, horse ancestors migrated back to Europe and filled new ecological niches, ultimately leading to modern horses. However, the proto-horses that these fossils represent are not the horses of today. Horse species existed in North America until approximately 12,500 years ago,[52] when they became extinct along with other large mammals and megafauna of the time during the Pleistocene extinction. Of horse species in North America, Leidy said:

> For it is very remarkable that the genus Equus should have so entirely passed away from the vast pastures of the western world, in future ages to be replaced by a foreign species to which the country has proved so well adapted.[53]

Like his many contemporary scientists, Leidy also admired and corresponded with Darwin and accumulated data and facts to support Darwin's work on evolution. Accordingly, the Academy of Natural Sciences in Philadelphia was one of the first institutions to recognize the importance of Darwin's theory.[54] Leidy was instrumental in Darwin's becoming a member of the Academy.

Modern horses and horse predecessors vary widely in size, number of toes, dentition (how their teeth are formed and arranged), and skull structure. How horse species and proto-horses evolved is a prime example of an organism changing and adapting to variations in the environment. The earliest recognized horse ancestors were small-dog-sized, five-toed browsers that dwelled in a woodland setting—called *Hyracotherium* (formerly; now, again, *Eohippus*, the "dawn horse"). For a time, *Eohippus* was not recognized as a valid genus, and all specimens were referred to as *Hyracotherium*. A recent study has resurrected *Eohippus* for a portion of the specimens assigned to *Hyracotherium*. In 1989, paleontologists reclassified the specimen of *Hyracotherium* described by Owen as a *Palaeotherium* (an extinct group of herbivorous mammals), related to horses but not a horse. It is the only species now referred to as *Hyracotherium*. The first ancestor of the modern horse is once again *Eohippus*.[55] *Eohippus* was first described by Marsh (of the Bone Wars) in 1876.[56] *Eohippus* had teeth with shallow roots adapted to eating the relatively soft leaves and shrubs of the forest environment, through browsing. From the early Eocene Epoch to the present, climatic conditions changed as the woodlands gave way to drier savannas. Grass replaced the trees and shrubs in the new settings and contained a much higher silica content than leaves, shrubs, or twigs. As the grasslands opened up in the Oligocene Epoch, horse ancestors emerged from the shrinking forests to find new sources of food. The animals that could adapt to these climatic variations passed on their genes and illustrate the principles of natural selection. The changes can be seen in the fossil record, in response to the food they ate,[57] with an increase in tooth length proportionally to the higher silica content in the grasses. Silica makes grass harder to break down and digest; thus, the teeth needed to be longer and ever-growing as they were worn down by the grasses. Horse ancestors with longer teeth for grinding in these conditions out-survived those that did not and were more likely to pass on their genes.

Around the same time, in the Oligocene Epoch, modern horse ancestors began to undergo adaptation from the five-toed horse ancestors of the woodlands to favor those with fewer toes in the savanna setting. More open areas required further travel to water sources with a drier climate. Weight was transferred to three toes and ultimately to the one toe or hoof we see today in modern horses. Leftover, or vestigial, toes are evident in the splint bones of the modern horse. Various species of horse ancestors are known to have lived at the same time and possibly interbred.[58] Horse fossils, while abundant, are not demonstrative of a straight-line progression from one form to the other, from five-toed to three-toed to one-toed. The history is much more complicated and branches widely across time and geography (color insert 7.1).[59]

However, some illustrations show horse lineage as a vertical progression, from *Hyracotherium* to modern *Equus*, promoting a misconception of the linear progression of life. Paleontologists are correcting these ideas slowly. The fossil record of the Equidae has allowed scientists to understand essential principles of how life changes on the Earth in response to the environment, given time.

In 2013, the oldest horse-ancestor DNA was sequenced from a 700,000-year-old leg bone from the middle Pleistocene,[60] preserved in the Arctic permafrost in the Yukon Territory of Canada. Although at the time it was the oldest DNA to be reconstructed, researchers believe that DNA might be able to survive for more than a million years under cold conditions, perhaps as long as 1.5 million years.[61]

Evolution in Action

One of the most famous examples of evolution (or natural selection) witnessed by humans is that of the peppered moth (*Biston betularia*) in England in the nineteenth century. These insects, also called "Darwin's moths," demonstrated changes in their coloration (melanism) in direct correlation with ash content, from the burning of coal, in their environments. Coal fueled the Industrial Revolution in the late 1800s in England, and many people burned it in their homes for heat, especially in the cities. London's population before the turn of the twentieth century reached 6.5 million. With a large population came significant problems, among them cholera and the "Great Stink," when pollution of the Thames from raw sewage and dead animals overwhelmed the city in the 1850s. The soot, along with the famous London fog, often turned into smog, marking it for decades as one of the most polluted cities in the world.

British ecologist H. B. D. Kettlewell (1907–1979) studied the peppered moth from collections and modern examples in the mid-1950s and conducted quantitative surveys.[62] He found that moths, inhabiting the city areas, were darker in color—form *carbonaria*—than moths in the countryside, which were the light-colored ones—form *typica*. In North America, the dark moth is form *swettaria*. Kettlewell's research showed that predation by birds increased with the *typica* individuals because they stood out and were more natural prey compared to the darker-colored moths, which were better able to survive in soot-polluted environments. Kettlewell called the changes "industrial melanism" spurred by a mutation switching on a gene that caused the expression of darker color in some of the moths. By the end of the nineteenth century, in many industrialized areas of southern England, the wild *typica* moths had completed died out. Natural selection may not have been the only factor, as later research in the 1990s confirmed,[63] but it points strongly to it as the primary driver of the rapidly induced evolution in the moth.

Paleontologists have located transitional fossils (those that exhibit characteristics of the ancestral form of the species and also newer traits) that provide further proof of the theory of evolution. These specimens are difficult to locate than more common organisms because, again, the process of fossilization is so rare. One case in point is *Archaeopteryx*, a birdlike dinosaur, evidence of which was reported in 1861 by the German paleontologist Christian Erich Hermann von Meyer (1801–1869), from the Solnhofen Quarry in the southern German state of Bavaria. Meyer described a single feather that was fossilized in the limestone dated to the Upper Jurassic Period, 150 million years ago. The debate centered on what creature to assign the feather to, but in 1863, the preeminent paleontologist at the time, Owen, after comparing the original specimen with a more complete Solnhofen specimen in the British Museum of Natural History, definitively identified it as an *Archaeopteryx* feather. These animals were more like birds (avian dinosaurs) than dinosaurs, with wings, feathers, and a large beak, along with reptilian features of bony tail, flat breastbone, and three claws emanating from each wing (figure 7.4). To date, there have been twelve significant finds of *Archaeopteryx*.

Although *Archaeopteryx* may be the most frequently cited transitional fossil, there are at least two other more recently discovered and reported transitional fossils. The first is *Tiktaalik rosae*, a Devonian fish/amphibian transitional fossil discovered on Ellesmere Island, Canada, in 2004.[64] Later, in 2014, the same three paleontologists who first found the fossil reported on the pelvic girdle of

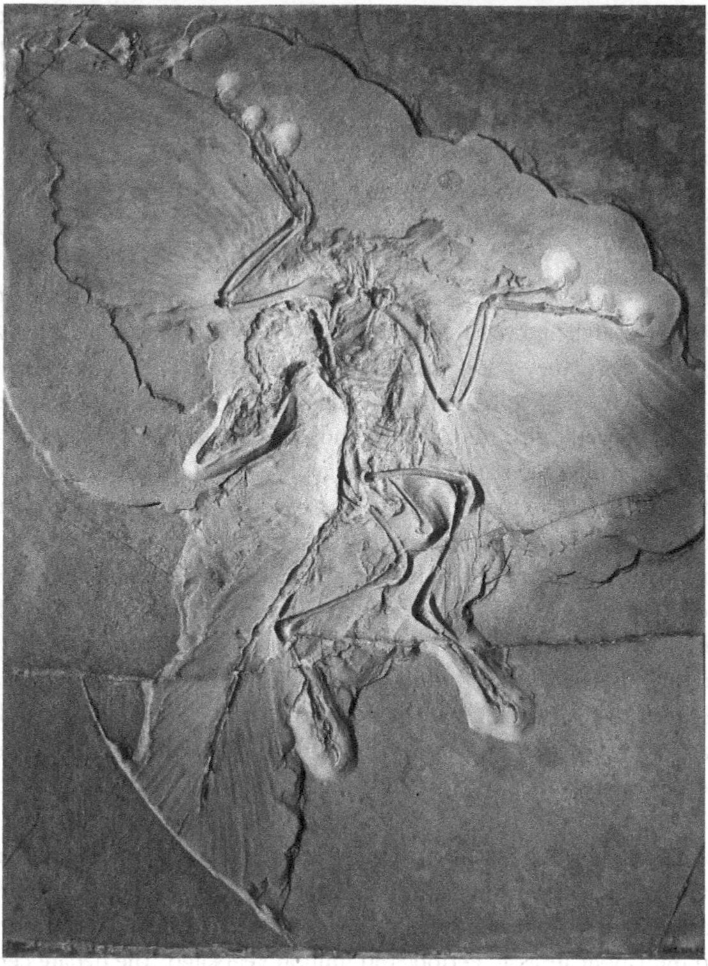

Figure 7.4 *Archaeopteryx lithographica*, original specimen displayed at the Museum für Naturkunde in Berlin. (Photo by H. Raab, 2009)

Tiktaalik. This discovery led to considering *Tiktaalik* an intermediate form with a reptile-like head, a fish-like body, and a tetrapod-like pelvic girdle—a "fishapod." A second possible transitional form reported recently is a nearly complete specimen of *Elpistostege watsoni*, from Quebec, Canada.[65] Australian paleontologist John Long (1957–) suggests that, based on the development of digits, this fish lies between *Tiktaalik* and the first amphibian known, *Acanthostega*. This organism is not newly known but a much more complete specimen than previously located. Together, these two fish/amphibians add significantly to the transition from fish to tetrapods.

Extinction Events and Their Impact on Evolution

Change in organisms through natural selection and evolution, including genetic drift, gene mutations, the impact of migration, population isolation, and the other mechanisms of evolution, has proceeded through the vast span of geologic time, particularly since the beginning of the Phanerozoic Eon. However, evolution has not proceeded without major and minor interruptions in the form of extinction events, or periods when species and genera died off from various factors. One of the first naturalists to recognize extinctions in the fossil record was Cuvier, working on fossil elephants in the Paris Basin.[66]

Extinctions have drastically altered the face of life on the planet and impacted the evolution of life forms. The range of extinctions, from more minor to major, exists on a continuum, and the definition for a "mass" extinction is fuzzy.[67] In general terms, a mass extinction is an event during which a significant number of living genera die out in a relatively short amount of time. The speed of such an event can occur in the blink of a geologic eye, which still means it could be millions of years. Reasons for extinctions vary from asteroid impacts to severe carbon-cycle fluctuations promulgating changes to the environment. Often a triggering incident will happen to drastically alter conditions on the planet, even the collapse of food chains, and life cannot adapt quickly enough.[68] Geologists correlate the rapidity of an extinction with a more significant loss in species and genera. The mass extinctions herald new eras of life after each event, though some life forms persist. An example of an extinction affecting evolution is that of the rodents and multituberculates (figure 7.5).[69] The little-known order of Multituberculata is the longest-lived order of mammals ever, extending from early in the late Jurassic to the late Eocene, more than 165 million years. The dominant theory for the extinction of such a long-lived taxon is that it never fully recovered from the Cretaceous-Paleogene extinction (K-Pg) and was eventually outcompeted by rodents.[70]

Mass Extinctions

The literature often discusses five mass extinctions in the Earth's biography, but this number neglects significant extinctions in the Precambrian Supereon. Evidence is more sparse going back in the geologic record, but paleontologists and geologists may have found indications of the extinction events before the Phanerozoic Eon. One of these occurred following the build-up of oxygen in the oceans and the atmosphere from the Great Oxygenation Event in the Proterozoic Eon. Cyanobacteria were harnessing the energy of the sun, the beginning of photosynthesis, utilizing carbon dioxide from the air, and changing the very nature

(a)

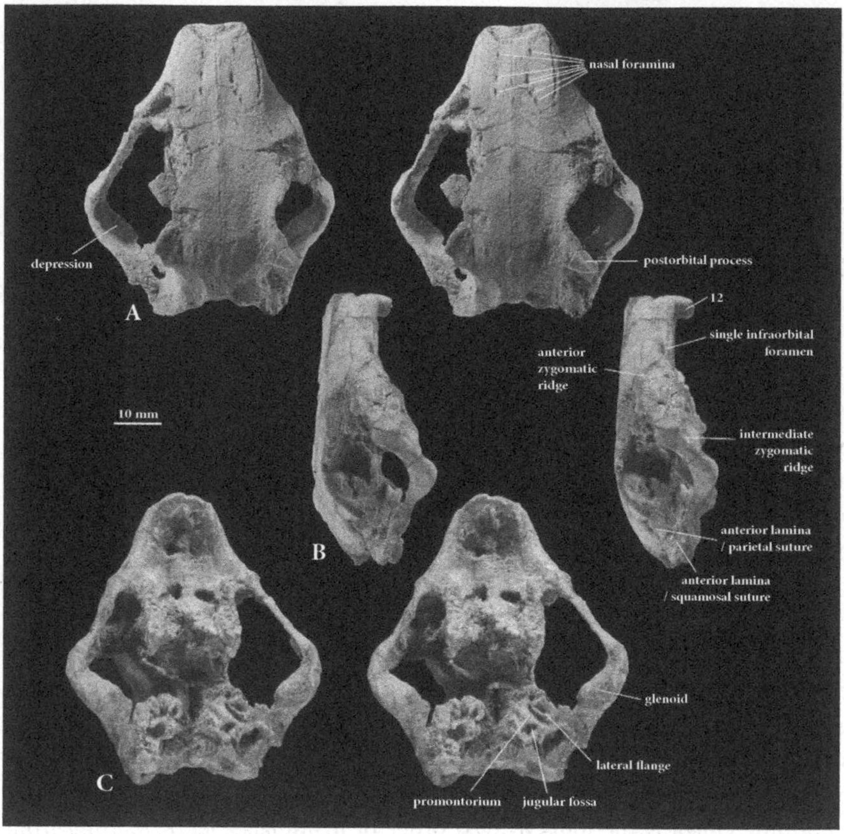

(b)

Figure 7.5 (a) Multituberculate *Catopsbaatar catopsaloides* skulls. (Kielan-Jaworowska, Hurum, and Lopatin, 2005) (b) Multituberculate life drawing. (Kielan-Jaworowska and Hurum, 2006; illustration by Bogusław Waksmundzki)

of the planet and the chemistry of the oceans by adding oxygen to the oceans and possibly reducing the nutrients in the waters. The theory is that in doing so, they killed off a large number of their cousins, the anaerobic bacteria, causing earlier "feast" conditions to devolve into "famine" for the next 1,000 million years.[71] Paleontologists log another major extinction wiping out much of the soft-bodied Ediacaran Period (Neoproterozoic Era) fauna and marking the end of the Precambrian.[72] The Cambrian explosion of life marks the beginning of the Phanerozoic Eon, when organisms diversified and increased in complexity.

During the Phanerozoic Eon, geologists track five significant extinctions on the Earth, and each time life has rebounded in new ways. The number of extinction events is in the scores, but only five have killed enough organisms to qualify as mass extinctions in the last 541 million years. As discussed in chapter 3, three mass extinctions mark the boundaries between major geologic eras of the Phanerozoic (between the Paleozoic and the Mesozoic and between the Mesozoic and the Cenozoic).

Among the five mass extinctions, most people are aware of the asteroid impact 65 million years ago, associated with the K-Pg event that ended the presence of the non-avian dinosaurs on the Earth and closed the Mesozoic Era. However, though this mass extinction is the closest to modern time and ushered in the Cenozoic Era, it was only the third-largest in the planet's history. The most significant mass extinction in Earth's biography occurred late in the Permian, 252 million years ago, called the end-Permian extinction or the "Great Dying." During this time, the climate changed dramatically, with severe greenhouse conditions. In a cascade of events, including the destabilization of the carbon and methane cycles, conditions on the Earth became inimical to life. Some paleontologists had estimated that more than 90 percent of the marine species on the planet became extinct. However, recent research contends that the number of losses in genera and families may be far less, around 81 percent in the oceans.[73] Steven Stanley, of the University of Hawaii, Manoa, the author of the research, argues that prior paleontologists overestimated the number of family and genera losses because background extinctions, tens of millions of years earlier, took out flora and fauna before the main event. Nevertheless, it was the most extensive mass extinction, and stratigraphers recognize its significance to geologic time. It ended the Paleozoic Era, and a new era commenced, the Mesozoic.

The Great Dying at the end of the Permian Period and other events, such as the Triassic-Jurassic Period extinction, are analogues for the modern climate. Another time of high greenhouse gases is called the Paleogene-Eocene Thermal Maximum (PETM) at the end of the Paleocene Epoch in the Cenozoic Era. This was a time of intense warming and related to a rise in carbon dioxide in the atmosphere and the ocean. Ocean circulation patterns changed dramatically, causing a depletion of oxygen in the ocean waters and a corresponding extinction of 30 to 50 percent of the marine foraminifera.[74] Dating of the PETM indicates that it lasted just 170,000 years, a relatively short time in the geologic record.[75] The data from the PETM have confirmed that rising carbon dioxide levels and methane are the cause of increasing global temperatures, including surface ocean temperatures, melting ice caps, and rising sea levels. It also most likely prompted the evolution of modern mammal forms.[76] Before the end of the Paleocene, most fossil mammals are unrecognizable when compared with subsequently evolved mammals.

The sources of these events were natural, unlike the current situation of humans burning fossil fuels. However, the original causes of the increased carbon and methane do not matter when it comes to the disruption of both cycles. Scientists now know that higher values of carbon dioxide and methane have significant impacts. Researchers point to evidence that the world currently is undergoing a sixth mass extinction event, a biodiversity crisis, one caused by pressure on species through habitat reduction, hunting, and environmental degradation, based on rapid global warming.

Biodiversity throughout the Geologic Record

Biologists measure the diversity of life by examining the number of species and their complexity. Biodiversity numbers correlate with the amount of land or area in which a species has to spread out; the larger the area, the higher the number of species. For paleontologists, however, it is much more difficult to calculate the area component in the fossil record. Fossils, as mentioned before, make up only a fraction of life that lived at a particular time, resulting in incomplete data from the fossil record; the incomplete fossil record leaves gaps in what organisms were alive at any given time. The rocks containing fossils are likewise limited in where they are exposed or where they are accessible. Paleontologists estimate the volume of rock or outcrop that contains the fossil to compensate for this difficulty. In a newer method, paleontologists back-calculate for the area effect based on interactive mapping of fossils in an area and use of software image analysis to create graphs relating paleospecies to area.[77] Computer models and simulations are employed as well to help sort out the number of species over time in the fossil record, and researchers need to consider the modeling process and how the models work in analyzing and correcting the fossil record.[78] This system results in an estimate at best, because, for example, organisms that did not live together may be fossilized together. Dead organisms can easily be moved by stream flow or floods to be thrown together on point bars or other areas of deposition.

Scientists at the Marine Biological Lab at the Woods Hole Oceanographic Institution in Massachusetts, undertook early work aggregating and quantifying species over geologic time in the 1970s, where renowned paleontologists including Gould, Thomas M. Schopf, Daniel S. Simberloff, and David M. Raup met, researched, and began working on computer models addressing paleospecies. One of Gould's graduate students was John J. Sepkoski (1948–1999), to whom Gould gave the arduous task of making a worldwide database of all marine fossils. Sepkoski labored for nearly ten years to create his *Compendium of Fossil Marine Families*, published in 1982, and he reported on his kinetic model of fossil families in 1984.[79] From this work came the now-famous "Sepkoski Curve," a graph of the prevalence of marine families over the past 600 million years. The graph shows an increase in fauna throughout the geologic record, with interruptions from extinction events. These events track the beginnings and endings of the three major Phanerozoic Eras.

In 1992, Sepkoski highlighted a total of 4,075 taxonomic families of marine animals.[80] Other paleontologists have made graphs for land animals,[81] plants,[82] and invertebrate forms of life. The Sepkoski Curve highlights the five mass extinctions of the Phanerozoic Eon, with the largest occurring at the end of the Permian Period. It also demonstrates that over the geologic record, biodiversity has increased, despite the setbacks of mass extinctions. Other researchers have statistically examined patterns in Sepkoski's data (color insert 7.2), finding a 62-million-year cycle in diversity.[83] Paleontologists continue to explore the reason for the cycle.

The unified theory of evolution and its history have been interwoven with the chronicle of geology and fossils. It is one of the three overarching principles of the Earth and holds a place among the giants of theories.

∞

Unifying Theories and the Future

These three overarching theories, in the fullest sense of the word "theory"—geologic time, plate tectonics, and evolution—have been refined and verified through testing of hypotheses, debate, argument, and used in predictive ways. This is the scientific method in action. All three hypotheses were initially discovered and elaborated through observation, reasoning, and posing questions. Each theory was subsequently confirmed with quantitative methods.

Further geologic research continues to add to the current understanding of each of the unifying geologic theories. Geologists are discovering more details about particular stratigraphic and geologic units, and their impact on geologic time, as geologists gather more data and information.

Scientists in the future will undoubtedly discover more about critical features on the ocean floor, deep-sea trenches, seamounts, and forces related to bathymetric changes in the oceans, which will help fill out and provide new insight for the science of geology and plate tectonics. The National Ocean Service notes that only 5 percent of the ocean has been explored as of 2014.[84] But oceans cover about 70 percent of the Earth's surface and are the location where a significant portion of plate tectonic activities occurs. GPS sensors are now used to track plate movements and provide refinement for the rates of movement of various plates. Vectors show the direction and magnitude of plate motion based on GPS sensors.

Theories, even those that define the most basic tenets of geology, will change and become more sophisticated over time as geologists gather data, more information is analyzed and tested, and further research is undertaken.

The key precepts of the Earth are geologic time, how the continents and oceans form and move, and how life changes over time in response to the environment. As rhythm, harmony, and melody are the elements of a song, so, too, are these geologic precepts the basis, foundation, and framework for the performance of the Earth as a whole.

8

The Biography of the Earth

The Precambrian Story

The song of the Earth, which began some 4,600 million years ago, may be best understood by examining its constituent parts. Much like the way music theorists may deconstruct a contemporary song in segments that include an intro, verses, a pre-chorus, a chorus, a bridge, and an outro, geologists arrange the order of the Earth's development from the immense periods known as eons, into smaller subdivisions, consisting of eras, periods, epochs, and stages (color insert 8.1). The Precambrian is an exception, being a supereon, and the topic of this chapter.

The Precambrian Supereon, spans the birth of the Earth from 4,600 million years ago, after the solar system and planets coalesced from a nebula formed after the Big Bang birth of the Universe (1,380 million years ago)[1] to 541 million years ago. The Precambrian represents an immense amount of geologic time, roughly 88 percent, or more than three-quarters, of the Earth's duration (figure 8.1). Stratigraphers reclassified the Precambrian, which is named for all of geologic time before the Cambrian Period of the Paleozoic Era, from an eon to a supereon.[2] It is the only supereon, thus far, in geologic time. Eons within the Precambrian are the Hadean (named for Hades) 4,600–4,000 million years ago, the Archean ("old") 4,000–2,500 million years ago, and the Proterozoic ("early" life) 2,500–541 million years ago. The Archean and Proterozoic Eons are further subdivided into eras.

Though the Precambrian makes up the bulk of the time scale of the Earth, it is the least known. Geologists have ascertained the approximate dates within it, through numerical age-dating, but what happened during that time (its record), and when, are less certain. Many rocks from the Precambrian Supereon are gone, recycled through the origination, movement, and destruction of tectonic plates. Only in the stable cratons that make up the unaltered areas at the centers of continents, or in slivers here and there on the surface of the planet, may geologists find these ancient rocks. In the western hemisphere, geologists find Precambrian-aged formations in Greenland, Canada, parts of the United States (Rocky Mountain region), and South America.

The Precambrian is the time in the Earth's biography when the five spheres of the planet became established. These consist of (1) the magnetosphere, the protective magnetic field around the planet; (2) the geosphere, the structure

Song of the Earth. Elisabeth Ervin-Blankenheim, Oxford University Press. © Oxford University Press 2021.
DOI: 10.1093/oso/9780197502464.003.0009

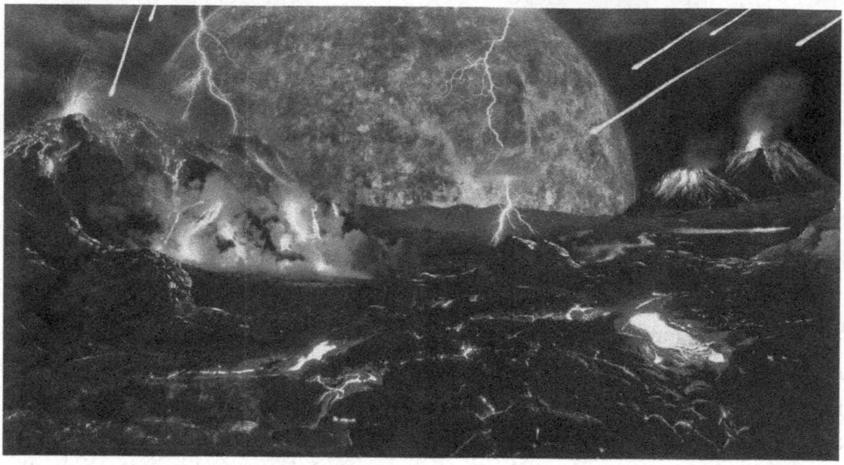

Figure 8.1 Hadean Eon, artist's impression. (Artist Tim Bertelink, 2016)

of the solid Earth, including the tectonic plates and all rocks and minerals, the continents, and the ocean floor; (3) the hydrosphere, all water on the Earth, including ice and liquid water on the surface; (4) the atmosphere, the air envelope around the Earth, its composition and climatic impacts; and (5) the biosphere, all life on the Earth. These spheres interact dynamically; changes in one sphere can have immense implications for the others. Indeed, life as we know it would not exist without all spheres of the Earth operating together. This interplay will be discussed throughout chapters 8–11 which address the Earth's biography.

One dynamic interaction of these spheres is evidenced by the presence of three very different atmospheres (color insert 8.2) that have occurred on the planet, the development of which spans the Precambrian Supereon and extends beyond it with the Earth's present atmosphere. The first atmosphere, which would have been toxic to life known today, formed during the Hadean Eon and consisted mainly of hydrogen and helium, gases derived from the solar nebula, and the protoplanetary disc. The sun and planets formed from a spinning cloud of particles and gases. On the Earth, these light gases were all eventually lost to solar winds because of the new planet's lack of fully developed gravitational and magnetic fields. The second atmosphere, established early in the Archean Eon, was made up of carbon dioxide, water vapor, methane, and nitrogen, from volcanic gases and the beginning of plate tectonics. This air, too, would have been toxic to life as it exists today. The Earth's current nitrogen, oxygen, and trace-gas atmosphere morphed from the gases of this second Archean atmosphere. Cyanobacteria, in the form of stromatolites, became more dominant over hundreds of millions of years and, in an elegant and almost incomprehensibly long transformation, gradually converted the atmosphere to oxygen from carbon dioxide through photosynthesis, forming the atmosphere we know. This third and current atmosphere began to appear in the late Archean Eon and continued to develop in the Proterozoic Eon. This atmosphere was and remains dynamic; the amount of O_2 and CO_2 has varied throughout time.

Hadean Eon, 4,600–4,000 Million Years Ago

The poet Dante Alighieri and various Renaissance artists imaged Hades as the bowels of hell at the center of the Earth. Indeed, early philosophers suggested that Hades was hell-like (figure 8.1). Nevertheless, as the study of the Earth proceeded, scientific discoveries replaced folklore. Instead of Hades at the planet's center, scientists discerned that the Earth had layers, and within were a core and mantle. They made these discoveries through the study of seismic waves that pass variably through materials, depending on whether they are liquid or solid.

Geologists know little about this earliest eon. Some think the Hadean was hot, with molten rock, lava bombs, and generally inhospitable conditions. In the hot-early-Earth version, the only water around would have been in the form of steam. Oceans consisted of hot magma, and no landmasses had emerged—truly a hell-like scenario. Others speculate that conditions were cooler and that liquid water would have been possible fairly early in the planet's history.[3] Evidence for a cold early Earth during the Hadean Eon includes geochemical models of global carbon dioxide cycles and data from oxygen isotopes. Under these conditions, life could have begun earlier, with liquid water to support it.

The processes of plate tectonics probably were not even a glimmer during the Hadean Eon under hot-early-Earth conditions, and thus, most likely, there were no continents. Particular thermal, compositional, and fluid conditions are needed to generate the dynamics to create the conditions necessary for tectonism, but as some scientists speculate, it was cooler during this time, and incipient plate tectonics may have begun.

During the first billion years, extending from the Hadean to the Archean Eon, meteors and asteroids pummeled the early Earth in a period called the Heavy Bombardment.[4] No impact craters were made on the Earth because there was no lithosphere to record the collisions. However, scientists have found analogue evidence from the impacts on the other inner planets, on our moon, and on other moons in the solar system.

Few, if any, rocks have been found, so far, from the Hadean Eon. In the 1980s, geologists (see chapter 4) found zircon crystals in detrital deposits from the Jack Hills in Southwestern Australia, age-dated to 4,400 million years ago.[5] Geologists speculate that these zircon crystals formed in ancient magma chambers. Rocks from the Acasta Gneiss in Canada have been dated near the upper boundary, 3,960 million years ago.[6]

An event early in the history of the Earth was the formation of the moon. At 4,500 million years ago, during the Hadean Eon, the moon was formed following a cataclysmic impact with a large protoplanet. In the most popular theory, astronomers propose that the moon was shaped in the aftermath of a Mars-sized object, which they named Theia, smashing into the Earth, creating a ring of particles that eventually coalesced to form the moon. They dub this event the "Big Splash" or the "Giant Impact" theory (figure 8.2).

How did scientists hypothesize such theories when the occurrence in question is so far back in geologic time? Of course, they had no direct evidence of what exactly occurred in the making of the moon, but instead, starting in the 1970s, they began to develop mathematical models (representations of conditions in the solar system) to test hypotheses about the moon's origin, including that of the Big Splash.[7] In more recent years, scientists created new models to refine the Big Splash origin of the moon. The newer data indicated that multiple impacts might have been necessary—the "mini-moons" hypothesis. Astronomers theorize that fragments from the impacts might have led to various moons that would have coalesced eventually into one lunar mass. The debate remains centered on the similar composition of the moon and the Earth and similar mixes of oxygen isotopes, which are like a fingerprint in detecting the age and history of a

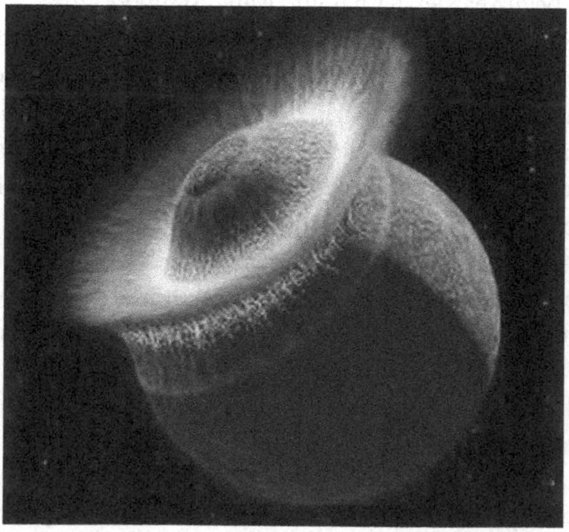

Figure 8.2 Formation of the moon: The Big Splash. (Artist Memomiguel, 2012)

particular rock. Another question is whether the collision between the Earth and Theia, if it occurred, was a glancing blow or a direct hit.

The Big Splash theory fits the majority of the data astronomers and physicists know about the moon, including the high-spin component (angular momentum) of both the Earth and the moon. Their initial thoughts were that Theia hit the Earth with a glancing impact, excavating 40 percent of the material that would make up the new moon. This theory accounts for the high-silica-rich, low-volatile (chemical compounds with low boiling points, such as nitrogen, water, carbon dioxide, ammonia, hydrogen, methane, and sulfur dioxide), and iron-poor content of the moon. In this view, the iron core of Theia melted when it hit and joined with the Earth's core. The difficulty with this model is that although it accounts for low concentrations of volatiles on the moon, it does not address the issue of the moon being composed of apparently the same isotopes as the Earth.

In 2014, physicists fine-tuned their models and came up with some exciting results. These recent calculations indicate that Theia hit the Earth in a high-speed impact that generated immense temperatures and pressures and completely demolished the protoplanet, which melted into the Earth. The immense impact created conditions that exceeded the "supercritical threshold" for materials. In such conditions, liquids and gases (and most solids) no longer exist. The lot is smashed together into a fluid substance with unique properties. Some of the properties are gas-like, allowing the substance to move through a solid, and others are liquid-like, able to dissolve materials. However, all operate together. Physicists pose that Theia merged with the Earth's mass, generating a supercritical fluid that spread from what had once been the Earth's core, generating a "cloud" that was neither solid nor liquid but a supercritical mass of stuff.[8] It was a continuous, seamless creation estimated to have a radius of 10,000 kilometers (6,200 miles). It would have been an uninterrupted structure, unlike the rings of Saturn. Astronomer call this new creation a synestia (from the Greek *syn* for "the same" and Hestia for the goddess of building and construction). One possible shape for the hypothesized synestia would have been doughnut-like, without the hole in the middle, but many other geometries could theoretically occur. The Earth and the moon alike condensed out of this supercritical fluid. Scientists have never observed such a phenomenon in the solar system, but new isotopic data exist to support it.[9]

In the meantime, chemists made advances in measuring isotopes, allowing for the detection of even smaller amounts than ever before. These new developments support the idea of a synestia, because scientists can now detect minuscule isotopic differences between the moon and the Earth. When the scientists analyzed moon rocks, they found potassium isotopes to be slightly heavier than those on the Earth. This difference, they theorize, can be accounted for by the cooling during the collapse of the supercritical fluid, as materials separated out of the mass. The slow-impact idea, as first proposed in the Big Splash, would not have generated enough energy to cause such a signal in lunar potassium. But the synestia will remain hypothetical until an actual synestia is observed in the universe.

As heavier atoms started to settle out on the Earth by gravity, the iron and nickel inner core was forged in the center of the planet, which was under so much pressure that it was solid. The outer liquid core separated from the inner core and, as it rotated, initiated the magnetic field that today protects the planet. Researchers theorize that this segregation and development of the planet's core, thus setting up the Earth's magnetosphere, occurred between 4,200 million and 3,300 million years ago during the Hadean and continuing into the Archean Eon. The data pertinent to the magnetosphere come from the isotopic analysis of minerals discussed above, the detrital zircons from the Jack Hills in Western Australia.[10] The challenge of dating minerals like the zircons is that they must be unaltered by metamorphic processes of temperature and pressure that would obliterate the original signature of the crystal.

At first, the magnetic field of the planet was not very strong, and it took time for it to develop fully. However, once the magnetosphere formed, its role in the history of the Earth and the viability of life on it cannot be overestimated, as it provided (and continues to provide) vital protection from harmful rays of the sun, the solar wind, and other kinds of radiation in space (figure 8.3). During the times when the magnetic field undergoes reversals, for a relatively short amount of time, it is weakened or possibly absent, producing fluctuation in the strength of the field. The last magnetic reversal was 773,000 years ago, and while some researchers think it took 22,000 years to complete,[11] others speculate that the reversal happened in 8,000 years. Magnetic reversals are not uncommon; in the last 20 million years, they have occurred every 200,000 to 300,000 years.[12] No one knows exactly what happens during magnetic reversals. However, the magnetic field weakens, and cosmic rays could bombard the Earth and impact life.

It is difficult to imagine the Earth without plate tectonics, the creation and destruction of crust and movement of the large lithospheric plates, as geologists now understand it. However, data indicate that temperatures and pressures did not favor the development of a mobile crust until much later in the Earth's biography, especially if warm- or hot-early-Earth conditions existed. Over the planet's first 2 billion years, scientists think that "stagnant lid conditions" occurred in the mantle, and one large plate of basalt covered the surface.[13] In this early thermal regime, a plethora of hot mantle plumes likely predominated, with possible recycling of the lid through mantle turnover. However, these conditions would have had to happen without convection cells (as are found with today's tectonic settings) because of the presence of immense heat.[14] Geodynamic models (those that predict the Earth's early temperature regime, thermal conditions, and mantle

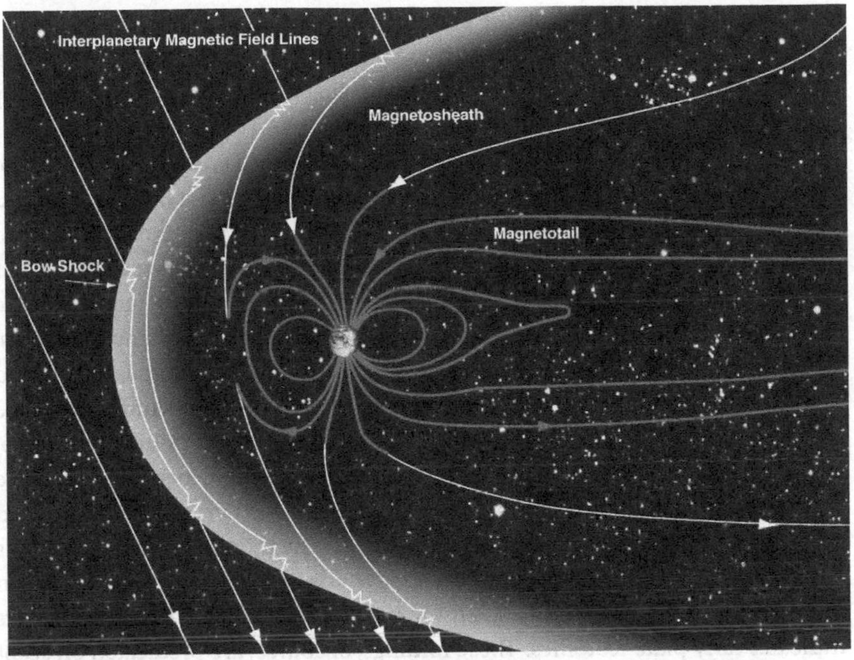

Figure 8.3 The Earth's magnetosphere. (NASA/Goddard, 2017; rendering by Aaron Kaase)

conditions) indicate that high temperatures in the mantle during this time of the Earth's early history have the main effect of decreasing convective stresses and mantle viscosity, which result in the mantle not supporting convection cells.

Planetary scientists and geologists theorize that Venus, which has a stagnant lid, represents similar conditions on the early Earth, with its lack of plate tectonics and one dominant plate that does not move. Another line of evidence is that the magnetic poles recorded in ancient rocks (see chapter 5) were vastly different once plate tectonics got under way in the mid- to late Precambrian Supereon. Rocks from the stagnant lid period do not show changes in pole locations over time, unlike rocks when plate tectonics became active. However, the debate is ongoing as to when stagnant lid conditions gave way to the modern plate tectonics, but pole positions indicate that the breakup of the lid occurred around 1,100 million years ago after the mid-Proterozoic Eon.

The first atmosphere on the Earth arose from the noble gases contained in the protoplanetary disc, which coalesced as captured gases from the nebula surrounded the incipient Earth. However, the energy flux from the sun was much higher than in the present day. Accordingly, this energy flux stripped the envelope of hydrogen- and helium-rich atmosphere from the planet, and it was lost to space.[15] In addition, in the early Hadean Eon, the gravitational field was weak or nonexistent and not strong enough to retain the Earth's first atmosphere.

One of the Jack Hills zircons offers a tantalizing clue to the potential for early life in the Hadean. Researchers sorted through thousands of these detrital zircons and found two containing inclusions (material contained within the mineral) of graphite, a pure carbon mineral. They rejected one of the crystals because it had a crack, which could have allowed extraneous material to penetrate it. However, the graphite in the remaining zircon was older than the crystal and dated 4,100 million years ago.[16] Graphite, depending on its isotopic signature, can be an indirect indicator of carbon-based life, depending on its isotopic analysis, and paleontologists classified it as a chemical fossil. Chemical fossils are created as an organism decomposes and leaves behind molecules that imply evidence of life by its isotopic signature. Not only could the graphite point to early life, but it also might elucidate planetary processes and the very beginning of plate tectonics with subduction zones. For such to occur, the Earth may have cooled more rapidly in the Hadean than scientists previously thought.[17] The researchers posit that carbon in the inclusion must have arisen from a continental origin and indicates that the planet cooled more quickly. The presence of continents indicates early plate tectonics. These findings, of course, are predicated on scant evidence, and the debate will continue about what the zircons mean for early life and what was going on in the Hadean, until the models are refined and if, with luck, other samples are found from the Earth's earliest time.

Archean Eon, 4,000–2,500 Million Years Ago

The following eon, the Archean, beginning roughly 4,000 million years ago, consists of four eras: Eoarchean (early), 4,000–3,600 million years ago; Paleoarchean (ancient), 3,600–3,200 million years ago; Mesoarchean (middle), 3,200–2,800 million years ago; and Neoarchean (new), 2,800–2,500 million years ago. Stratigraphers have not defined a firm date for the beginning of the Archean Eon. As geologists find and identify minerals and rocks farther back in geologic time, the Eoarchean boundary is moved back earlier and earlier.[18] The eras of the Archean Eon are defined chronometrically, by the use of numerical age-dating of minerals or rocks from that time.

The early Archean is marked by (and roughly corresponds to) the end of heavy bombardment of the Earth by asteroids and comets, though a period of increased activity occurred from 3,900 million to 3,800 million years ago in a period called the Late Heavy Bombardment. Physicists and astronomers attribute this time of increased barrage to asteroids and comets related to the movement of the gas giant planets, Jupiter and Saturn, from outer to inner orbits and back again.[19] This era also has evidence of the first crust on the Earth. Rocks from western Greenland, called the Isua Greenstone Belt, are 3,800 million years old. Geologists infer that a crust must have existed and that plate tectonics was beginning to be active, because these rocks were created in a tectonic setting. As the planet continued to cool, it began to differentiate into the various layers seen today. These developments led to "modern plate tectonics" in that the crust and mantle moved by convection, which requires a rigid lithosphere and lateral motion of the cratons. The crust started to separate into the lighter rocks, such as granites, and the denser material—the basalts. From 3,500 million to 2,700 million years ago, the formation of landmasses occurred by aggregation of the crust into the light, silica-rich material, massing into protocontinents, and dense, iron-rich material making up the oceanic crust.

The Archean is known for the start of the Earth's interacting spheres, the first supercontinent, the development of ocean water chemistry, and the transition from the second atmosphere to the third (modern) atmosphere, through the agency of early life. The second atmosphere, composed mainly of carbon dioxide, water vapor, methane, and nitrogen, was the first to stick around for some time, but it contained little or no oxygen. Thus, there was no protective ozone layer in the stratosphere during that time. Scientists think that once plate tectonics was under way on the early Earth, large amounts of nitrogen were released from the mantle through fissures and where tectonic plates collide and dive down.[20]

Scientists are not in agreement, and research continues, about the source of water and when it first was present on the Earth. As noted, many think conditions in the Hadean Eon would have been too hot for liquid water, and it could have existed only once the Earth cooled during the Archean Eon. In this scenario, they speculate that water on the Earth originated from asteroids hitting the planet during the time of the Late Heavy Bombardment, which carried the molecule along with them as they hit the Earth.[21] Modern asteroids contain little water, but it is most likely that ancient asteroids carried more water in the earlier history of the solar system.

More recent studies, based on the analysis of hydrogen isotopes of asteroids and meteorites, demonstrate that the water on the Earth has the same isotopic signal as the hydrogen isotopes in meteorites from the asteroid Vesta, located near the Earth in the inner part of the solar system. The geochemical data suggest the planet could have accreted under "wet" conditions, with water, as far back as the birth of the Earth 4,600 million years ago.[22] In this version, the water-rich meteorites provided as much as 30 percent of all the water on the Earth from the very beginning of the planet, with the later asteroids bringing additional water.

Not only were asteroids, and perhaps meteorites, the likely source of water on the early Earth, but volcanoes began to emit water vapor along with carbon dioxide and hydrogen sulfide. As the Earth's second atmosphere developed, geologists cite volcanic eruptions as a further source of water on the planet. The Earth is situated in a unique position in the solar system, a place where water can exist in its three forms—gas, solid, and liquid. Without liquid water, life as we know it would not have developed.

Around 3,800 million years ago (although some scientists suggest it could have been earlier), the Earth cooled below the boiling point (212 degrees F), initiating condensation of water to create the early ocean.[23] Scientists speculate that it may have rained for millennia during the Archean. The water cycle during the time of the second atmosphere was initiated with evaporation over the young ocean. Meanwhile, as the outer core of the planet now protected the Earth's second atmosphere with the magnetosphere, volcanic activity released carbon dioxide and methane.

The first life on the Earth may have been non-oxygen-consuming bacteria, extremophiles, from deep-sea hydrothermal volcanic vents called black smokers. However, planktonic life, photosynthesizing organisms, may have existed far earlier than indicated by the geologic record. Geologists have discovered biogenic isotopic markers in the rocks of the Greenstone Belt of the Isua Province in southwest Greenland, dated as 3,700 million years old, providing evidence of photosynthesis and single-celled life.[24] Several hundred million years later in the geologic record, a new life form evolved, single-celled cyanobacteria, 3,400 million years ago. These cyanobacteria were the first verified photosynthesizers and gave rise to structures known as stromatolites as they processed carbon dioxide to create oxygen through photosynthesis. At first, all of the oxygen the stromatolites produced dissolved in the oceans, which were abundant in reduced, dissolved iron, making the waters green in color. As the oxygen in the waters increased, the iron precipitated in the ocean basins, and the oceans turned blue. Geologists like to say the oceans "rusted" as a result of this process, around 2,600 million to 2,400 million years ago, right at the end of the Archean and the beginning of the Proterozoic. The iron deposition on the seafloor created the vast banded iron formations that make up the source of most of the iron mined today.

Stromatolites are the Earth's oldest fossils. For many years, the most ancient stromatolites were dated at 3,500 million years, from the Pilbara Province in Australia. However, in 2016, geologists found much older layers of stromatolites, also from the Isua supracrustal belt rocks in southwest Greenland, dated around 3,700 million years ago.[25] Stromatolites are created by cyanobacteria. Most lived and still thrive in shallow, clear water or within intertidal flats, built up layer by layer in rounded humps of colonies, with trapped particles between their layers. Geologists have found some stromatolites (lithoherms, a domed stromatolite) in deep water, up to 700 meters (2,300 feet) in the Florida Straits.[26] They may not be the earliest types of life on the Earth, but they were the first photosynthesizers. They have the critical ability to absorb carbon dioxide and release oxygen as a byproduct and, as discussed above, had an enormous impact on the development of the Earth's second atmosphere. Modern stromatolites grow and can be seen in the Indian Ocean at Australia's Shark Bay and elsewhere.

Recent research has shown that early life, first forming in the Archean Eon, had no protection from the sun's ultraviolet rays, but some of the first bacteria may have developed their own sunscreen.[27] These bacteria, called iron-oxidizing bacteria, lived in the iron-rich oceans and converted the iron from one form to another, providing a barrier from the sun's rays for themselves. Not all bacteria in the Archean had that ability; others were photosynthesizing around the same time.

Ultimately, enough oxygen was built up in the air once the oceans had absorbed all of the initial oxygen, causing them, as noted above, to rust over a long time span during what is called the Great Oxygenation Event (GOE). Scientists think the GOE may have started at the end of the Archean, 2,650 million years ago, but that the full oxidation event, resulting in the development of the third atmosphere, most likely occurred in stages—as many as seven phases, spanning from the end of the Archean Eon into the Phanerozoic Eon. Ultimately, the GOE enriched the oxygen in the atmosphere to the current oxygen content of 21 percent.[28]

Researchers think the initial rise in oxygenation during the Archean Eon also was related to the formation of the start of the supercontinents and landmasses.[29] As the continents aggregated, mountains rose, and weathering and erosion began to affect the high peaks. The transport of the weathered material carried sediments and nutrients into the oceans, providing beneficial conditions for life in the seas. This development underscores the harmonious interaction of all the Earth's spheres. In the geosphere, tectonics and weathering influenced the biosphere by providing nutrients for colonies of stromatolites. The life forms subsequently influenced the atmosphere through photosynthesis and the release of oxygen, which in turn influenced more growth in types of life forms.

The free oxygen in the air was bombarded by ultraviolet particles, making the two oxygen atoms split apart and combine with a third oxygen atom, forming protective ozone (pure O_3) in the stratosphere. Ozone was and remains vital for the continuation of biological development. Once the Earth's organisms were protected from ultraviolet rays by ozone, they were able to exploit other environments.

Proterozoic Eon, 2,500–541 Million Years Ago

The last eon before the Phanerozoic is the Proterozoic. The Proterozoic Eon began 2,500 million years ago, and within it are three eras: Paleoproterozoic (old),

2,500–1,600 million years ago; Mesoproterozoic (middle), 1,600–1,000 million years ago; and Neoproterozoic (new), 1,000–541 million years ago. Again, the boundaries between these eras are defined chronometrically. Unlike the Hadean Eon, with no eras, and the Archean, with eras but no geologically defined periods, the Proterozoic contains further subdivisions of the three eras, or geologic periods, like those of the Phanerozoic. Furthermore, while geologists name these times for general geologic processes typically occurring during a particular time, they are not uniquely "diagnostic" of the time unit.[30] For example, the Cryogenian Period of the Neoproterozoic Era, from approximately 720–635 million years ago, is named for the recurring snow and ice of the Snowball Earth times.

The Earth began, in the Proterozoic Eon, to bear more resemblance to its appearance today. As mentioned during discussion of the Archean Eon, now there were continents and plate tectonics and an atmosphere that was more hospitable to life forms as they developed and took root. Geologists have analyzed Precambrian rocks from Greenland and have discovered, based on their isotopic signatures, that incipient plate tectonics occurred as far back as 3,500–3,200 million years ago in the Archean Eon.[31]

Paleomagnetic data place the initial movement of cratons to 1,880–1,110 million years ago in the Proterozoic Era.[32] However, the lack of data in the form of rocks from that time and a poor understanding of the magnetic field before the Proterozoic limit the ability to recognize events farther back in time. Given the advances in paleomagnetism research and the discovery of other, older rocks, the record of early modern tectonics may ultimately reach back to the early Proterozoic and even the Archean Eras.

In the Paleoproterozoic Era, cataclysmic events were triggered by the movement of lithospheric plates, as they aggregated from the embryonic continental masses and oceanic plates some 2,500 million years ago. These events marked the beginning of the second phase of plate tectonic actions and the rise of modern plate tectonics. No longer was there one continuous crust, the stagnant lid. The planet had cooled, and the geothermal gradient (the increase of temperature with depth in the Earth's interior) increased.

Mantle processes changed with the difference in the heat regime, thus creating new structures in the crust. Pieces of landmasses were smashed together, forming mountain ranges, trenches, and suture zones (areas where tectonic plates have been stitched together by shearing and compressive forces). Immense and impactful events triggered by the movement of the great lithospheric plates created and aggregated the continental masses in the early history of the planet. Landmasses, island chains, and protocontinents were smashed together, some forming mountain ranges. One example of early tectonics was the Grenville orogeny—mountain-building event—which occurred during the

Mesoproterozoic, 1,500–1,000 million years ago, as the protocontinents assembled into the large landmass Rodinia. The mountains of the Appalachian Blue Ridge, as well as the Adirondacks and New Jersey Highlands, were created in the Grenville period.[33] Areas associated with the Grenville orogeny are also known in Texas, Mexico, and Canada. During this time, an ocean to the southeast of Laurentia (what would become North America) closed. Stresses from the merging of the landmasses extend as far north as Vermont and Canada and south to Texas, where the Llano uplift is related to the episode, and portions of Central America.[34] Rifting with back-arc spreading created basins, like the Sea of Japan. Thrust faults were formed along with a metamorphic belt of rock and exotic terranes.

These masses, over time, aggregated into even more massive continents, fused through suture zones where the material is sheared together, buckled, and stretched, under monumental stress. An instance of these deep shears still in evidence today can be observed in the Cheyenne Belt, located along the northeast axis of the Medicine Bow Mountains and the Snowy Range in Wyoming, among other locations.

In the Cheyenne Belt, Proterozoic island arcs collided 1,780–1,750 million years ago with an older Archean region of the Wyoming Province. Geologists agree that the Cheyenne Belt is one of the most impressive suture zones in rocks of the Precambrian in the western United States. This seaming together of lands did not occur in an instant of geologic time. Instead, it took millions of years of inexorable grinding and shearing of ocean sediments and island arc materials against the young continent, creating heat, temperature, and pressure changes that cooked and deformed these materials but did not melt the rocks trapped in between.

Geologists have discovered unusual rocks called migmatites, along with ancient stromatolites, in and near the Snowy Range in Wyoming's Cheyenne Belt. Located next to these migmatites are deformed rocks with tight chevron folds, such as the French slate (figure 8.4a).

Before the shearing event, an ocean once existed on the edge of Wyoming, where now there are mountains, during the Proterozoic Eon. The quartzite and other metamorphic rocks of the Medicine Bow Mountains at Lake Marie originally were formed from sands, reefs, and near-shore clays in shallow oceans, before lithification, including cementation, made them into sandstone, limestone, and shale, respectively. The ocean environment lasted long enough for sediments to accumulate and lithify, at least until the shearing event as the tectonic plates smashed and slid past each other. Moreover, before the shearing, the limestone further morphed into a rock called dolomite as magnesium infiltrated its matrix by means of circulating water. The collisions brought about metamorphic forces of compression and shear and transformed the sandstone into the quartzite seen

Figure 8.4 (a) Chevron folds, French slate, Cheyenne Belt, Wyoming. (b) Stromatolite from Nash Fork Formation, Snowy Range, Wyoming, in a meta-dolomite. (Photos by the author, 2017)

today, the dolomite into meta-dolomite, and the shale into slate. Ripple marks from ocean currents and bedding planes are still visible in the quartzite.

The meta-dolomite, called the Nash Fork Formation, is slightly younger than the quartzite and contains vast numbers of stromatolites (figure 8.4b). These Wyoming stromatolites, 2,000 million years in age, are among the largest and best known in the world. Geologists have noted three ancient reef structures in the Snowy Range area that measure from 55 meters (180 feet) to 966 meters (more than 3,100 feet) long. Also found were more than 150 bioherms (mounds of algal mats) resulting in stromatolites 1–33 meters (3–108 feet) wide and 3–33 meters (10–108 feet) long. Some geologists think the Proterozoic stromatolites are exceedingly large because no burrowers or other creatures existed to disturb their growth.

The stromatolites of the Snowy Range, preserved within the meta-dolomite, were likewise squeezed and deformed during compression and shearing, but their essence, shapes, and structures still can be seen. The changes to the environment of these earliest life forms was a calamity to them. These creatures, of course, were not cognizant of their fate, but by being incarnate in the rock, they have sent a message forward in time to tell of an Earth time long gone.

In the late Proterozoic Era, starting between 800 million and 750 million years ago, the supercontinent Rodinia split-rifted apart into smaller landmasses, forming three continents: North Rodinia, South Rodinia, and Congo. Just 150 million years later, in an event called the Pan-African collision, a new supercontinent was formed when these three landmasses fused back together to become what is called Pannotia (Greater Gondwana). The continents rotated and collided between 650 million and 560 million years ago. Nevertheless, then, nearly as soon as Pannotia came together as a supercontinent, it broke up into the four continents of the Paleozoic Era: Laurentia (North America), Baltica (Europe), Siberia, and Gondwana. Pannotia existed for about 50 million years.[35] The Pan-African collision creating Pannotia gave rise to mountains, and the climate was colder, which dropped the sea levels. Ice sheets were present at both poles, though paleoclimatologists believe that the ocean was free of ice in the region of the equator—called "Ice House" conditions.

Geologists theorize the existence of Pannotia based on how the continents, their margins, and their edges fit together, like pieces of a jigsaw puzzle. Though evidence is scarce, they find matching sedimentary and igneous rocks of the same age and composition across the once-separated continents. Geologists also verified the location of Pannotia during the Neoproterozoic Era by using paleomagnetic pole data from the same time frame.

Continuing from the Archean Eon, stromatolites kept producing oxygen, with the excess oxygen filling the Proterozoic atmosphere in pulses. More recent research pins down the dates of the GOE to the beginning of the Paleoproterozoic,

2,426 million years ago, and continuing to 2,060 million years ago.[36] This may have led to the first mass extinction, that of the anaerobic bacteria. Scientists speculate that the oxidation events were interrelated with the development of large igneous provinces on the landmasses as well as the beginning of glaciation.

Several feedback mechanisms among carbon dioxide, methane, weathering of silicate rocks, and the composition of the ocean waters were initiated, causing the lowering of global temperatures. This eon is known for the first of its three "Snowball Earths," when the entire planet froze, resulting from carbon dioxide and methane decreases in atmospheric concentrations, triggering worldwide episodes of cold, snow, and ice. Once the temperature plummeted and ice formed over half of the Earth's surface, the white of the ice reflected the energy of the sun, causing even colder conditions. This time, known as the Huronian glaciation, consisted of at least three periods of glaciation and occurred from about 2,400 million to 2,250 million years ago.[37] The most recent may be known only from South Africa.

Snowball Earths and Emergence of More Complex Life

In the late Proterozoic Eon, the Neoproterozoic Era hosted several additional Snowball Earth episodes, when the entire planet froze and glaciers and ice dominated. The first of these occurred 720 million years ago and may have lasted until 660 million years ago, when it suddenly ended, known as the Sturtian snowball period. Geologists note the possible existence of the earlier Kaigas snowball period (80–735 million years ago), but the rocks used to define it may not be glacially derived, and the period is questionable. Another episode happened 637 million years ago and lasted for 2 million years. Both of these times of near, if not complete, freezing of the Earth occurred during a time called the Cryogenian Period of the Neoproterozoic Era, so named for the two ice ages.[38] Geologists know the planet was mostly frozen because deposits from the glaciers and ice are found in southern Australia, South Africa, and other continents that can be dated by isotopes to those times.[39]

Along with glacial deposits—some found at what is now the equator—dropstones from the Neoproterozoic have been discovered in numerous locations, confirming the presence of ice sheets. Such dropstones occur when the ice meets an ocean. After that, melting and releasing rocks entrained in the ice sheet become embedded in the material below and then encased as the surrounding sediments become rock.

Scientists continue to debate whether ice completely covered the Earth or if there was an open area of water near the equator ("Slushball Earth").[40] No matter which concept scientists find to be more accurate, either situation points to an intensely frigid time for the planet multiple times. The Snowball Earth events discussed here are only those that occurred during the Precambrian. They represent the cryogenic time.

The intricate balance of gases in the atmosphere, the chemistry of the oceans, erosion of the landmasses, and the activities of organisms have always had an enormous effect on the Earth's climate. The content of carbon dioxide in the atmosphere during the late Proterozoic may have been low, but it was, as it is today, a greenhouse gas. This gas, along with methane, created a blanket of warmth around the entire Earth. Nevertheless, the ratio of carbon dioxide, oxygen, and other gases is a delicate dance that sometimes reaches a tipping point, and the cascading effects could be catastrophic. Paleoclimatologists think that, in the Neoproterozoic Era, the average temperature lows reached those of today's Antarctica, about -20 degrees C (approximately -4 degrees F). Snow and ice pervaded the planet, with intense winds and a hostile environment. At the equator, open water existed, the extent of which is unknown. Nevertheless, life persisted in these areas, and in these times, around warm geothermal vents.

Up to the time of the first Snowball Earth, cyanobacteria were the primary life forms, called prokaryotes because their cells are not organized with a nucleus. However, suddenly, in the latest Proterozoic Eon, there were new signs of a different "take" on life from what had previously been on the planet. In essence, cells with a nucleus, or eukaryotes, developed. The rise of the eukaryotes represents an increase in the complexity and intricacy of life. Eukaryotes do not appear as fossils in the rock record until 1,500 million years ago, just before the episodes of Snowball Earth. These creatures were the foundation of all future complex organisms, and from this point in the geologic record, new organisms developed rapidly. They emerged as Snowball Earth and the grip of the massive ice and glacial times began and then melted away. Some geologists think that the extreme conditions of the glacial times acted as a bottleneck in the evolution of the eukaryotes and may have encouraged a radiative burst in the next era, the Paleozoic, when life burgeoned in the Great Cambrian Biodiversification.

The Ediacaran biota are large, soft-bodied animals, the oldest animals found so far on the Earth. They appear in geologic time before fossils that mark the terminal Proterozoic, "small shelly fossils," evolved and before the explosion of multicellular life in the Cambrian. The Ediacaran fauna are named for a locality in south Australia where their fossils are located in rocks dated to a time just after the first Snowball Earth.[41] The Ediacaran organisms were found in preserved reef deposits, not only in Australia but also in Namibia, Newfoundland, and Africa. Many consisted of bubble-like structures 3 centimeters (1 inch) across, with interlinking networks between them, and they are thought to be either protosponges or complex microbial colonies. Some of these multicellular animals were like modern jellyfish and worms. Another was *Kimberella quadrata*, identified initially as a jellyfish but recently reclassified as a mollusk-like bilateral organism (figure 8.5). Sponges generally are recognized as the first complex animals on the Earth.

Figure 8.5 *Kimberella quadrata*. (Royal Belgian Institute of Natural Sciences; photo by Eduard Solà Vázquez, 2009)

After the Ediacaran fauna emerged, at the boundary between the Precambrian Supereon and the Phanerozoic Eon, small shelly fossils are found. These are the fossils of a plethora of tiny creatures, only millimeters long, containing an external hard shell often made of phosphatic material, the first shells recorded in the geologic record. Paleontologists have discovered their fossils in Siberia,[42] Namibia, and South China.

Because the Precambrian Supereon is so far removed in time, it is a challenging period to study. Debates are ongoing as to whether life became more complex and specialized before the Snowball Earths or after. What is inarguable, though, is that the development of multicellular animals was related to the worldwide ice and glaciers of the Snowball episodes.

Research shows that as the ice sheets melted and the world returned to a warmer state, icy waters and sediments carried large amounts of phosphorus to the oceans, where it precipitated the flourishing of sea life. The losers of the Snowball Earth episodes may have been early life forms, but the winners, as will be seen, are those creatures in the next eon as the Cambrian explosion of life ensued.

Color insert I.1. Earth limb with cloud cover, as seen from space, Apollo 10, May 1969. (Image Science & Analysis Laboratory, NASA Johnson Space Center, 1969)

Color insert 1.1. Siccar Point, location of Hutton's missing geologic time, an unconformity. (Modified by the author and R. Gary Raham, 2020, based on a photo by Anne Burgess, Geograph.org.uk)

Color insert 1.2. William Smith's geological map of England and Wales, with part of Scotland, British statute miles, 30[= 155 mm (Smith, 1815; British Library)

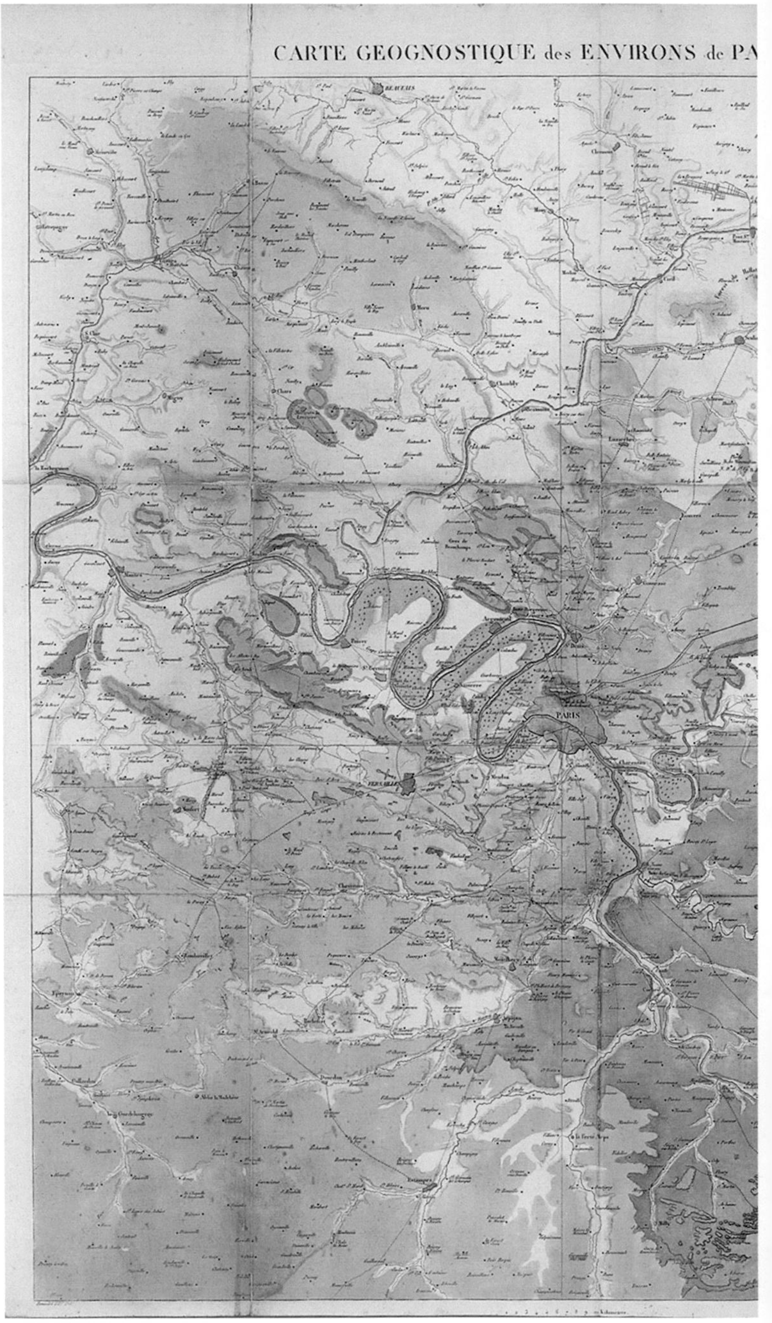

Color insert 1.3. Geologic map of the Paris Basin, 1:200,000. (Cuvier and Brongniart, 1810; courtesy of David Rumsey Map Collection, David Rumsey Map Center, Stanford Libraries)

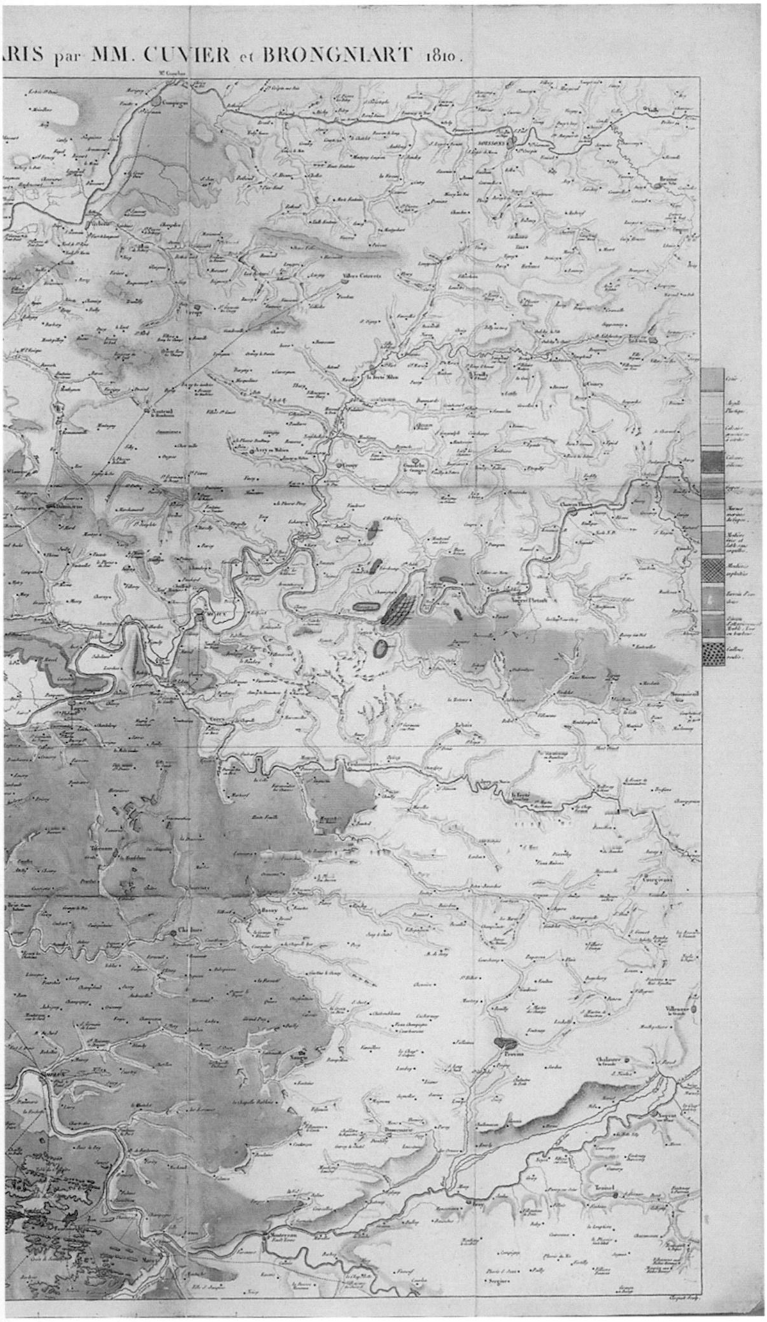
ARIS par MM. CUVIER et BRONGNIART 1810.

MOODIE PLATE 9

Hylonomus lyelli Dawson. 1, maxillæ and skull bones; 1*a*, sternal bones; 2, mandible; 3, humerus, ribs, and vertebræ; 4, posterior limb; 5, pelvis; 6, caudal vertebræ. Nearly natural size. Erect tree, Coal formation, South Joggins, Nova Scotia. Photograph by Dawson, published through the courtesy of Dr. Arthur Willey. Original in the British Museum.

Color insert 2.1. *Hylonomus lyelli* Dawson (some assembly required), Joggins Fossil Cliffs, first land-dwelling animal, found by W. Dawson. (Moody, 1916, plate 9; photo by William Dawson, 1916)

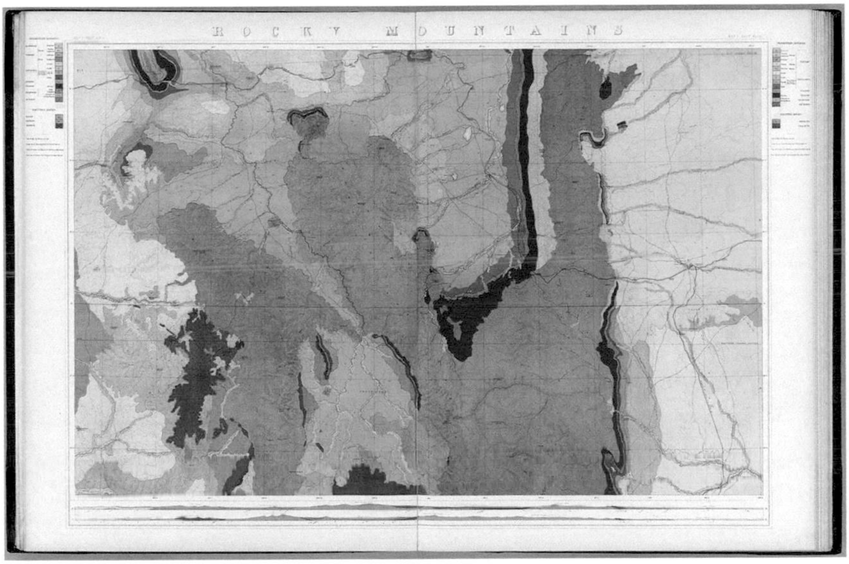

Color insert 2.2. Rocky Mountains, Wyoming, and Colorado, atlas map, scale 1:253,400, 1876, by Clarence King. (King, 1878; courtesy of David Rumsey Map Collection, David Rumsey Map Center, Stanford Libraries, https://purl.stanford.edu/tw016xh4452)

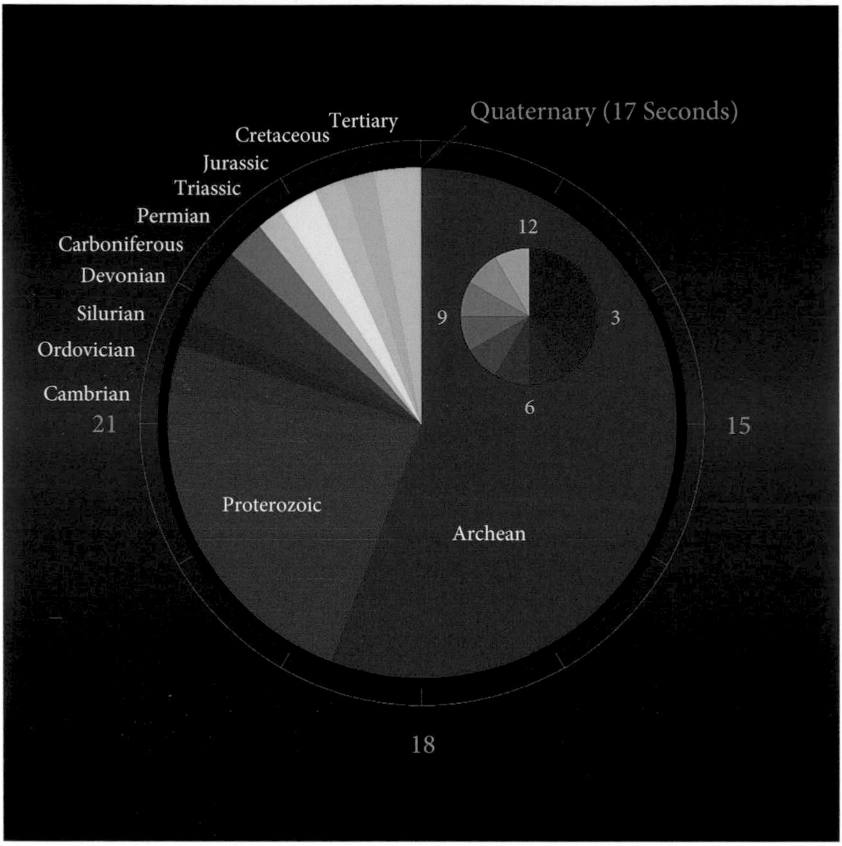

Color insert 3.1. The twenty-four-hour clock representing the span of geologic time. (Woudloper, 2010; derivative work: Hardwigg; James.mcd.nz, 2007)

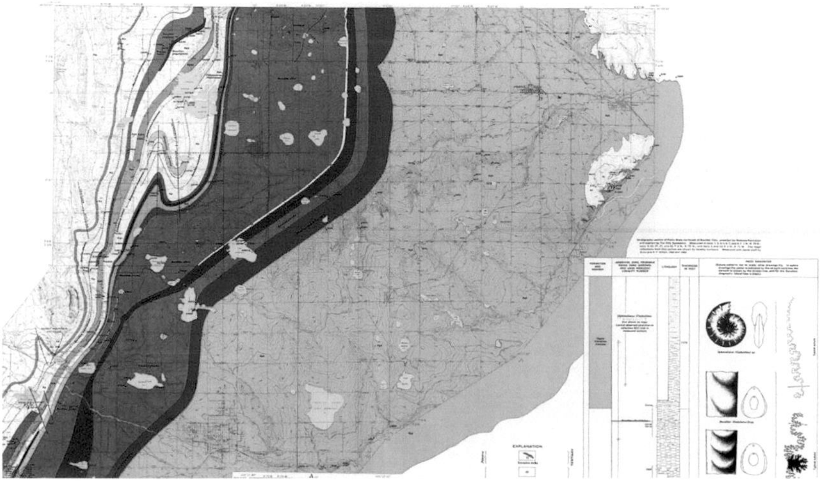

Color insert 3.2. Ammonite zones near Loveland, Colorado. (Scott and Cobban, 1965, cropped from original)

Simplified Geological Timescale				
Supereon/Eon	**Era**	**Period**	**Epoch**	**Date (Million years ago)**
Phanerozoic Eon	Cenozoic	Quaternary	Holocene	0.01
			Pleistocene	2.58
		Neogene	Pliocene	5.3
			Miocene	23
		Paleogene	Oligocene	34
			Eocene	56
			Paleocene	66
	Mesozoic	Cretaceous		145
		Jurassic		201
		Triassic		252
	Paleozoic	Permian		299
		Carboniferous		359
		Devonian		419
		Silurian		444
		Ordovician		485
		Cambrian		541
Precambrian Supereon: Proterozoic Eon				2,500
Precambrian Supereon: Archean Eon				4,000
Precambrian Supereon: Hadean Eon				4,600

Color insert 4.1. Simplified geological time scale. (Illustration by R. Gary Raham, 2020, with input from author, based on the International Commission on Stratigraphy,https://stratigraphy.org/ICSchart/ChronostratChart2020-03.pdf).

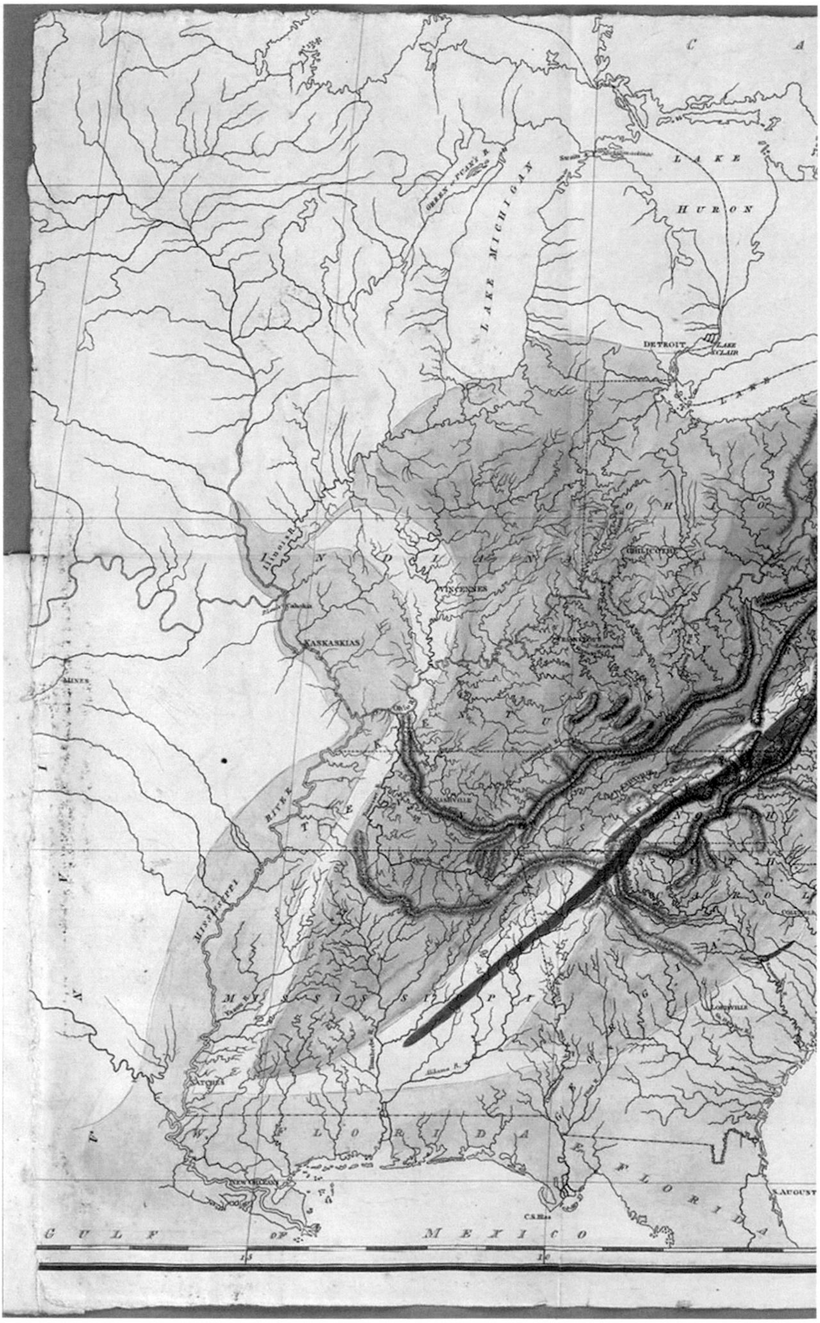

Color insert 4.2. First formal geologic map of the United States, scale of miles 69½ to a degree (1:4,942,080; 1 inch = 78 miles). (Maclure, Tanner, and Lewis, 1809; map made by Samuel G. Lewis; courtesy of David Rumsey Map Collection, David Rumsey Map Center, Stanford Libraries, https://purl.stanford.edu/bs213mj3674)

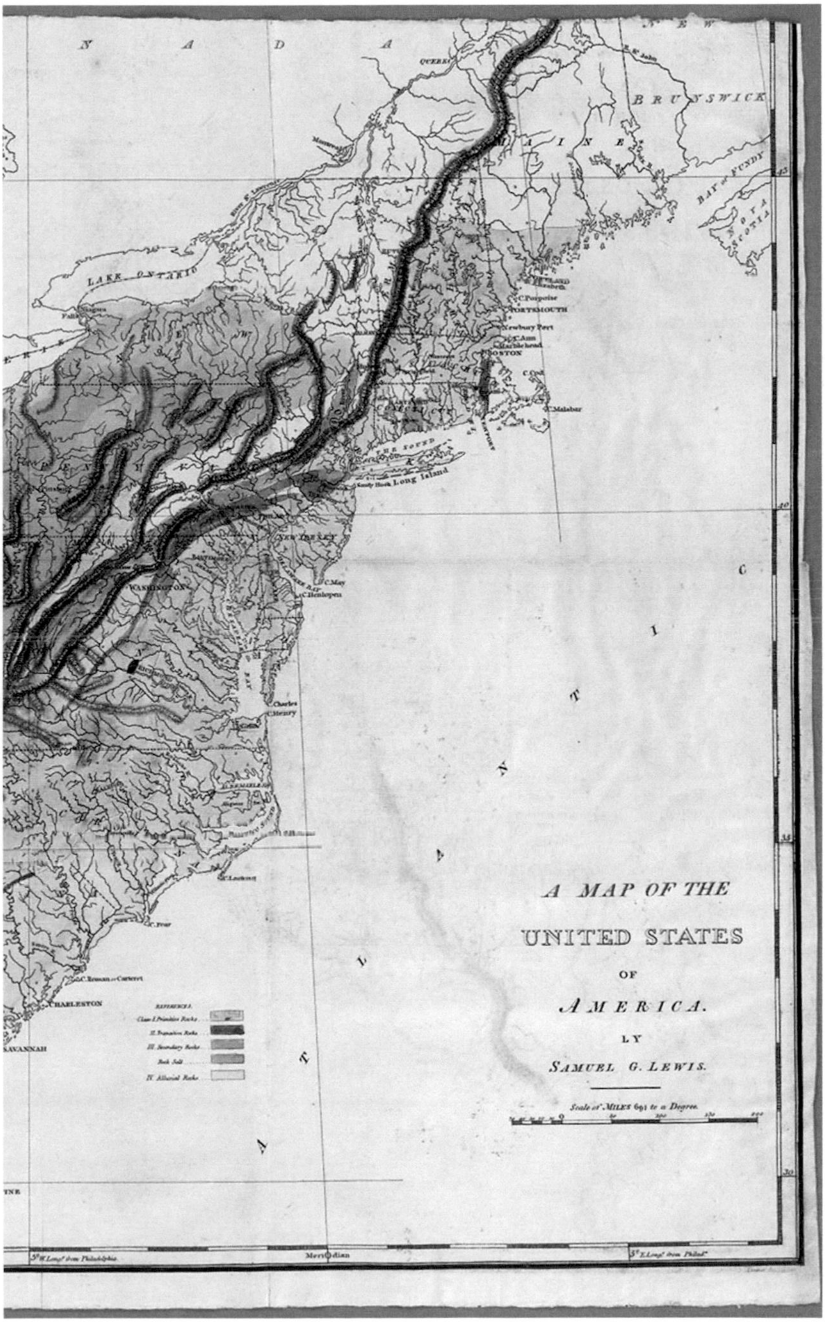
A MAP OF THE
UNITED STATES
OF
AMERICA.
BY
SAMUEL G. LEWIS.
QUEBEC
BRUNSWICK
MAINE
BAY of FUNDY
LAKE ONTARIO
C.Porpoise
PORTSMOUTH
Newbury Port
Marblehead
BOSTON
C.Malabar
THE SOUND
Long Island
NEW JERSEY
C.May
C.Henlopen
WASHINGTON
C.Charles
C.Henry
C.Lookout
C.Fear
CHARLESTON
SAVANNAH
Meridian

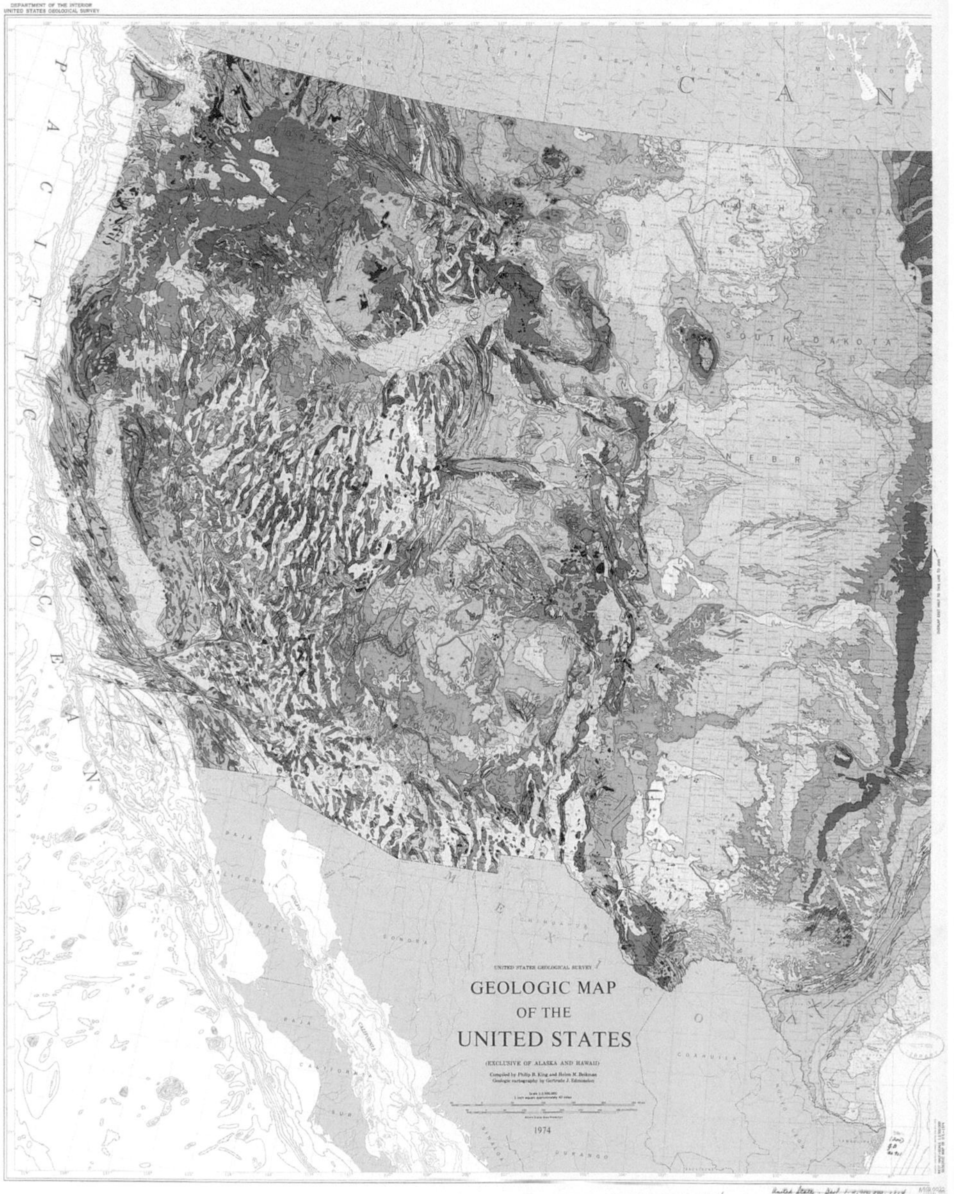

Color insert 4.3. Geologic map of conterminous United States, scale 1:2,500,000. (King, Beikman, and Edmontson, 1974)

GEOLOGIC MAP OF THE UNITED STATES

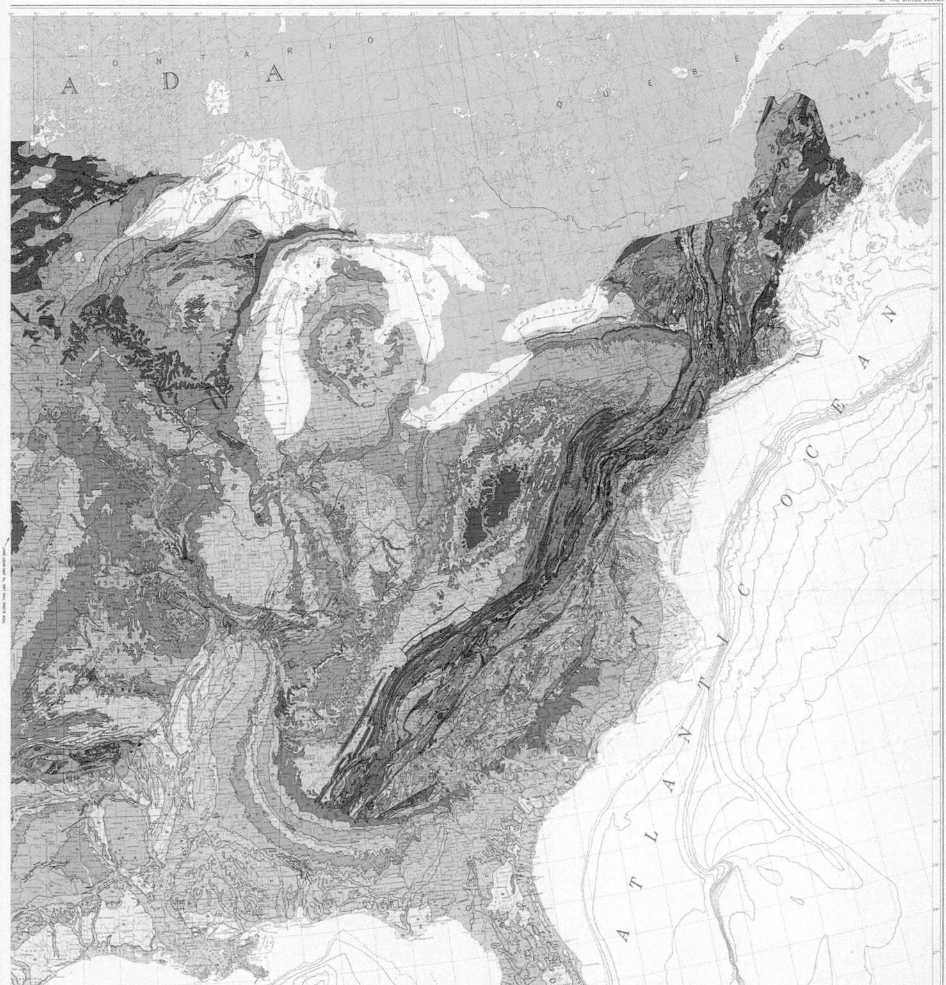

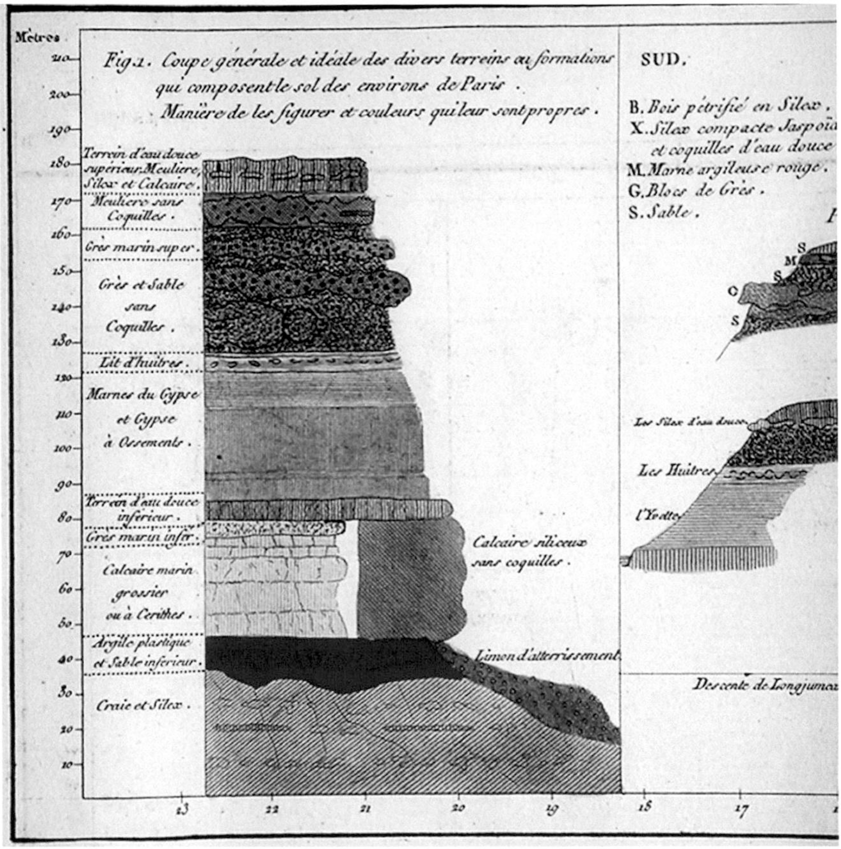

Color insert 4.4. Stratigraphic cross section. (Cuvier and Brongniart, 1811, plate I; courtesy of David Rumsey Map Collection, David Rumsey Map Center, Stanford Libraries)

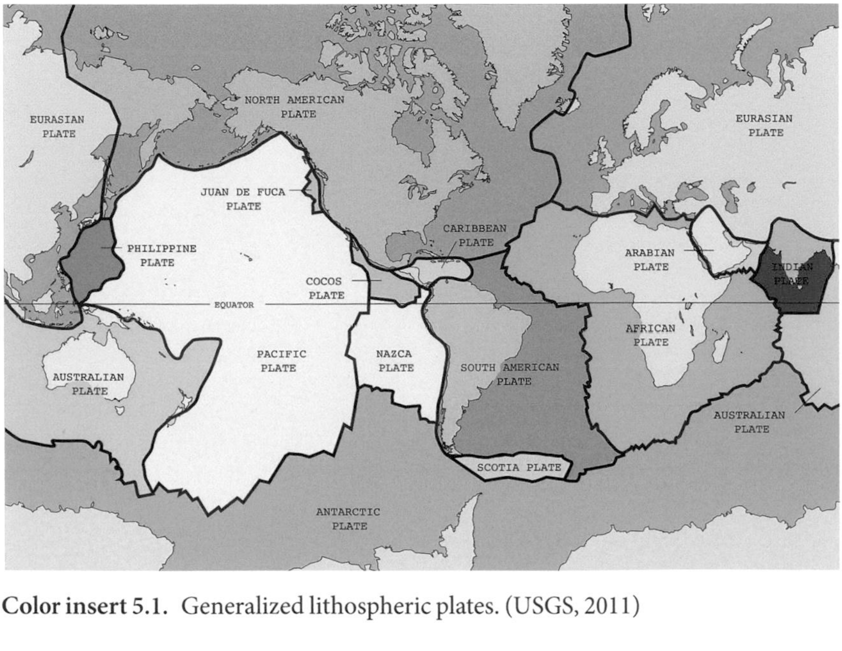

Color insert 5.1. Generalized lithospheric plates. (USGS, 2011)

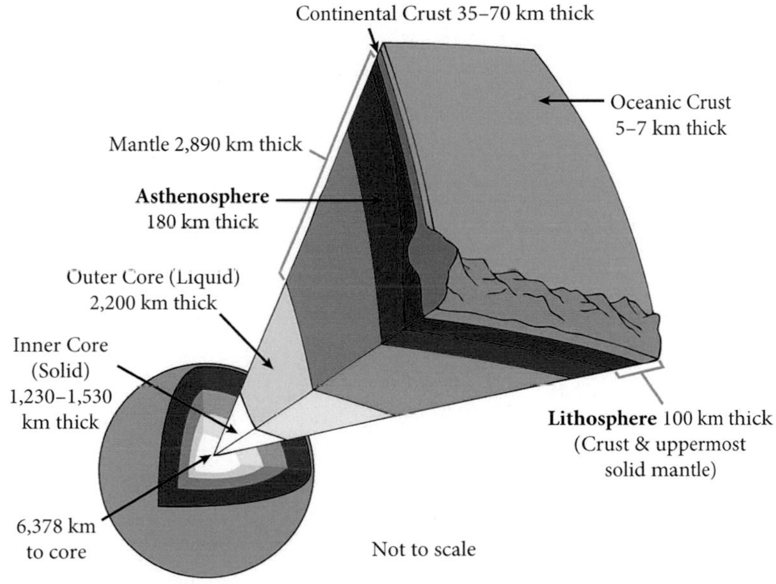

Color insert 5.2. Layers of the Earth. (Illustration by R. Gary Raham, 2020, with input from the author)

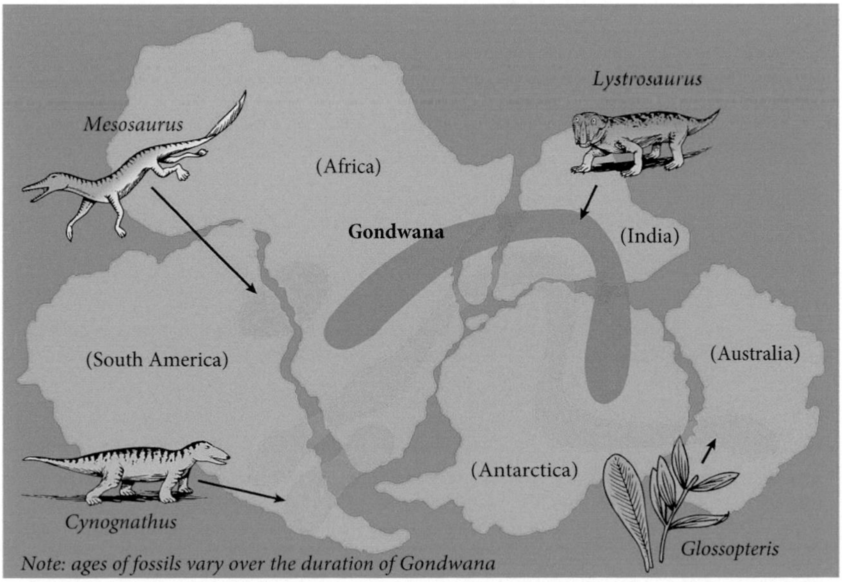

Color insert 5.3. Gondwanaland reconstruction with *Lystrosaurus*, *Cynognathus*, *Mesosaurus*, and *Glossopteris* distributions. (Illustration by R. Gary Raham, 2020, adapted from Snider-Pellegrini and Wegener map, with input from the author)

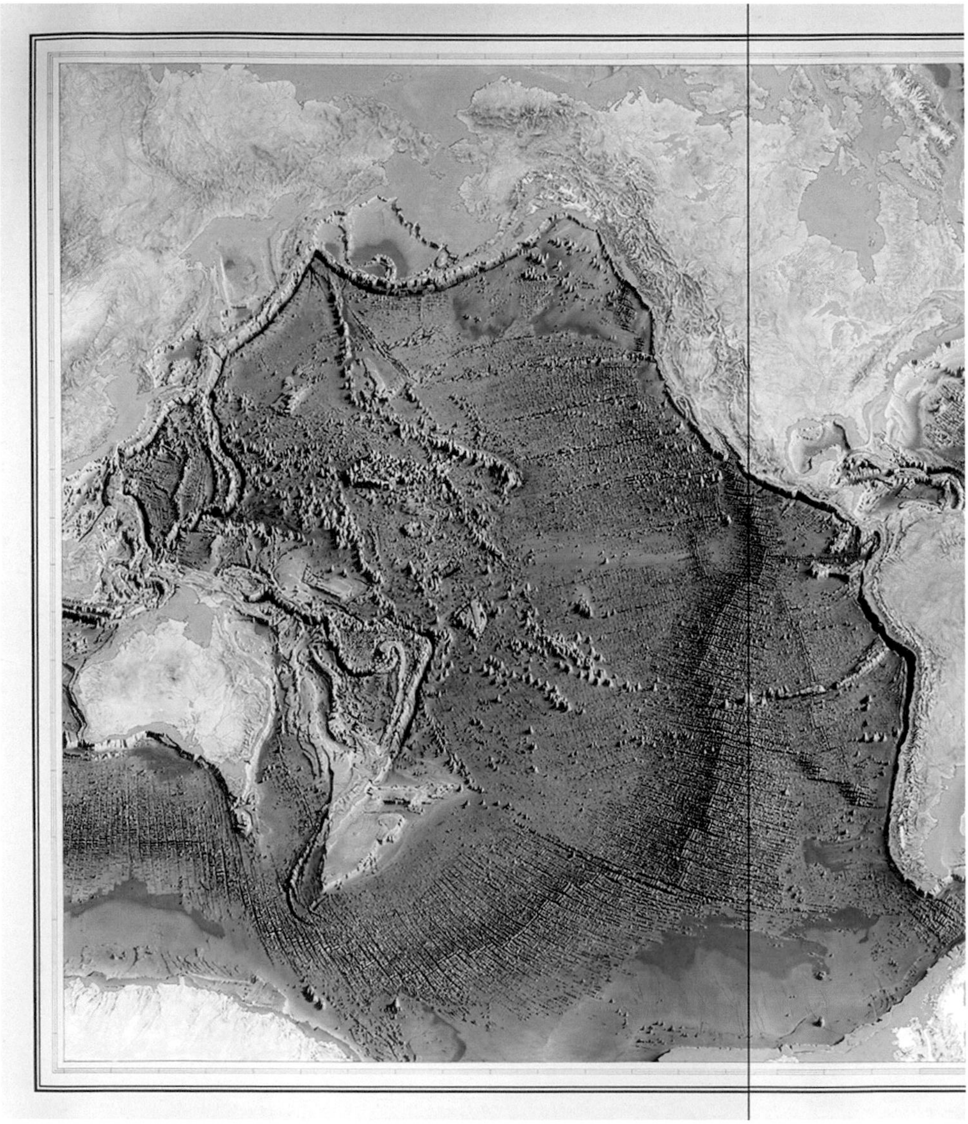

Color insert 5.4. World ocean floor map. (Heezen and Tharp, 1977, map painted by Heinrich C. Berann; Library of Congress, 1977)

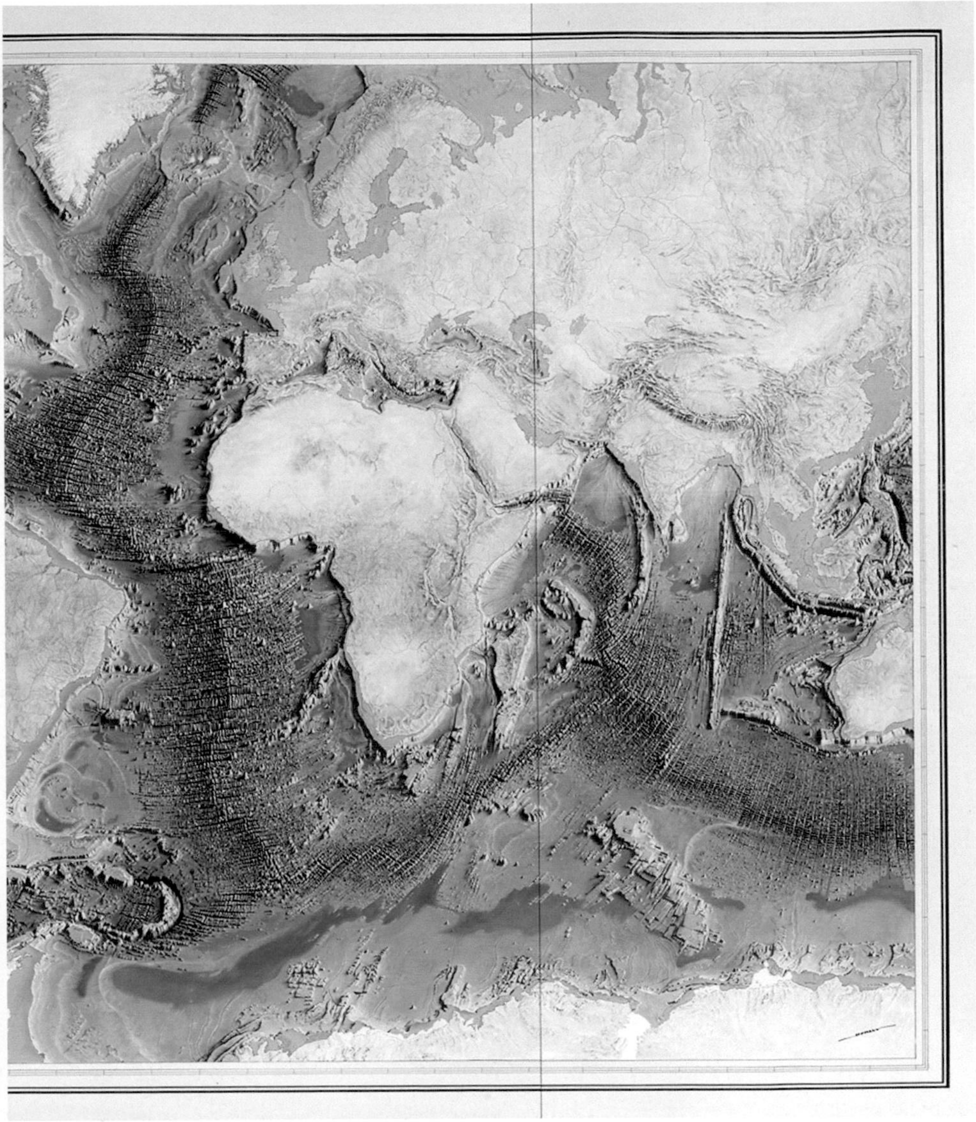

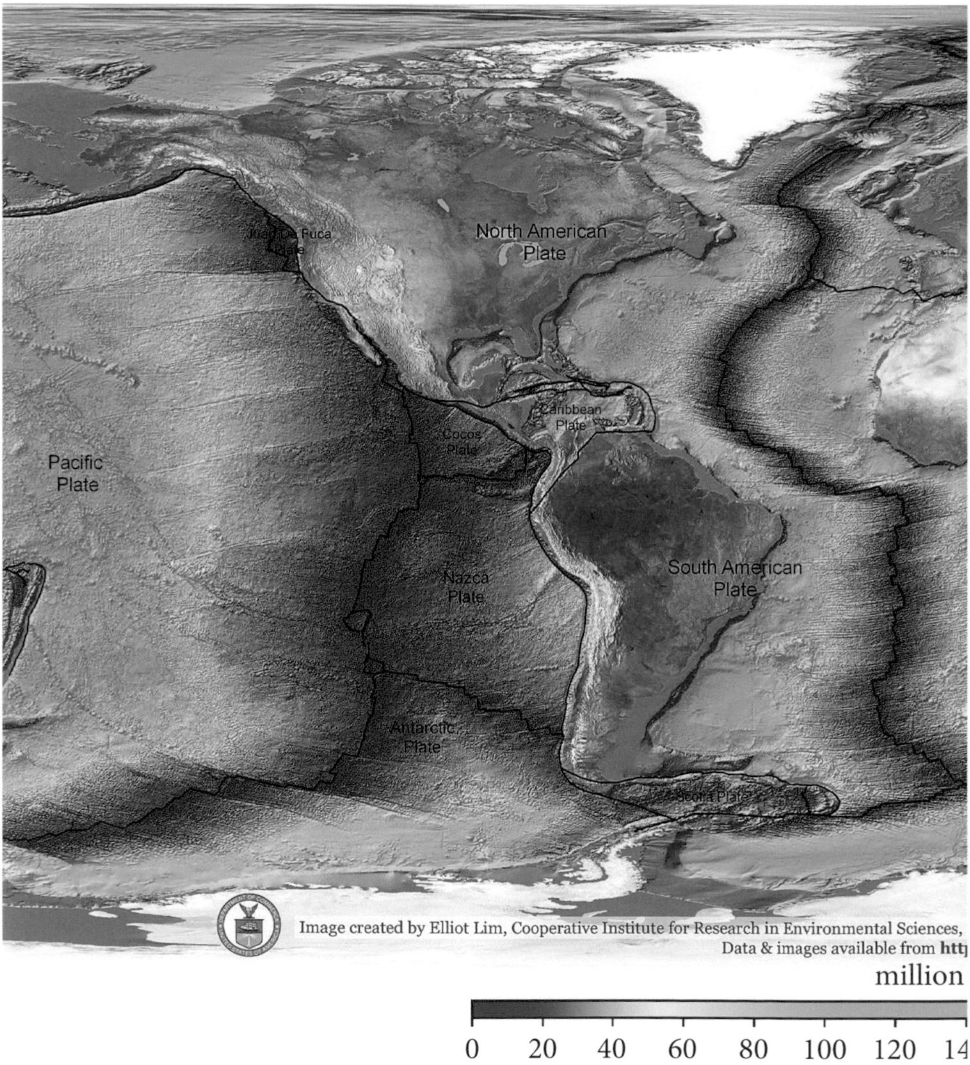

Color insert 6.1. Age of the ocean floor in millions of years, with age ranges in color. (Müller, R.D., Sdrolias, M., Gaina, C., and Roest, W.R., 2008)

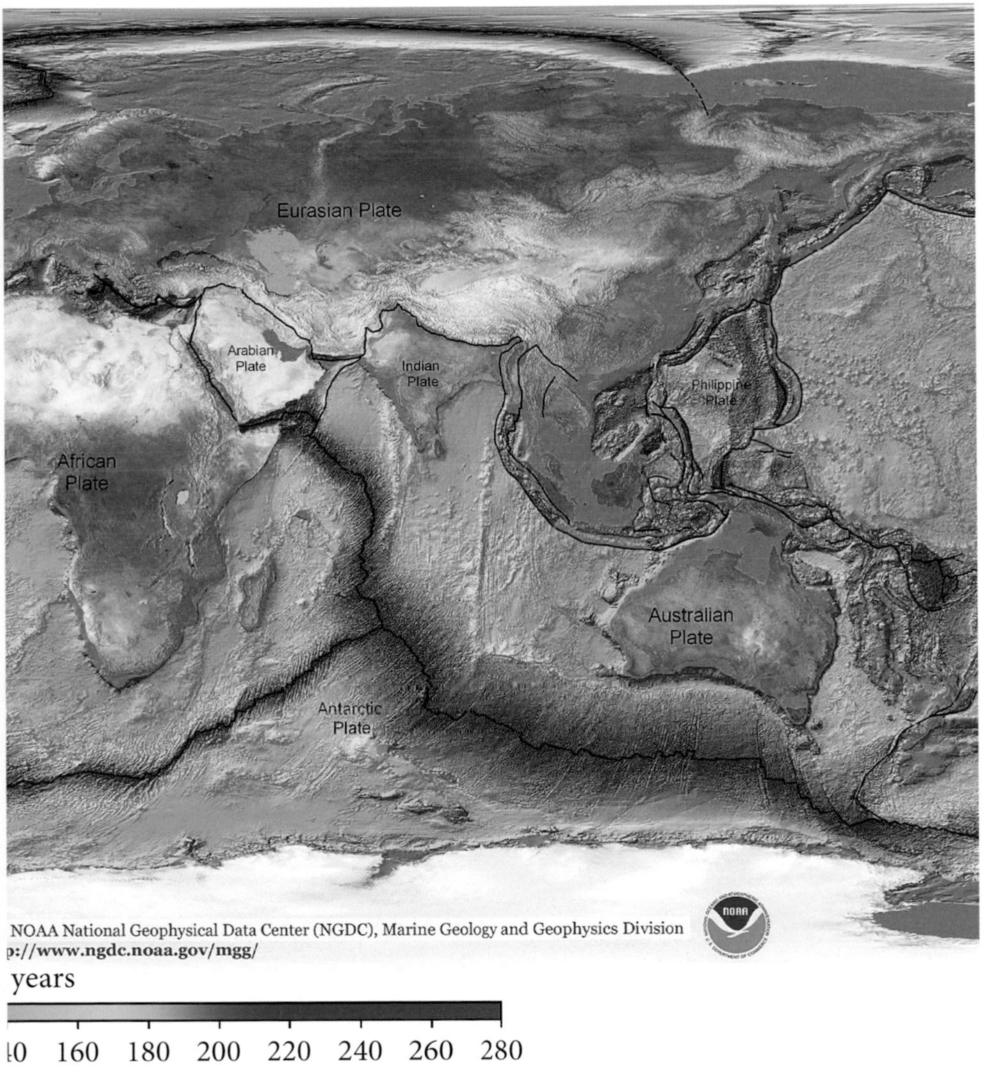
Eurasian Plate
Arabian Plate
Indian Plate
Philippine Plate
African Plate
Australian Plate
Antarctic Plate
NOAA National Geophysical Data Center (NGDC), Marine Geology and Geophysics Division
p://www.ngdc.noaa.gov/mgg/
years
ı0 160 180 200 220 240 260 280

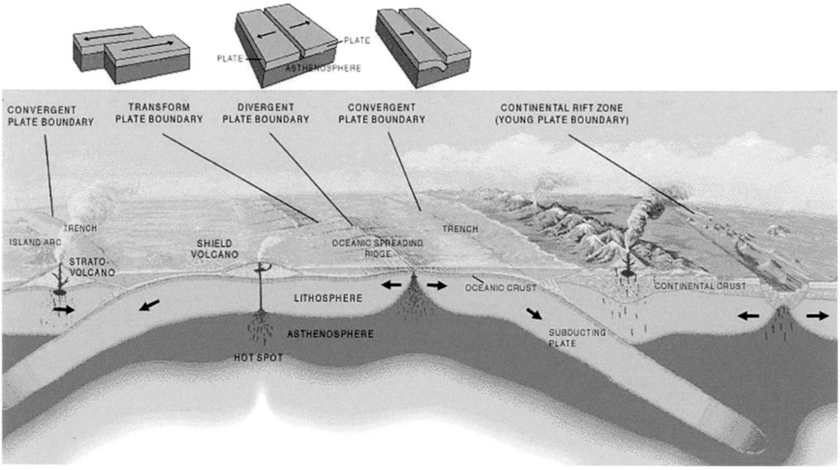

Color insert 6.2. Faults and specific plate boundaries. (Cross section in Vigil, n.d.)

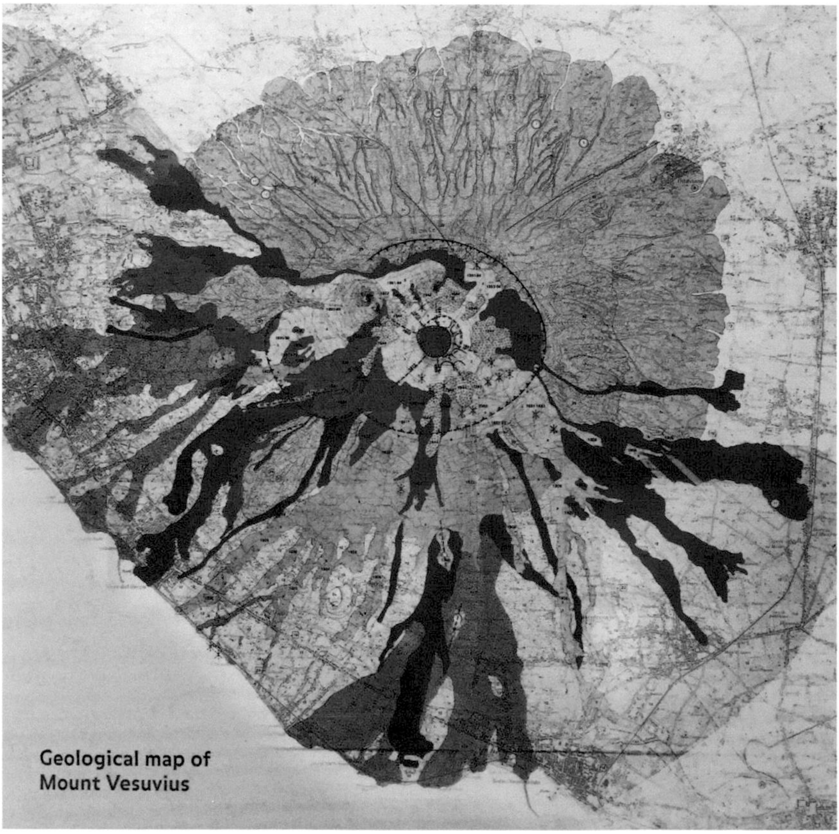

Color insert 6.3. Eighteenth-century geologic map of Mount Vesuvius. (Displayed at the David S. and Ruth L. Gottesman Hall of Planet Earth in the American Museum of Natural History; photograph by SingALittle, 2018)

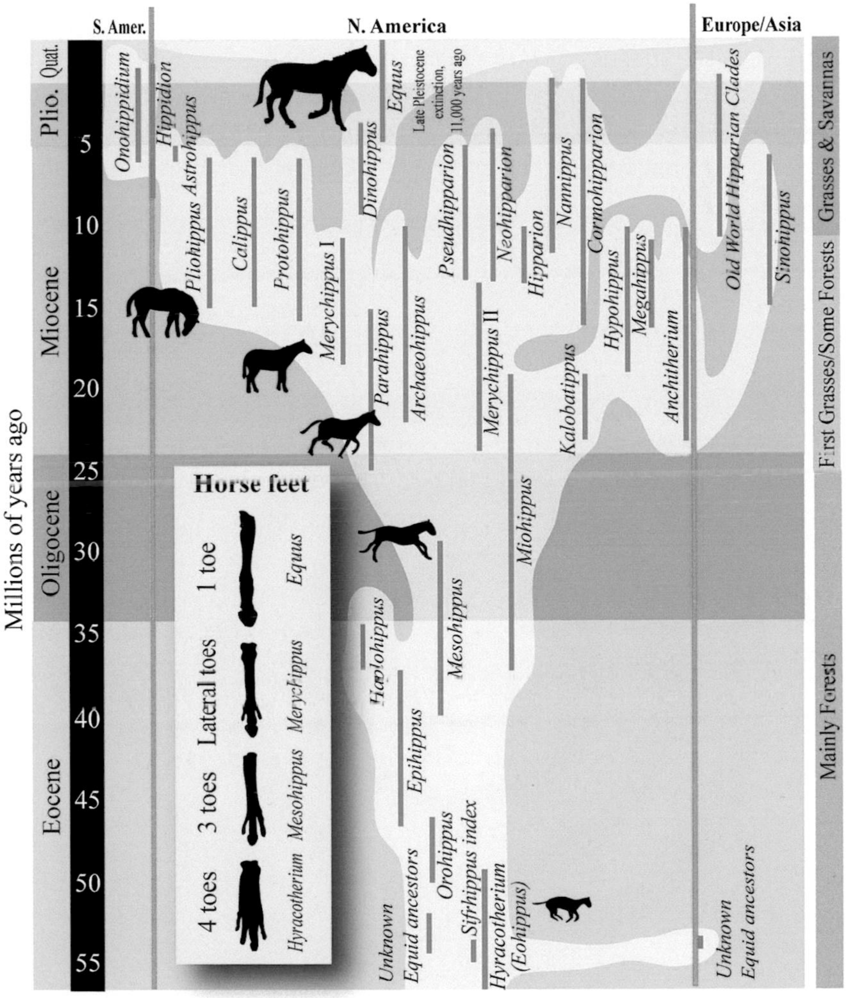

Color insert 7.1. Horse evolution and ecosystem changes in the Cenozoic Era. (Illustration by R. Gary Raham, 2020, with input from the author; adapted from MacFadden, 2005)

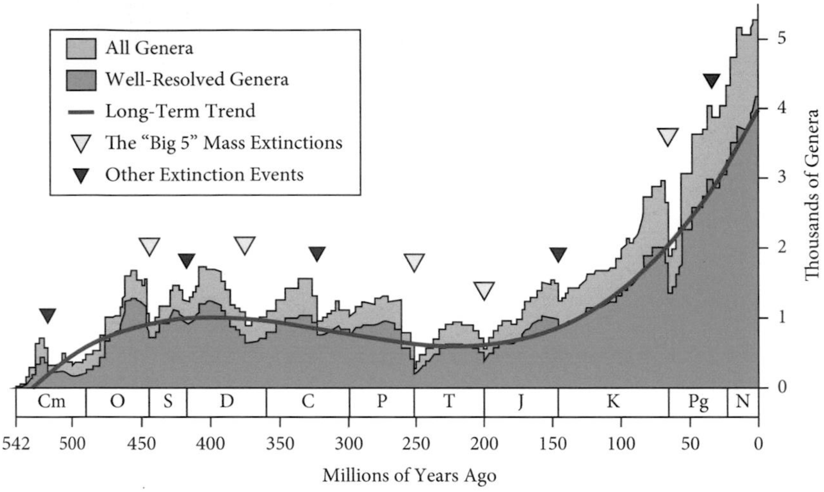

Color insert 7.2. Phanerozoic biodiversity. (Rohde and Muller, 2005, based on Sepkoski, 1992)

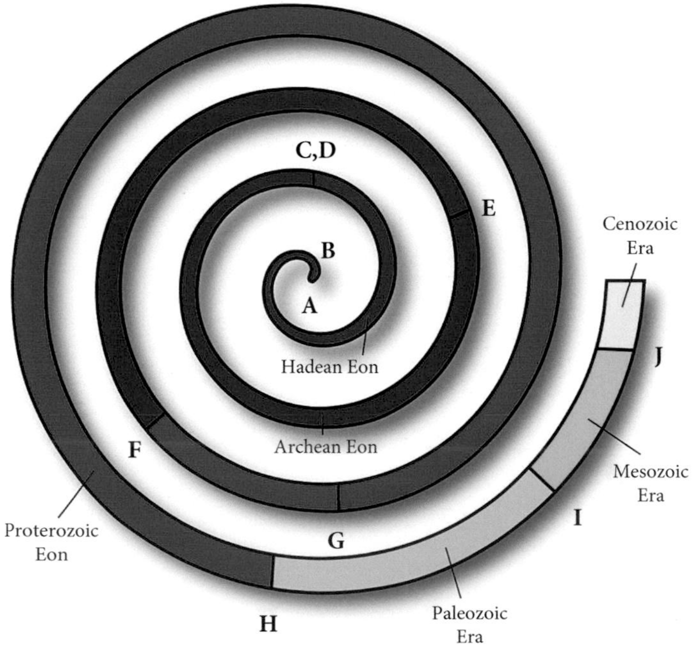

LEGEND

A 4,550 Ma: Birth of Earth; Start of Hadean Eon
B 4,527 Ma: Formation of Moon
C 4,000 Ma: End of Heavy Bombardment
D 4,000 Ma: Start of Archean Eon
E 3,200 Ma: First Photosynthesis
F 2,500 Ma: Start of Proterozoic Eon
G 2,300 Ma: First Oxygenated Atmosphere
H 541 Ma: Start of Paleozoic Era
I 251 Ma: Start of Mesozoic Era
J 66 Ma: Start of Cenozoic Era

Ma = millions of years ago
Timeline colors adhere to International Commission on stratigraphy (ICS) color classification system
Timeline graphic shows correct relative lengths of Eras & Eons

Color insert 8.1. Proportional geologic time scale with major events. (Concept and initial diagram by the author, 2017; final illustration by R. Gary Raham, 2020)

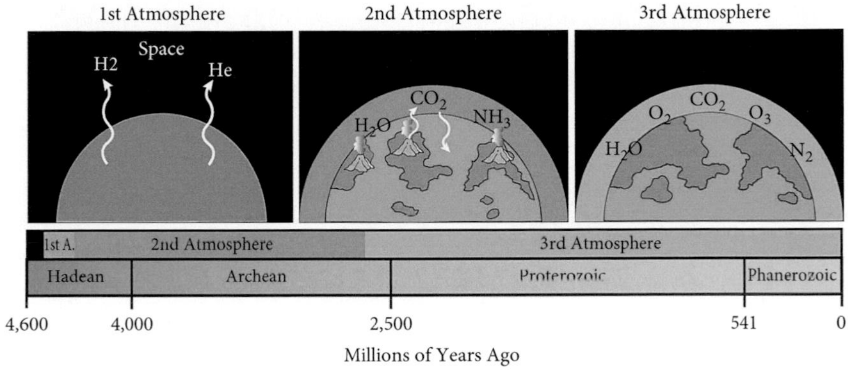

Color insert 8.2. Evolution of the Earth's atmosphere. (R. Gary Raham, with input from the author, based on images from NOAA, 2021, "How Did Earth's Atmosphere Form?")

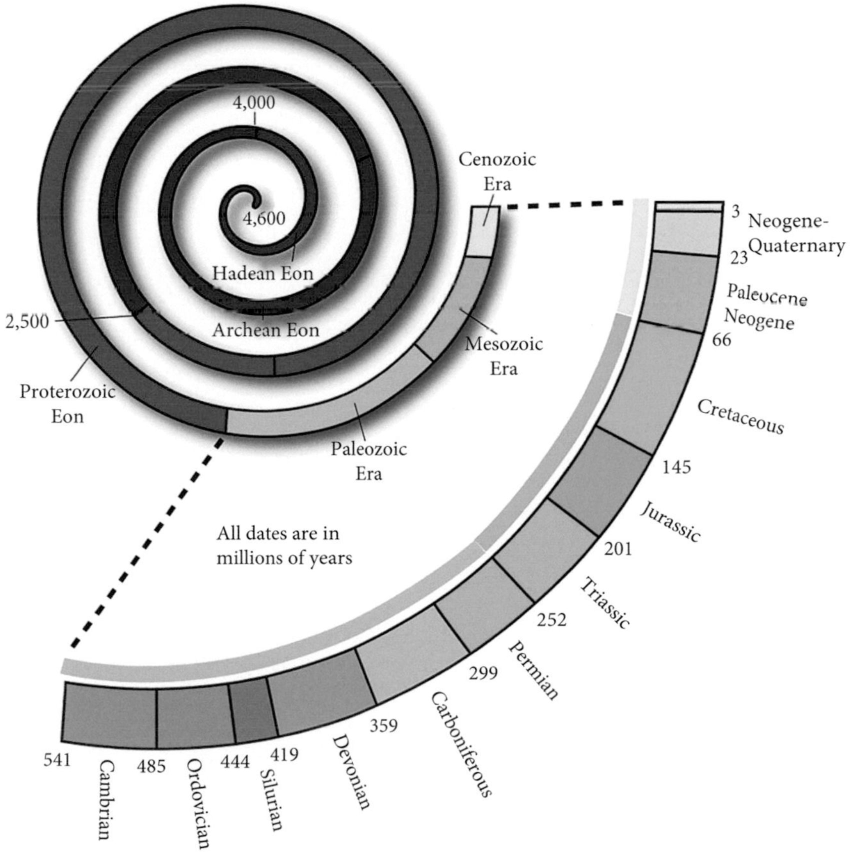

Color insert 9.1. Expanded Phanerozoic Eon, geologic time scale. (Designed by the author; final illustration by R. Gary Raham, 2020)

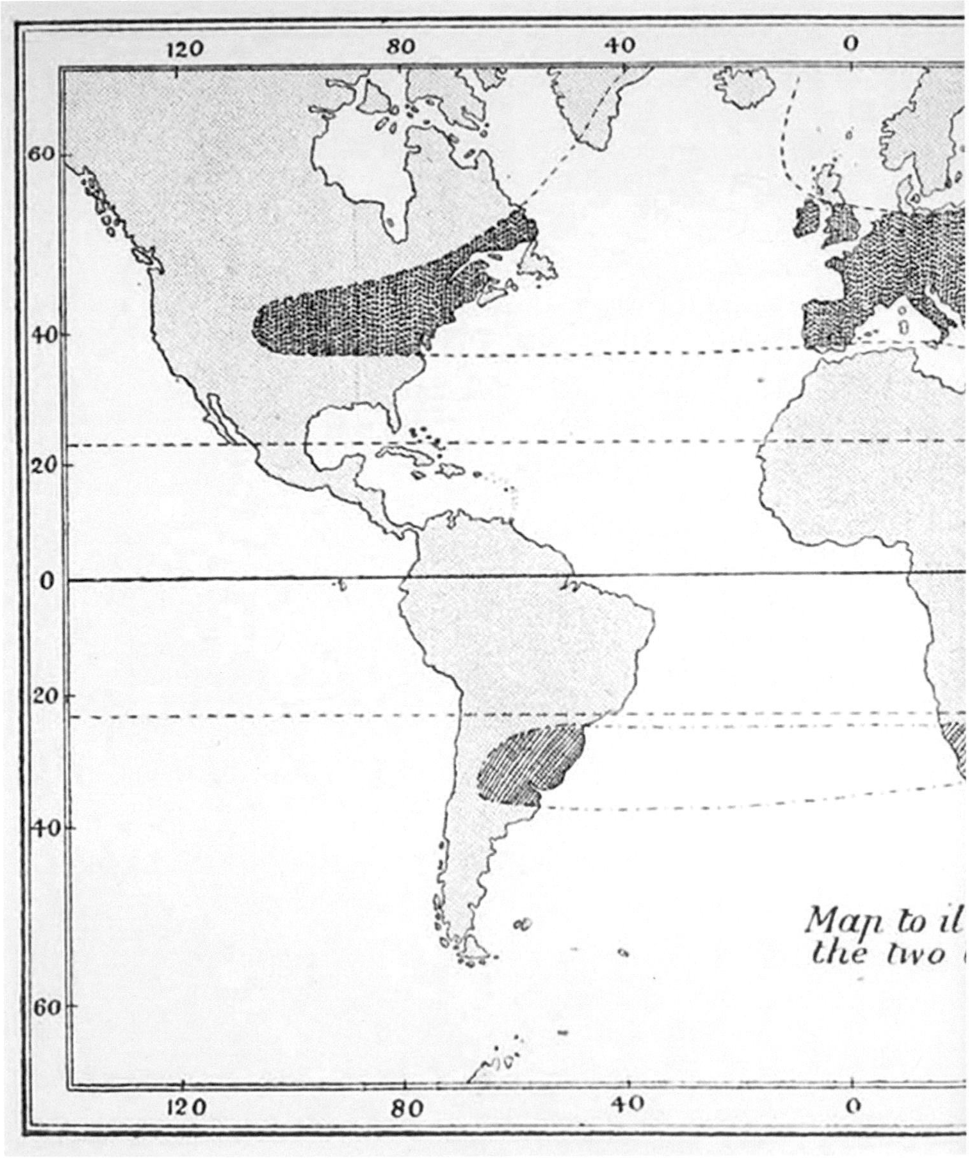

Color insert 9.2. Distribution of northern and southern *Glossopteris* fossils. (Arber, 1905, p. xix)

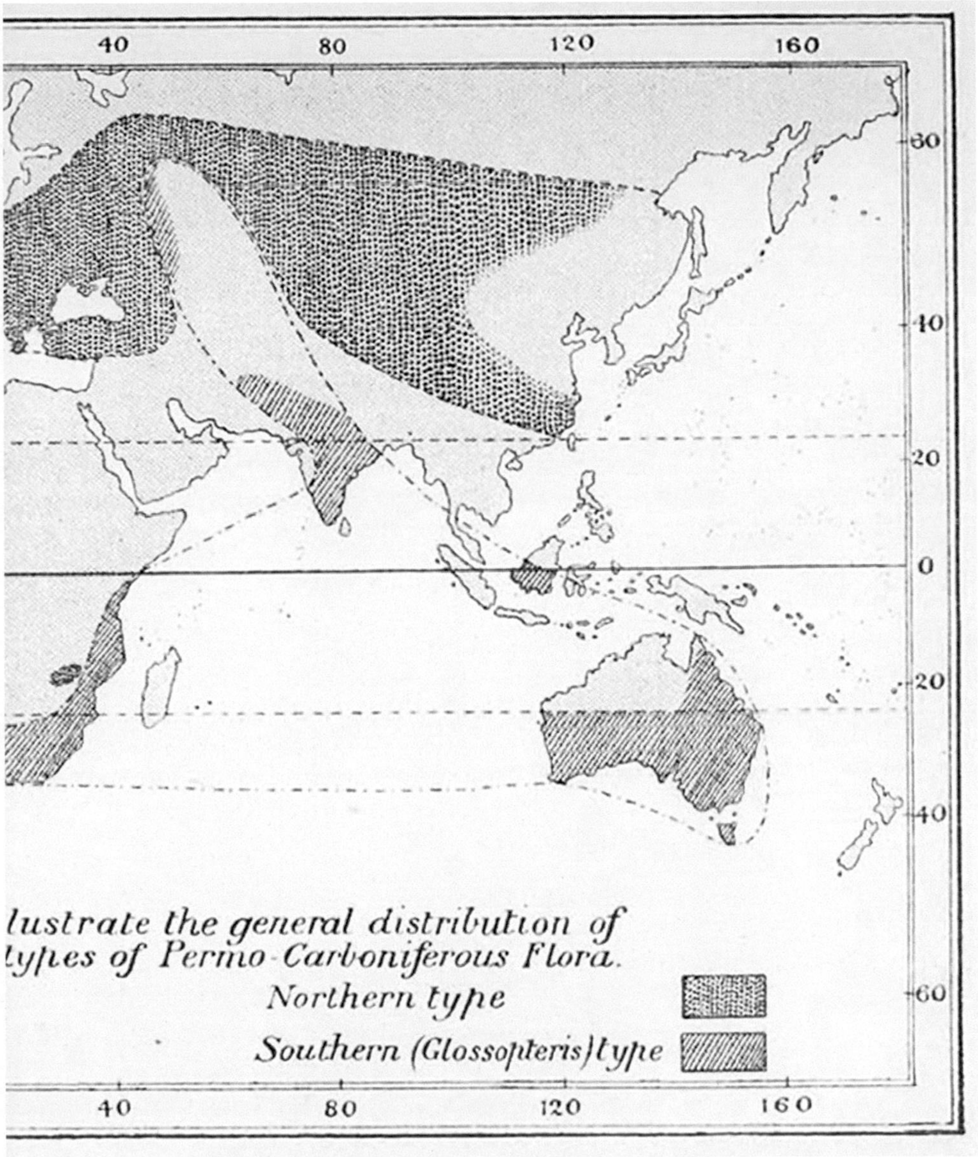
40
80
120
160
60
40
20
0
20
40
60
lustrate the general distribution of
types of Permo-Carboniferous Flora.
Northern type
Southern (Glossopteris) type

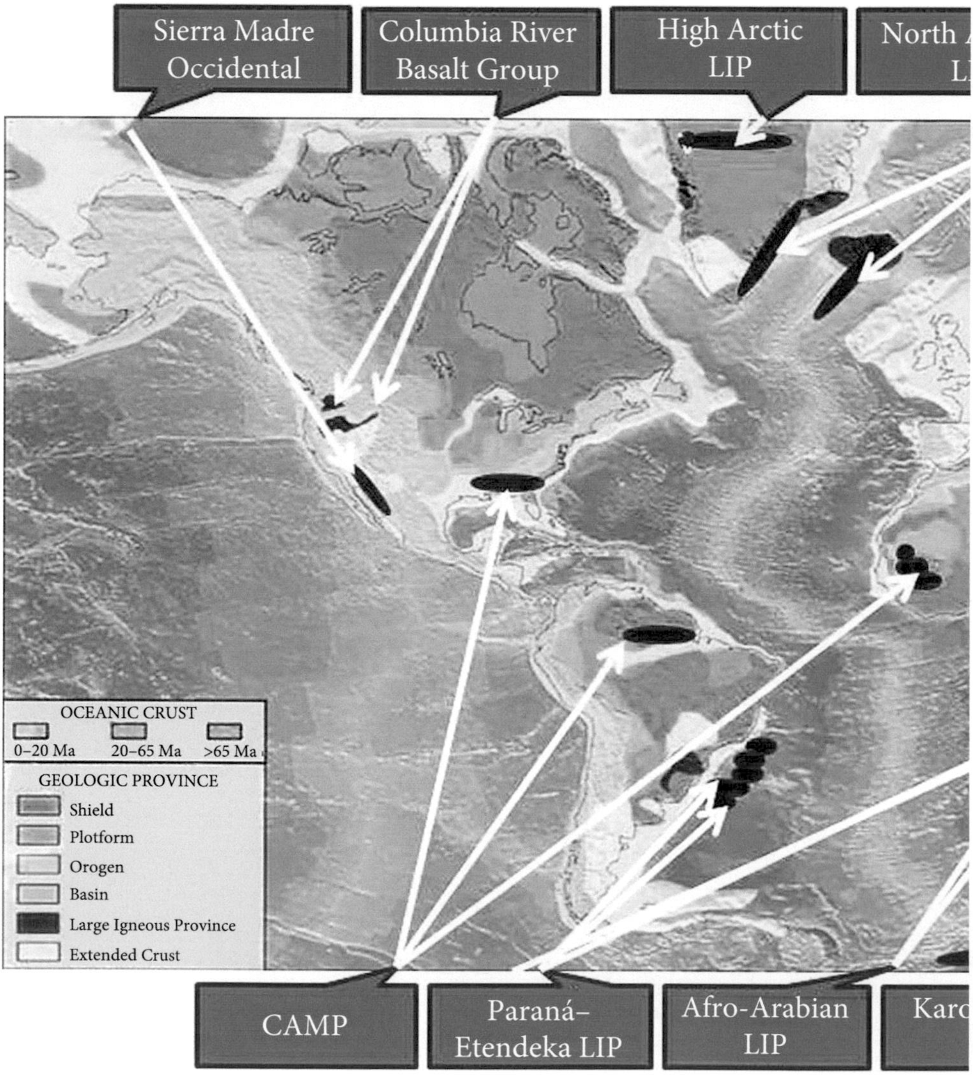

Color insert 9.3. Flood basalt map. (Adaptation of USGS and NOAA map by Williamborg, 2011)

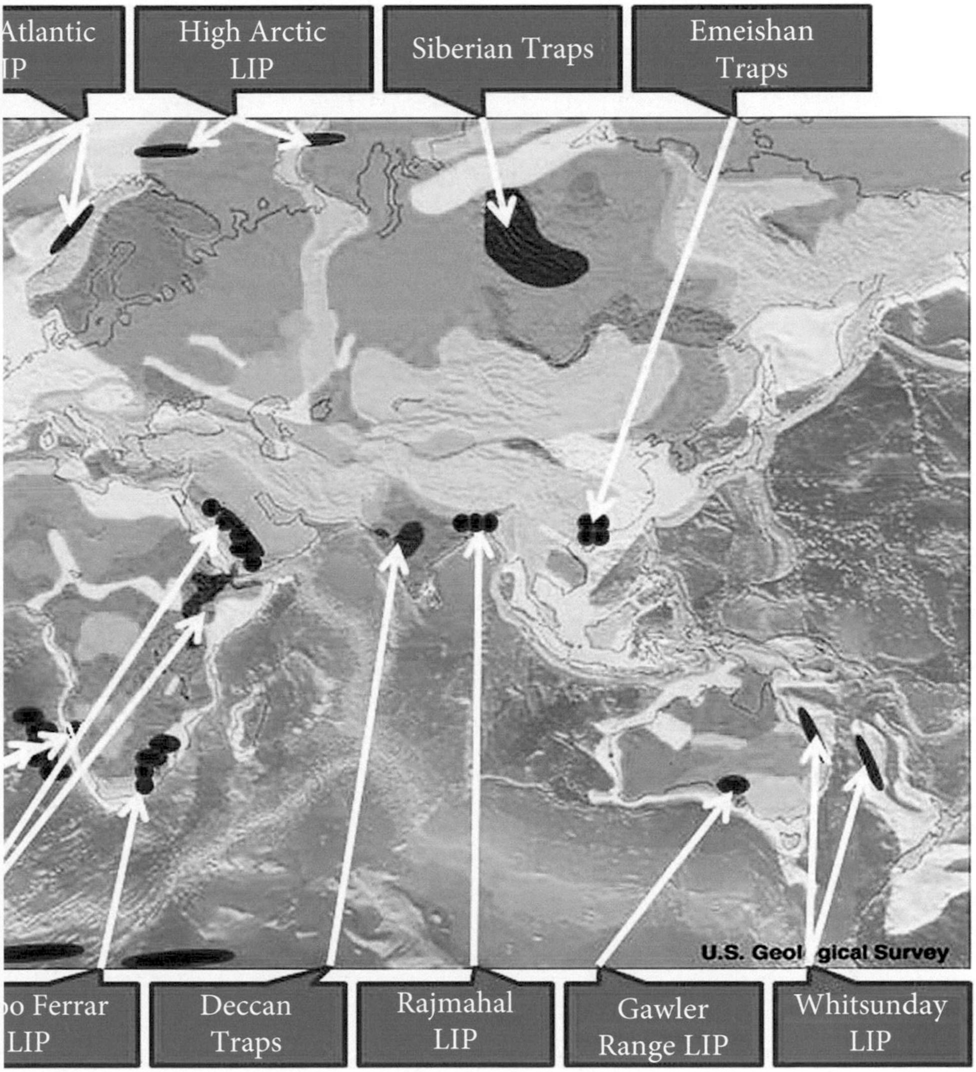

Atlantic
IP
High Arctic
LIP
Siberian Traps
Emeishan
Traps
U.S. Geol gical Survey
oo Ferrar
LIP
Deccan
Traps
Rajmahal
LIP
Gawler
Range LIP
Whitsunday
LIP

Color insert 10.1. Dinosaurs of the Hell Creek Formation. Back (larger), left to right: *Ankylosaurus*, *Tyrannosaurus*, *Quetzalcoatlus*, *Triceratops*. Front (smaller), left to right: *Acheroraptor* (formerly *Caenagnathid*), *Pachycephalosaurus*, *Struthiomimus*, and *Anzu* (formerly classified as a dromaeosaurid). (Durbed, 2012)

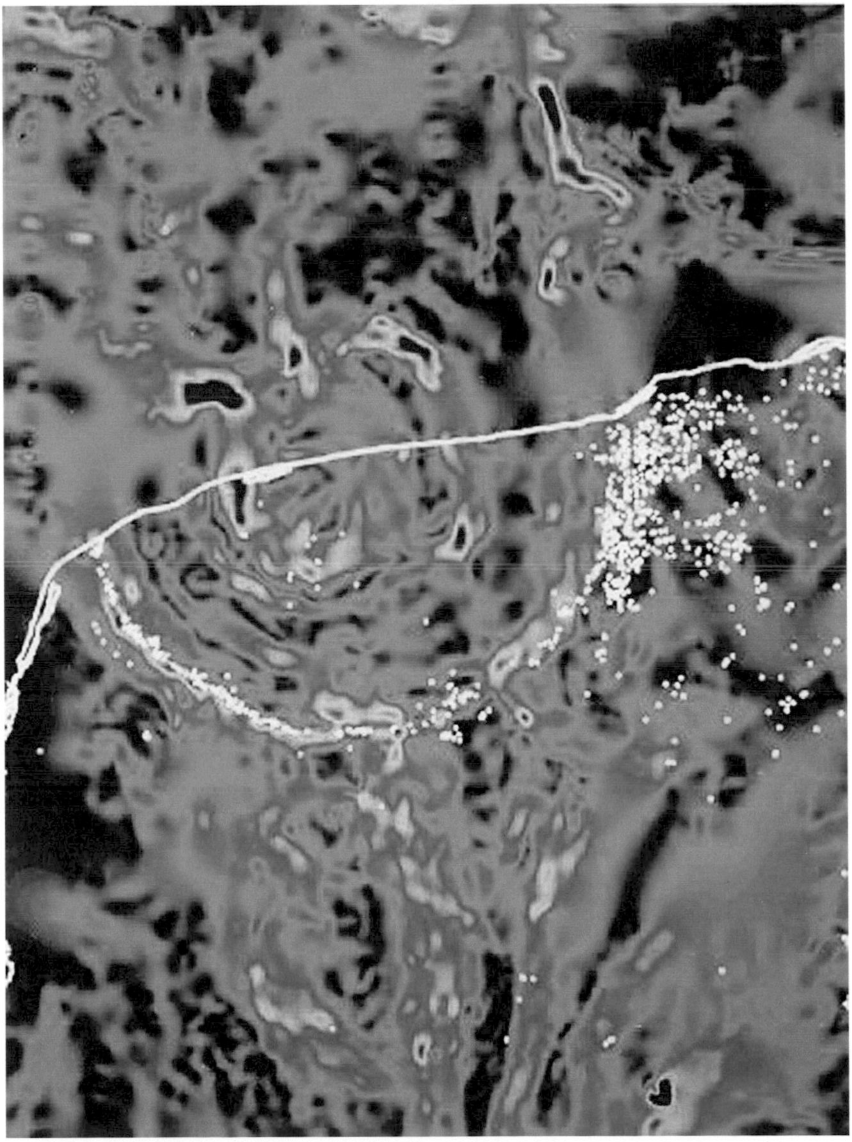

Color insert 10.2. Gravity anomaly map of the Chicxulub impact crater, with the coastline shown as a white line, white dots representing water-filled sinkholes called cenotes after the Maya word. (USGS, 2019)
USGS, 2019, Chicxulub Anomaly, https://commons.wikimedia.org/wiki/File:Chicxulub-Anomaly.jpg.

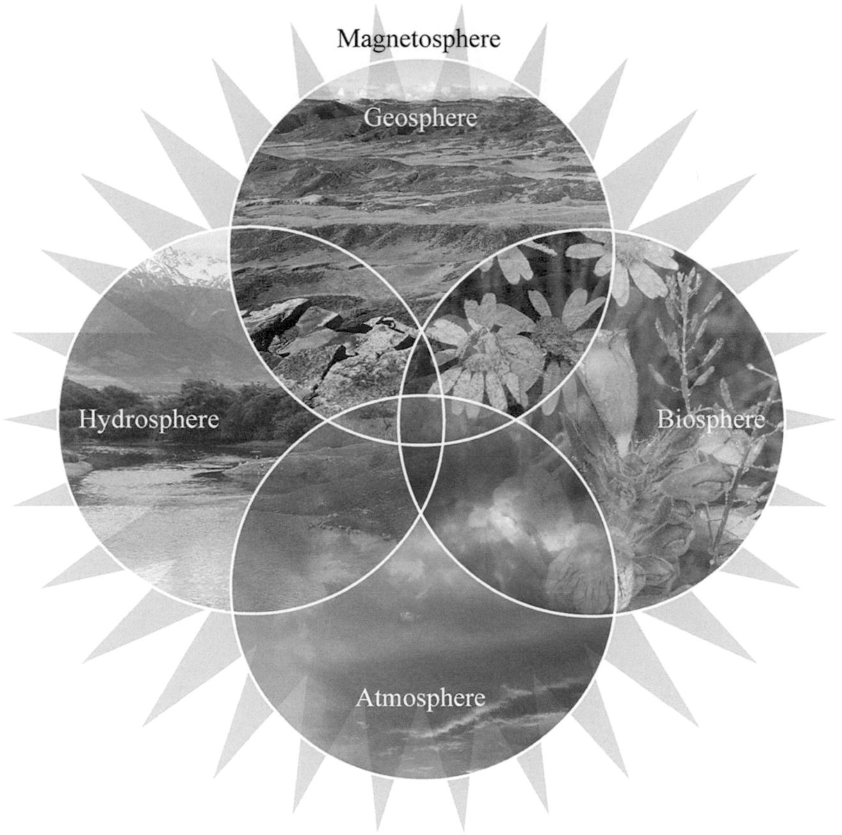

Color insert 12.1. The five overlapping spheres of the Earth. (Concept and initial diagram by the author, 2020; final illustration by R. Gary Raham, 2020)

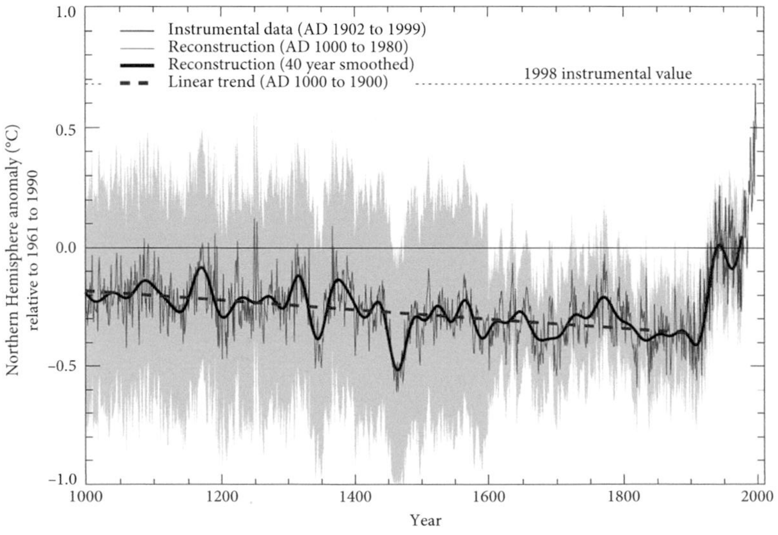

Color insert 12.2. Millennial Northern Hemisphere (NH) temperature reconstruction (blue) and instrumental data (red) from AD 1000 to 1999, adapted from Mann et al. (1999). Smoother version of NH series (black), linear trend from AD 1000 to 1850 (purple-dashed) and two standard error limits (grey shaded) are shown. (Folland et al., 2001, p. 134, Intergovernmental Panel on Climate Change, 2001)

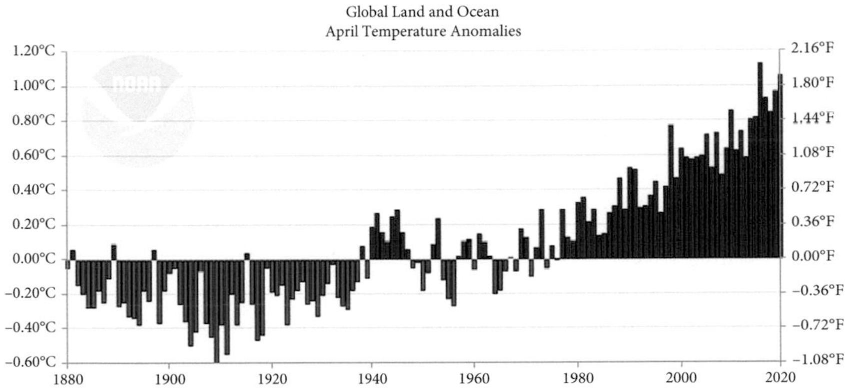

Color insert 12.3. Global land and sea temperatures, departures from average, 1880–2020. (NOAA, 2020)

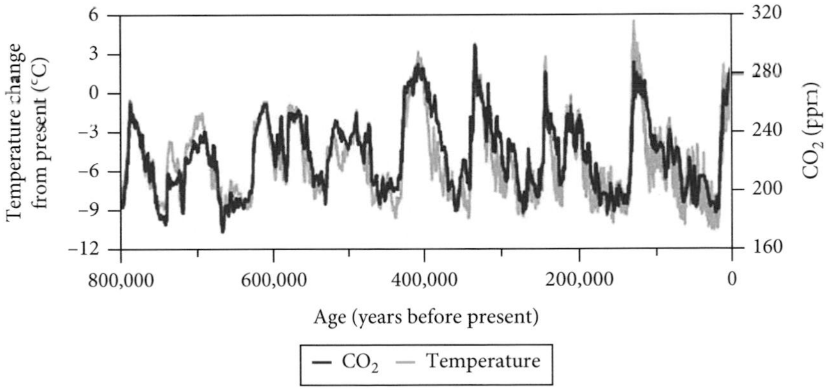

Color insert 12.4. Temperature and atmospheric carbon dioxide correspondence through the last 800,000 years. (NASA, 2008)

(a)

(b)

Color insert 12.5. Geologic time as a vessel, with Greek amphora and the sailing vessel *Rembrandt* in Greenland. (Amphora photo by Dennis Jarvis, 2010; ship photo by Vaido Otsar, 2016)

9

The Biography of the Earth

Paleozoic Era

As we have seen, the Precambrian Supereon encompasses the bulk of geologic time—more than three-quarters of the biography of the Earth. Now we head into the last 12 percent of the story: the Phanerozoic Eon (the divisions of geologic time scale are rounded to the nearest million) (color insert 9.1). The term means "visible life," and life surely is varied and evident during the Phanerozoic. Though the shortest of the eons, it is an exciting one, because it is here that life takes hold on the planet in its myriad forms, alongside such major events as extinctions, floods, massive fires, the formation and demise of supercontinents, ice ages, and droughts. The story begins 541 million years ago and continues to the present day.

The Paleozoic ("old life") Era is the first of three eras in the Phanerozoic Eon. The longest, the Paleozoic lasted from 541 million to 251.9 million years ago, a span of 289.1 million years. It is marked on the lower boundary by the appearance of complex life after the times of Snowball Earth and on the upper boundary by one of the largest mass extinctions ever recorded on the planet, at the end of the Permian Era. This era has six periods, the Cambrian, Ordovician, Silurian, Devonian, Carboniferous, and Permian.

During the Paleozoic, life as we know it began to rapidly diversify on the Earth. In part, this burgeoning of life forms was an effect of the many ecosystems created at that time by the shallow, epicontinental ("upon the continent") seas. These seas extended inland over continental interiors when sea levels were high and continental elevations were low, influenced by the actions of plate tectonics. By contrast, "shelf seas" are shallow seas that occur when ocean waters flood the continental shelf areas, and they also are rich environs for life.

In the early portion of the Paleozoic, the landmasses that had been assembled into the supercontinent Pannotia at the end of the Proterozoic Eon continued to break apart, a process that had begun during the latter part of the Neoproterozoic 550 million years ago. In some locations, forces rifted and tore Pannotia asunder, and in other areas, great landmasses collided. Four continents created from the ultimate split of the supercontinent are known as Laurentia, Baltica, Gondwana, and Siberia. Additionally, a microcontinent called Avalonia became a crucial

Song of the Earth. Elisabeth Ervin-Blankenheim, Oxford University Press. © Oxford University Press 2021.
DOI: 10.1093/oso/9780197502464.003.0010

player in the process, as it was located near the edge of Laurentia and was frequently smashed as the other continents rammed into each other.

The continents, riding upon massive tectonic plates, jockeyed for position, resulting in great mountain-building episodes during the Paleozoic. The late Cambrian time recorded the beginning of the Caledonian orogenic phases that created still-extant ranges, including the Scottish Highlands and the mountains of Norway, Ireland, and Wales. This event lasted through the Devonian Period—over 200 million years—and is related to the closing of the Iapetus Ocean as Laurasia, Baltica, and Avalonia collided. The earliest of the Caledonia phases was the Penobscotian, evident today in unconformities in the rocks of northern Wales, southeast Ireland, and eastern Maine.[1] Geologists have recorded many other phases of the Caledonian event, with some spanning extensive areas in their influence and others being more local in their effects.

Cambrian Period, 541–485.4 Million Years Ago

The first of the Paleozoic periods—the Cambrian—is subdivided into four epochs, the Terreneuvian, Series 2, Miaolingian and Furongian, based on the development of organisms from the simple single-celled to the more complex. This epoch tracks the continued evolution of shells as an external means of protection, which began in the late Precambrian Era. Creatures such as trilobites and other arthropods were among the first predators in the history of the Earth. Paleontologists have associated every single event in this change in life forms with significant fluctuations in the carbon cycle.

The Cambrian Period carried on the late Precambrian Supereon of mineralized shells, known as small shelly fossils (see chapter 8), a fossil assemblage. What differed, however, is seen in the evidence of complex trace fossils—impressions of burrows and feeding tubes—which appeared for the first time in the early part of the period. Scientists used to think that the development of mineralized shells marks the boundary of the earliest Cambrian, where life forms were significantly different from the prior Precambrian fauna, the Ediacaran, with their soft, shell-less bodies. Later, more refined data proved that the age of the small shelly fossils began before that time. With a duration of 55.6 million years, the Cambrian has four epochs (from oldest to youngest)—the Terreneuvian, Series 2, the Miaolingian, and the Furongian—and ten stages based on the distribution and biodiversity of small shelly fossils, trilobites, brachiopods, and ocean-dwelling arthropods.

Vast changes in diversity and type of life forms during the Cambrian period have been known as the "Cambrian explosion of life." However, the term explosion is a misnomer as it did not occur in a flash but extended 50 million years or more from the Cambrian Period until the start of the Ordovician Period. Even before the Cambrian, life had started diversifying in the late Precambrian. A new name is proposed for this time of burgeoning life, the "Great Cambrian Biodiversification."[2]

Paleontologists view this time more in terms of the Cambrian radiation of life, which originated from an increase in oxygen levels in the oceans, allowing for the development of the first predators.[3] This development spurred more adaptations in creatures to avoid being eaten, including burrowing into the ocean floor, hiding with better camouflage, and increased swimming abilities.

Fossils of the Burgess Shale of British Columbia, for example, illustrate the magnitude of species that proliferated during this time, as animals expanded into all ocean areas. One of the most famous deposits in the world, the Burgess rocks, preserved even the soft parts of the animals. Richard McConnell (1857–1942) of the Geological Survey of Canada examined the first of the Burgess Shale fossils in 1886 after laborers contacted him about "fossil bugs" on the slopes of Mount Stephen. These fossils turned out to be trilobites from a different formation located near the famous shale deposit. However, it set the stage for the critical discovery by Charles Walcott (1850–1942) in 1909. Walcott had visited the area to view the trilobite beds in 1907 and stumbled upon the fossils of the Burgess Shale.

Walcott described the Burgess Shale fossils in several publications and assigned many of them to the four known groups of arthropods (invertebrate organisms equipped with exoskeletons, jointed legs, and segmented bodies), basing his thoughts on what were then contemporaneous ideas. However, he was unable to classify some of the fossils, even as far as determining their families, orders, or classes.[4] Some, like twentieth-century paleontologist and popular writer Stephen Jay Gould, criticized Walcott's classification of the fossils.[5] Others acknowledge Walcott's contribution in collecting thousands of fragile fossils under challenging conditions and his intellectual effort of classification with the tools he had at hand.[6] In the late 1960s, Italian biologist Alberto Simonetta of the University of Florence began the task of reinterpreting some of the species.[7] In more recent years, British paleontologists Harry Whittington and Simon Conway Morris and Irish paleontologist Derek Briggs started to examine the fossils of the Burgess Shale. They began the arduous work to reconstruct, redescribe, and reclassify them.[8] Many of the creatures were flattened, with their remains spanning more than one bedding plane in the shale. At the time, paleontologists discovered that a significant number of the Burgess biota did not even fit into any existing phylum of animals.[9] Ultimately, Gould wrote the story of the reinterpretation and significance of the reclassified Burgess organisms in his book *Wonderful Life: The Burgess Shale and the Nature of History*, published in 1989, which included recognizing and viewing them in a three-dimensional sense.

The Burgess deposit is a record of a particular type of fossil preservation that captures soft parts of organisms along with the exoskeletons, in the Burgess Shale case, or skeletal material, in other locations, called *Lagerstätten* (German for "mother lode" or "lode place"). *Lagerstätten* deposits are relatively rare because they only form under unusual environmental conditions. Shale is the dominant rock of the formation at this location and is produced when clay, over geologic time, hardens and becomes lithified (made into rock). It is a product of quiet water environments—deep ocean basins, near-shore clays, bay muds, or lakebed deposits.

The arthropods of the Burgess Shale probably lived in shallow-water mud deposits near a large, vertical ledge of limestone called the Cathedral Reef Escarpment.[10] Geologists think an underwater landslide occurred, carrying the animals from their original location to the depths, with rapid burial and preservation of their tissues and outer skeletons. The Walcott Quarry, which holds the fossils encased in rock, is hardly a traditional quarry, being located on the side of a mountain approximately 8,000 feet in elevation. The size of the outcrop is about 1.8 meters (6 feet) wide by 61 meters (200 feet) long, a relatively small exposure that has produced an exceptional amount of Cambrian fossil material.

One Burgess Shale representative, *Anomalocaris*, was one of the fiercest predators ever born (figure 9.1). The now-extinct arthropod had grasping, armlike appendages that latched onto prey, then stuffed it into its mouth. *Anomalocaris* grew to 3 meters (10 feet) in length and was one of the biggest animals among the Burgess Shale biota.

Paleontologists have identified the first animals with backbones in the Burgess Shale deposit. An ancestor of all living vertebrates, *Pikaia gracilens* (figure 9.2), an early chordate, had a notochord (similar to cartilage). The chordate family includes all fish, amphibians, reptiles, birds, and mammals. This earliest human ancestor was only 5 centimeters (2 inches) long, with a flattened body, and swam by flexing its skeletal muscles from side to side. Researchers have found myomeres—chordate blocks of skeletal muscle—preserved in the shale. It is rare to discover such a trove of soft structures as has been located in this deposit.

The Burgess biota represents particular creatures that no one ever dreamed of: the "weird wonders" of Gould. Gould based his categorization on differences he saw between the Burgess fossils and morphologies of known fossils and their taxonomy. Even before Gould, Morris and Whittington found that a number of the creatures did not fit into the taxonomy of known phyla,[11] an idea Gould also advocated.[12]

The third wave of Burgess Shale fossil interpretation introduced new methods to understand and study relationships between flora and fauna. One was cladistics, the study of how closely groups of animals and plants are related based on creating phylogenetic trees measuring characteristics and their

(a)

(b)

Figure 9.1 (a) Fossil *Anomalocaris canadensis* grasping claw (~ 8.5 centimeters long). (YPM 35138, Yale University's Peabody Museum, New Haven; photo by James St. John) (b) *Anomalocaris canadensis* life drawing. (Nobu Tamura, 2014)

(a)

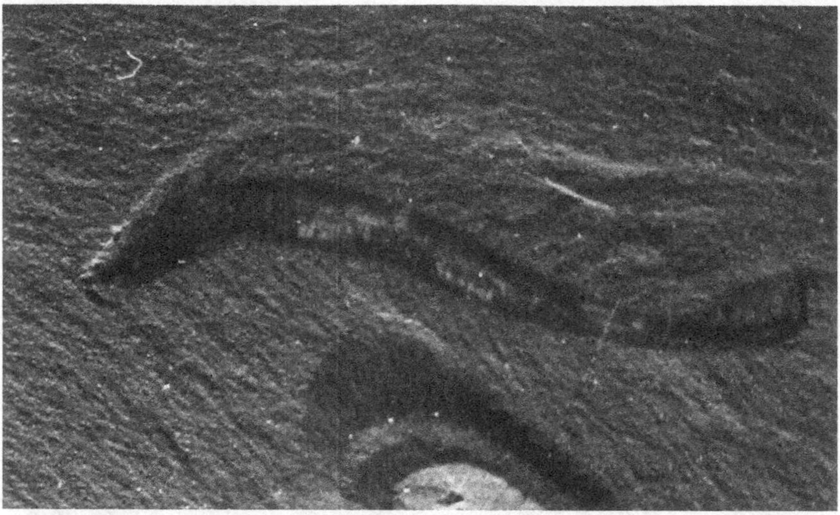

(b)

Figure 9.2 (a) *Pikaia gracilens* fossil. (Smithsonian Institution; photo by James L. Stuby, 2009) (b) *Pikaia gracilens* life drawing. (Nobu Tamura, 2016)

relationships relative to a common ancestor. The new methodology replaced the traditional methods called evolutionary systematics (the study of biological diversity). The first to write about cladistics was German zoologist Willi Hennig (1913–1976), who published a critical work on the topic in 1950, translated into English in 1966.[13] Along with cladistics was another technique called the stem group concept, a method of sorting out phylogenetic relationships by tracing the common ancestor and all its descendants through related groups and graphing them.

These new analyses led to the hypothesis that many of the Burgess fauna and flora belong to known phyla. For instance, *Anomalocaris* is now thought to belong to a stem group with the arthropods. While many of the Burgess biotas fit into existing phyla, such as the Arthropoda (arthropods) and Porifera (sponges), some may be considered as "other," yet to be further classified or belonging to a never-before-seen phylum. These animals are an example of a branch of the evolutionary vine that consisted of early "experiments." Paleontologists think they may have led to other clades (groups of organisms arising from a common ancestor), only to be found millions of years later, almost by chance. Time and testing of hypotheses based on cladistic analysis will shed further light on where to place the Burgess creatures.

Ordovician Period, 485.4–443.8 Million Years Ago

The Ordovician, which follows the Cambrian Period, lasted for 41.6 million years and is composed of three epochs—the Lower, Middle, and Upper—with seven stages based mainly on graptolite and conodont groupings and first appearances of those various species. The Caledonian orogeny continued during the Ordovician with the Grampian phase[14] of the event, building mountain ranges whose remnants exist as the Highlands of Scotland. Several stages of mountain building, including the Humberian (Newfoundland) and the Taconic (New England), occurred around the same time.

Graptolites are the hard shells of creatures that dwelled in floating marine colonies found in clays and other fine-grained ocean deposits. Individuals, zooids, lived in the segmented portions of the shell, like apartment dwellers, and were connected by a tube to each other, as categorized in the *Hemichordata* phylum. They range in age from the mid-Cambrian through the Carboniferous Periods. In the Ordovician, graptolites increased in diversity at a fast pace until 470 million years ago, when their numbers dramatically declined toward the end of the period.

Henry De la Beche, founder of the Geological Society of London, and Andrew Ramsey of University College, London, were reported to have found the first graptolites near St. David's, in the Ordovician strata in south Wales in 1841.[15] Nevertheless, stratigraphers did not utilize graptolites to solve stratigraphic relationships until the work of English geologist Charles Lapworth in 1878.[16] Lapworth started sorting out the ages of various strata in the southern uplands of Scotland and then extended his work to the rest of Britain and internationally, ultimately developing twenty graptolite zones from the late Cambrian to the Silurian. His work has stood the test of time, even as more recent stratigraphers have subdivided and refined the Ordovician.

Conodonts are microfossils less than a millimeter long. Their "teeth" and venomous structures are shown in figure 9.3a,[17] with a life drawing in figure 9.3b. These are "index" fossils—species easy to identify and which lived for a relatively short time before changing in shape or structure, found in a wide variety of deposits. Conodonts are excellent index fossils because, though small, the material of which they are made—calcium phosphate—persists and survives in the geologic record. It can be easily removed from samples of rock and studied to determine the specific time in which they lived.

Conodonts were initially recognized in 1856 in a clay bed near St. Petersburg by C. H. Pander, a Russian paleontologist who thought they were unusual fish teeth and gave them their name.[18] Not much notice was taken of these microfossils until 1926, when American geologists E. O. Ulrich and R. S. Bassler found they could be used as a biostratigraphic marker to identify rock layers and sort out stratigraphy.[19] By the 1960s, researchers realized they were among the most critical microfossils and biostratigraphic markers from the Paleozoic to the Upper Triassic, when they became extinct.

Paleontologists speculate that fossils of conodonts, sometimes called "conodont elements," are the toothlike feeding structures of long-extinct jawless vertebrates that lived in the oceans. Moreover, geologists have discovered that particular conodonts were the first venomous animals ever found in the middle Ordovician Period. The teethlike structures are curved and have a grooved track through which they theorize the venom was delivered.

The middle of the Ordovician witnessed a substantial increase in marine biodiversity when the number of species tripled in what is known as the "Great Ordovician Biodiversification."[20] Geologists think this proliferation is a result, in part, of sea-level incursions onto continents, increasing the niches for life in the shallow seas. Also, toward the end of the Ordovician, two glacial periods occurred, causing related extinction events. Not only were these the first of five significant extinction episodes of the geologic time scale, but more than 60 percent of the marine genera were decimated.

(a)

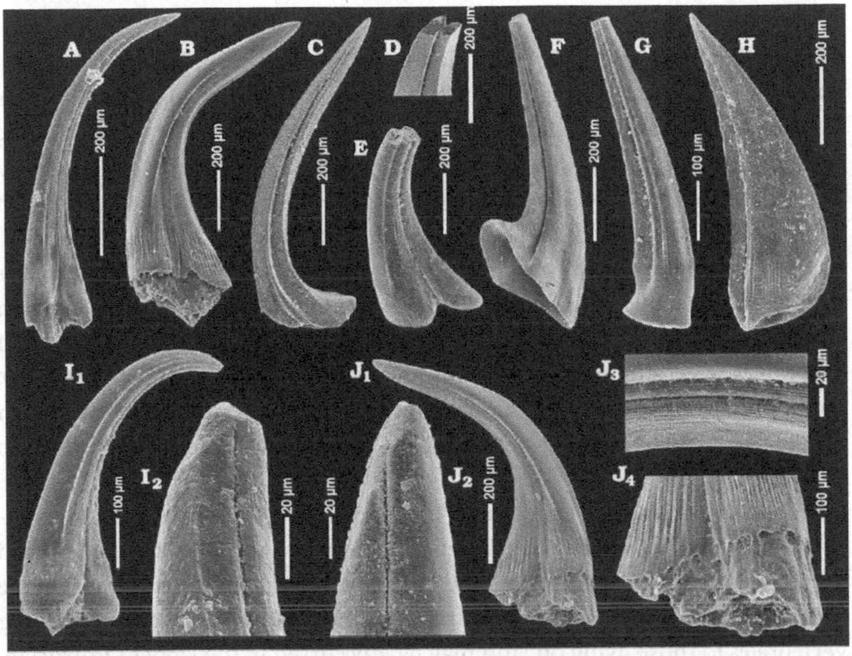

(b)

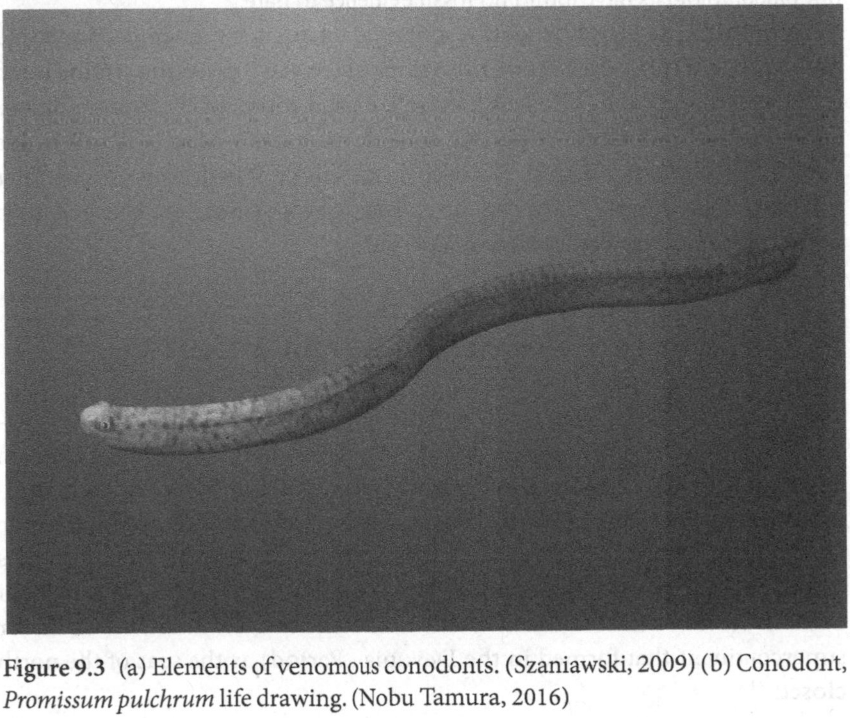

Figure 9.3 (a) Elements of venomous conodonts. (Szaniawski, 2009) (b) Conodont, *Promissum pulchrum* life drawing. (Nobu Tamura, 2016)

Silurian Period, 443.8–419.2 Million Years Ago

The Silurian has four epochs—the Llandovery, Wenlock, Ludlow, and Pridoli—and within the lower three epochs are seven stages based on changes in graptolites and conodonts. The period extended for 24.6 million years. The Silurian of the current geological time scale is that defined by Scottish geologist Roderick Murchison in 1839, the "Upper" Silurian.[21] Lapworth reclassified the beds of Murchison's "Lower" Silurian to be part of the Cambrian Period in 1879.[22]

Fauna in the Silurian was recovering from the extinction events in the Ordovician related to the two glacial periods. Graptolite species rebounded in numbers and served to distinguish the series and stages within this period. If graptolites were king of the fossils in the clays and shale deposits of this time, conodonts were primary in the limestones and carbonate-rich rocks. Jawless fish first appeared, including *Loganellia*, a Theolodonti (feeble-toothed) jawless fish (figure 9.4).

In addition to jawless fish, during the Silurian Period, jawed fish and freshwater fish species evolved. Coral reefs became more complex and widespread during this time, with the first life on land, as shown in the fossil record, in the form of early vascular plants and some insects. Movement from ocean-dwelling vertebrates to those living on land may have occurred in the Upper Ordovician, but paleontologists have found no fossil evidence to date.

At this time, sea levels rose in response to fluctuations in temperatures and melting ice sheets in the Upper Ordovician. There was a general warming trend in the early part of the period and, overall, a stabilization of climatic conditions compared to prior times. A period of black shale deposition occurred in the early Silurian, and this deposit has been a vital source of hydrocarbons that later migrated to sandstones. Curiously, no episodes of volcanism have been recorded in the Silurian; it was a relatively calm period.

Devonian Period, 419.2–358.9 Million Years Ago

The Devonian Period has three epochs—the Lower, Middle, and Upper—and within those three epochs are seven stages, categorized by changes in conodonts and ammonites. This period lasted 60.3 million years. Mountain-building episodes during the Devonian consisted of the Acadian phase of the Caledonian orogeny.[23] Evidence of these events can now be seen in breaks in the geologic record in the northern Appalachian Mountains and much of Britain. Pieces of the continent Gondwana collided with Laurussia (a minor supercontinent that formed in the Devonian Period) as the seas of the north closed.[24]

(a)

(b)

Figure 9.4 (a) Fossil *Loganellia*, an extinct fish (Thelodonti) from the Silurian. (Museon, The Hague; photo by Ghedo, 2011) (b) *Loganellia* life drawing. (Nobu Tamura)

During the Devonian, the continents existed in one hemisphere, and there were extensive shallow seas in the tropical latitudes, leading to the formation of thick reef deposits and, later, limestone sequences. The rise of fish species characterized life during this period. In England and on the European continent, geologists cite the development of the Old Red Sandstone as a characteristic rock of the period. A striking feature of these sediments, now lithified into rock, is their red color, which results from oxidation of iron. The sediments had a continental source and were oxidized subaerially (under the air).

Murchison discussed the Old Red Sandstone fishes in his report on the Silurian (and younger) systems.[25] One fish of particular note, with its prominent shield head, is an early Devonian *Agnatha*, a jawless fish, shown specifically is the *Agnatha* fish *Cephalaspis lyelli* (figure 9.5). The specimen came from Charles Lyell's collection. Jawed fish species were prominent in Lower Devonian rocks, including armored fish (*Placodermi*), bony fish (*Osteichthyes*), and lungfish (*Dipnoi* group of *Sarcopterygii*). Lungfish have a swim bladder that allows them to breathe air in dry periods when they burrow in mud to survive. The class of *Sarcopterygii*, or lobe-finned fish, includes the *Dipnoi* as well as tetrapods, which used the bony, muscular lobes of their fins to expand their territory onto the land, the first vertebrates to venture out of the oceans. Tetrapods, the four-legged land dwellers, constitute a biological superclass and are the ancestors of all living vertebrates with legs.

The Devonian is not only the "Age of Fishes" but also the age of the land plants, which arose in the Silurian Period. Paleontologists find them in abundance in the Devonian fossil record. Ferns, horsetails, and early seed plants colonized the land and built up the earliest forests. Soils developed on terra firma for the first time, and many were rich in oxygen, giving many Devonian deposits their red color, including the Old Red Sandstone.

As this period tracks the rise of fish genera and movement of life onto the continents, even more complex and varied organisms appear in the fossil record. These developments do not suggest a linear, straight-line trend of the development of life. Instead, as many researchers have noted,[26] extinctions of a significant number of species—mass extinctions—have occurred five times from the Paleozoic to the Cenozoic. The analogy of a branching tree may not encompass

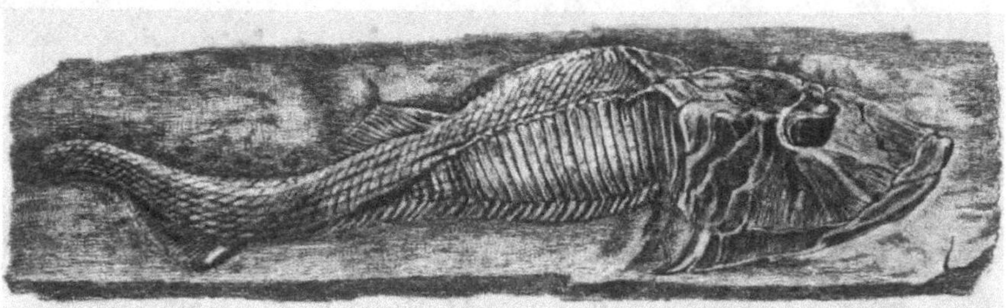

Figure 9.5 *Cephalaspis lyelli* from the Lower Devonian Old Red Sandstone. (Murchison, 1839, plate I)

the diversity formed after an extinction, as many lines may have moved into the niches created, and experiments with different anatomical forms may have occurred. Some of these experimental forms were not successful and not able to persist.

Toward the middle to late Devonian, the second of the five major extinction periods occurred in fauna—the "Hangenberg Event"—with half of the marine genera becoming extinct, including vertebrates, ammonites, and trilobites and some of the early terrestrial animals. Graptolite species had died out in the latter part of the early Devonian before the first appearance and fluorescence of ammonites. The Hangenberg Event was not just one extinction phase, according to recent authors, but up to twenty episodes of extinction.[27] Paleontologists think these losses were a result of changing environments and global climatic shifts when sea level rose and repeatedly fell, with rapid cycling from warm to cold temperatures. In particular, sea incursions onto land during times of warming brought oxygen-starved clastic sediments (sands, clays, and gravels) in floods, covering up limestone reefs and killing corals and other reef dwellers.

Carboniferous Period, 358.9–298.9 Million Years Ago

The Carboniferous Period, extending 60 million years, is known as the first significant coal-forming time on the Earth and contains two sub-periods, the Mississippian and Pennsylvanian, both with three epochs, the Lower, Middle, and Upper. There are three stages within the Mississippian Sub-period and four stages within the coal-bearing Pennsylvanian Sub-period. The stages and epochs of the Carboniferous Period are based on marine organisms—conodonts, foraminifera, ammonites, brachiopods, and benthic (living on the seafloor or in shallow marine sediments) fusulinids (an extinct order of foraminifera). Fusulinids, appearing for the first time in the early Carboniferous, are tiny grain-sized foraminifera—an important index fossil for correlating beds and strata. Index fossils are used for stratigraphy because they lived for a relatively short time over extensive areas, changing and evolving rapidly.

Atmospheric oxygen levels soared and allowed the further development of large plants, ferns, and other seed-bearing vegetation. The modern oxygen concentration in the atmosphere is 21 percent; however, in the Carboniferous, oxygen levels rose to 26 percent, based on the studies of charcoal concentrations in peat and mires from that time.[28] The increase in oxygen levels allowed large plants along with supersized insects and other creatures to grow and proliferate. Mega-flora during this period consisted of *Lycopsids* (seedless plants, the most common coal-forming plant), *Sphenopsids* (horsetail ferns), *Neuropteris* (extinct

Figure 9.6 Carboniferous Period *Neuropteris*, extinct seed fern. (Exhibit in the Houston Museum of Natural Science; photo by Daderot, 2014)

seed ferns; figure 9.6), and early forms of trees, including *Lepidodendron* (an extinct scale tree) and *Archaeopteris* (a treelike plant with leaves similar to ferns).

Paleontologists note a 25-million-year apparent break in the fossil record of the Lower Carboniferous, "Romer's Gap,"[29] from 360 million to 345 million years ago. However, work in Nova Scotia on the Carboniferous outcrops of Horton Bluff, Blue Beach, near the town of Hantsport, indicates a wide variety of tetrapod bones during that very time.[30] Some postulate that the high-energy environment of deposition and lack of finds are responsible for the seeming gap, rather than a reduction of species. Some paleontologists speculate that tetrapods were not impacted by the late Devonian extinction event and instead were present throughout the Romer's Gap time.

During the Carboniferous Period, tetrapods and reptiles filled (and in some cases refilled) ecological niches on the land. Tetrapods developed the amniotic egg, which could survive dry periods because of an outer shell or leathery covering. Laying eggs on the shores in protected areas gave tetrapods an edge in survival, and the extra casing on the eggs enhanced the chances that their young would survive. Tetrapods with their amniotic eggs diverged into two major groups, representing a critical divergence in the development of groups of tetrapods: the synapsids and the sauropsids. Paleontologists classify synapsids as animals whose skulls had one temporal (near the eye socket) opening, thought to be ancestral to mammals. The sauropsids (animals whose skulls had two temporal holes, the Sauropsida were the ancestors of the diapsids)[31] were more akin to reptiles and led to the dinosaur and bird lines. Two holes, in the case of sauropsids, allowed for more significant musculature of the jaw and increased crushing power.

Plants during this period grew up to 30 meters (100 feet) tall, among them the lycopods, clubmoss trees, tree ferns, giant horsetails, and other seed plants. Forests abounded, and those located near shorelines, lakes, and the oceans had copious plants that thrived in the warm, humid conditions. When they died, their tissues sank into the swamps and decayed into peat, and with time and burial, they formed vast coal deposits. These coal deposits range from low to high grade, interbedded with clays. The deposits contain coal balls, round ball-like structures found in coal beds made up of peat and plant materials. In the lab, paleontologists slice the coal ball and make an acetate peel, which is examined under the microscope to study the plant structures within. Detailed morphology of the plants can be seen, down to the cellular level, making identification of the Carboniferous plants possible. Plant communities were different in various areas of Pangea and Laurentia and in the southern part of Gondwana, with species of the northern *Glosopteris* living in Laurentia and southern *Glossopteris* living in Gondwana (color insert 9.2).[32] Paleobotanists also have found some intermixing of the two plant communities.[33]

Additionally, geologists have discovered specific sedimentary structures, such as filled stream channels and bedding planes (figure 9.7), as seen, for example, in Carboniferous deposits in the Stellarton Formation, Nova Scotia, to contain dark black layers of coal. Excavated areas of this outcrop in the Colburn Pit are where coal was extracted by hand in minor amounts to burn. The lower part of the Carboniferous, designated as the Mississippian Sub-period, is primarily composed of limestones. Accordingly, these formed when the sea levels rose and created ideal conditions for carbonate reefs.

Indeed, life in the oceans was recovering from the extinction events of the Devonian. However, oddly, the number of fish species during the Carboniferous, contrary to this trend, dropped considerably. The sizes of vertebrates in the oceans also generally decreased, for still-unexplained reasons.

Permian Period, 298.9–251.9 Million Years Ago

The last period of the Paleozoic Era is the Permian, enduring for 47 million years, and it has three epochs, the Cisuralian, Guadalupian, and Lopingian—as well as nine stages. Epochs and stages within the Permian are based on marine biostratigraphy and changes in conodonts, ammonites, and fusulinids, which disappeared at the end of the Permian.

Figure 9.7 Channel body in the Stellarton Formation. (Pennsylvanian Period), Coalburn Pit. (Photo by Michael C. Rygel, 2001)

On land, the synapsids and the sauropsids dominated. A group of the early synapsids were the pelycosaurs, among them *Dimetrodon* (sail-backed creature), which walked the shores and wetland areas in the Permian.

A disruption in the land-animal fossil record was known as "Olson's Gap." Some scientists call this break in the record "Olson's extinction," because it may represent an extinction event. The gap occurred for 5 million years, starting in the early Permian around 273.5 million years ago.[34] In part, because no similar interruption occurs in the marine record, geologists debate whether this gap is a result of extinction, poor preservation, sampling bias in locations of the sites where fossils have been found, or a real discontinuity.[35] Spencer Lucas (1976–), an American paleontologist and stratigrapher from the New Mexico Museum of Natural History, who had previously named this break in the record based on extensive analysis of marine fauna biostratigraphic correlation to land animals, reaffirmed in 2013 his commitment to Olson's Gap.[36] Some of the dispute is a result of the difficulty in correlating strata, fossils, and species between deposits critical to related fossils in Russia and those in the United States. However, Lucas performed an extensive analysis to back up his claim, and other authors have accepted it as of 2016.[37] The reason this break or gap is of particular note is that it spans the time from when pelycosaurs dominated in the Permian to when the therapsids were the main synapsid land animal—a critical pre-transition period for the mammal lines to come. Therapsids, such as *Scymnognathus* (figure 9.8), emerged from pelycosaur ancestors (synapsids) mainly through the Sphenacodontia clade (see chapter 5).

Paleontologists first identified a therapsid called *Cynognathus* as a "mammal-like reptile." Further research has shown that synapsid lines coevolved along with reptiles but were more mammal-like. Paleontologists today categorize them as protomammals.[38]

In the Permian, the landmasses grouped into the massive C-shaped supercontinent Pangea. Deserts and arid conditions were frequent across this immense continent, which was more than 130 million square kilometers (50 million square miles), because the continental interiors were far removed from the moderating climatic influences of the oceans. Mega-dunes, which abounded in the inland areas, are recorded in the rock record. These dunes become preserved in sandstone and demonstrate their original dune structure, with horizontal beds on the bottom and angular beds representing the faces of the dunes. These formations are often red from the oxidation of small amounts of iron in the sand-dune layers. The part of Pangea that was near the equator was tropical and had swamps producing coal deposits. Other areas in the tropical region developed reefs and corals, which later formed limestone.

The third great extinction in the history of life occurred around 250 million years ago, at the end of the Permian. Geologists think this event was the largest in the Earth's history, with up to fully 95 percent of marine genera being wiped out along with 70 percent of all land species.[39] At this time, tectonic forces began

Figure 9.8 Fossil of an extinct therapsid, *Scymnognathus*. (Museum of Paleontology, Tübingen; photo by Ghedo, 2013)

to shift their alignment, breaking apart the large lithospheric plates that made up the Pangea supercontinent. This process was not gentle; volcanic activity dramatically increased in the area that is now Siberia. During this time, rift valleys were created, which opened great "traps" (volcanic chasms from which voluminous amounts of basaltic lava flows, creating layer upon layer of deposits) called the Siberian Traps.[40] These massive lava deposits, also known as flood basalts, covered 5 million square kilometers (1.9 million square miles) (color insert 9.3). The layers of lava from the Siberian Traps would have buried an area the size of modern China 13 meters (40 feet) deep. Other flood basalt locales, called large igneous provinces, exist on the Earth and have been associated with mass extinctions, but none with a role as significant as the Siberian Traps.

Nevertheless, the flood basalts alone were not what produced the mass extinction. Geologists, in addition, speculate that magma pushed into the preexisting rock surrounding the rifts, kicking off subsequent eruptions. It appears that the volcanic deposits, especially the materials creating sills (horizontal deposits injected into surrounding rock, like the sill of a door), were the triggering mechanisms that ultimately led to the environmental catastrophe.[41] The volcanic materials were so hot that they caused the solid rock to vaporize gases that escaped into the atmosphere. Moreover, these layers into which the volcanic materials were injected were made of limestone and gypsum, consisting of calcium carbonate and calcium-magnesium carbonate, which resulted in the release of greenhouse gases such as carbon dioxide, methane, and sulfur dioxide.

Coal seams caught on fire. Think of Centralia, Pennsylvania, and its underground burning coal seams but magnified a million times. As the coal seams burned, more carbon dioxide was pumped into the air at a high rate, along with the gases from the volcanic sills. Ash and sulfur dioxide, as well as the other gases, spewed into the atmosphere. Sunlight was eventually blocked as the impacts of volcanism took hold.

Recent research has shown that, along with an increase in methane, carbon dioxide, and sulfur dioxide levels, there was a corresponding upsurge in microbes, which in turn generated more methane, along with much higher temperatures in the ocean and atmosphere, and amplified the increase of carbon dioxide in the atmosphere. It was global warming on a massive scale. Ocean acidification and rates of erosion ramped up. As more sediment and nutrients poured into the oceans, conditions were created in which there was no longer dissolved oxygen in the seawater. The change in the climate and warming produced wildfires. Geologists find excess metals in the rock record from this time, from the great amount of ash produced, along with evidence of increased and blooms of cyanobacteria because of nutrients released into the oceans.[42] Daniel Rothman, co-director of the Lorenz Center at MIT, has investigated with colleagues this methane-producing mechanism unleashed during the late Permian and related extinction event.[43] In essence, the global carbon cycle was disrupted, and a cascade of events took place, which ended most life on the Earth. Their work—to be discussed later—has implications for the current conditions on the Earth.

As the coal seams burned, more carbon dioxide was pumped into the air at a high rate, along with the gases from the volcanic sills. Ash and sulfur dioxide, as well as the other gases, spewed into the atmosphere. Sunlight was eventually blocked as the impacts of volcanism took hold. Acid rain precipitated from the skies to add to the misery for organic life. The food supply was dangerously impinged. To make matters worse, widespread forest fires were prevalent and added even more carbon dioxide to the air as the trees were incinerated. There was a spike in fungus at the Permian–Triassic boundary, which indicates to scientists that most trees died out quickly, and rocks from this time record the ash and charcoal from the wildfires.

These factors disrupted the carbon cycle, creating rapid global warming on a colossal scale because of the elevation of temperatures related to rising greenhouse gases. Climatic shifts impacted the weather patterns; parts of Pangea started to dry out and become more arid than ever. Deforestation occurred as growing zones shrank. Because the land surface was exposed, rainwater eroded the denuded ground much more than if vegetation had protected it. Much of the material was eroded by streams, wind, and sometimes ice and was deposited elsewhere.

The oceans were not safe for life at this time, and even more species extinction occurred there than on land. The deep water in the oceans had little oxygen in the Permian Period; most sea life lived in the shallow waters near the edges of the continent, but these, too, become stagnant as ocean circulation patterns were disrupted when carbon dioxide levels soared. Small organisms at the base of the food chain soon succumbed, and their loss propagated up the food chain, impacting the rest of life.

This disruption of the carbon cycle was global warming on such an enormous scale that animal and plant life could not adapt quickly enough. The extinction took place, possibly in several pulses of intensity, over 200,000 years,[44] a blink of the eye compared to the length of geologic time.

The severity of the Permian–Triassic extinction was significant enough that stratigraphers use it to mark both the end of the Paleozoic and the beginning of the Mesozoic, when the way was paved for the rise of the dinosaurs.

10

The Biography of the Earth

Mesozoic Era

The Mesozoic ("middle life") Era is the second within the Phanerozoic Eon and spans from 251.9 million to 66 million years ago. The Mesozoic has three periods, the Triassic, Jurassic, and Cretaceous. The era is bounded on each of its ends by major extinctions, several of the largest in the Earth's history. The first is known as the Great Dying, and the second marked the end of the era when the asteroid that collided with the Earth created the Chicxulub crater.

Pangea still was a supercontinent from the beginning of the Mesozoic Era to the late Triassic, enduring for 100 million years, but it continued to rift and split apart. By the Jurassic Period, the basin between Laurentia (North America) and Baltica (Europe) had opened up as the Mid-Atlantic Ridge formed—the beginning of the Atlantic Ocean. In the valleys made by the rifts, as ocean waters filled the fissures, sometimes the basins made in the rifting dried out and salts built up, creating thick evaporite deposits. The old convergent margins where mountains, such as the Appalachians, had formed on the east coast of Laurentia were transitioning to passive edges of the continents as the new ocean plates created at the mid-ocean ridge separated them farther and farther away.

On the western side of Laurentia, tectonic conditions favored convergence and trenches. California was built mainly by "exotic terranes" (pieces of other continents and ocean floor smashed onto continental cratons by tectonic forces) aggregating eastward to the craton at this time. Along with this increased tectonism on the west coast of Laurentia, mountain building occurred, first with the Sevier event and later, toward the end of the Mesozoic Era, with the Laramide orogeny. The Sevier orogeny started 170 million years ago when a volcanic arc, called the Sierran Arc, crashed along the western coast of Laurentia and created a series of low-angle thrust faults that broke and folded the brittle rocks, forming the Canadian Rockies and the mountains of western Wyoming. The exotic terranes also piled up to the west of the Sierran Arc, building up the western edge of the continent.

The Laramide mountain-building event was responsible for the most recent version of the Rocky Mountains, occurring between 70 million and 40 million years ago. The Rocky Mountains are far from any plate boundary, approximately 600 kilometers (1,000 miles) from the compressive forces on the west

Song of the Earth. Elisabeth Ervin-Blankenheim, Oxford University Press. © Oxford University Press 2021.
DOI: 10.1093/oso/9780197502464.003.0011

coast. Geologists speculate that instead of diving deep into the asthenosphere as most plates do at their convergent edges, the subducting plate, called the Farallon Plate, encountered a discontinuity. This barrier caused the plate to flatten out, reaching far inland, away from plate boundaries, and thus building the young mountain range. The compressive forces were so strong that some areas underwent shallow thrust faulting (low-angle reverse faults, related to convergent plate boundaries). Other regions had much steeper reverse faults (steeper-angle faults related to convergent plate boundaries) reaching into the crystalline basement rocks that later were uplifted and eroded, exposing the ancient strata.

The climate in the early Mesozoic was warm, but conditions cooled in the late Jurassic and early Cretaceous, the era's middle and last periods. Sea levels had begun rising in the early Jurassic Period and formed incursions on the continents. One such was an interior seaway along what is now the Front Range of the Rocky Mountains.

Triassic Period, 251.9–201.3 Million Years Ago

Mass extinctions bookend the Triassic Period at both of its boundaries: the Permian–Triassic extinction event in the late Permian, previously discussed, and the Triassic–Jurassic extinction event. Both extinction periods coincide with massive flood basalts emerging from the rifting of Pangea. The Triassic extended for 50.6 million years and is composed of three epochs—the Lower, Middle, and Upper—and seven stages. However, in the last decade, stratigraphers have hotly debated the number and placement of stages.[1] Divisions within the Triassic are based on marine ammonites, conodonts, land trace fossils, and tetrapod evolution. The latter divides this period into eight intervals based on land-vertebrate faunochrons (a division of geologic time based on classification of the fauna of the period), tying tetrapod evolution to time.[2] The climate in the Triassic was hot and dry, with no ice at the poles.

The Triassic was a critical time for the development of the tetrapod, which led to the salamanders, frogs, turtles, crocodiles, lizards, dinosaurs, and mammals. The oldest dinosaur fossils ever found are from the Upper Triassic Ischigualasto Formation of northwestern Argentina on the Chilean border. These rocks date from 231 million years ago and are made of floodplain deposits of sandstones, overbank muds, and ancient soils, in thickness of 400–700 meters (1312–2296 feet) over 7 kilometers (4.3 miles), which were located near an active volcanic area. The region was extremely arid, contained badlands, and is known as the Valley of the Moon.[3]

British paleontologist Richard Owen defined the term *Dinosauria* (see chapter 1) in his 1841 *Report on British Fossil Reptiles*. It was during the Triassic

Period that dinosaurs started to emerge and diversify into different types. Between 1887 and 1888, British paleontologist Harry G. Seeley of Cambridge University classified them into two groups: ornithischians and saurischians.[4] The Ornithischia were defined as "bird-hipped," with backward-oriented pubic bones. The Saurischia were classified as "lizard-hipped," with forward-oriented pubic bones. Paleontologists realized there was a need for a new, third group: the theropod dinosaurs. The Theropoda (animals with hollow bones and three-toed limbs, "beast-footed") were proposed for the carnivorous dinosaurs by American paleontologist Othniel Marsh in 1881.[5] The theropods, which, like the saurischians, had lizard-like hips, evolved to include fierce predators, among them *Tyrannosaurus rex*. Primitive theropods were carnivorous, but with changes over time, some later theropods survived on vegetation, others on fish, and yet others on insects.

Instead of formal groups, another change to classification occurred when paleontologists reassigned the groups to clades, with the common ancestor *Dinosauria*. Seeley's overall classification stood the test of time for 130 years, until 2017, when paleontologists, including Matthew Baron, also at Cambridge, upended the traditional dinosaur clades, which had been based on hip configuration.[6] With stunning new research, Baron and other paleontologists discovered that the prior classification did not match their latest analysis. They found that the ornithischians and the theropods, despite different hip styles, were more alike based on twenty-one similar anatomical features in an extensive review of seventy-four dinosaur taxa for 457 characters, reviving the name "Ornithoscelida" as a clade for both. This new theory explains why theropods and the ornithischians had feathers, and saurischians did not. It appears now that the avian dinosaurs emerged from the theropod lines, later in the middle Jurassic, 165–150 million years ago.[7]

Despite the many species of dinosaurs identified by Cope and Marsh of "Bone Wars" fame (see chapter 7), finding good specimens has always been a challenge. The ornithischians are relatively well represented in the fossil record. However, others, such as the theropods, were not discovered often—at least, not until the fossil finds in the Ischigualasto Formation in 1958, a joint Argentinian-U.S. paleontological expedition. Among the participants was Harvard paleontologist Alfred S. Romer, who noted the richness of fossils from the beds. Romer packed up and attempted to take some of the specimens back to Harvard when he left the site. However, they were impounded at the border and remained there for several years until Harvard could convince custom authorities to release them.[8] Paleontologists have located some of the earliest theropods in the Ischigualasto Formation, which, as earlier noted, was created as Pangea split apart. The discovery of the back half of a *Herrerasaurus* skeleton, one early theropod, was

Figure 10.1 *Herrerasaurus* skull, one of the earliest dinosaurs, part of the Ischiqualasto fauna, Triassic Period. (Joerim, 2016)

made in 1961 by Argentinian rancher Victorino Herrera, for whom the creature was named (figure 10.1). Later finds included the skull of the creature.

Another spectacular specimen discovered in the Ischigualasto Formation is *Eoraptor*, found in 1993.[9] The Ischigualasto Formation is in the Upper Triassic Epoch, Carnian Age, approximately 227.8 million years old. To date, it has yielded three or four taxa of basal theropods.[10]

Paleontologists found an additional important dinosaur site in the Chinle Formation of the Colorado Plateau in Utah, Upper Triassic Period in age. These rocks are slightly younger than the Ischigualasto Formation. The Chinle also represents floodplains and overbank deposits, much like the Ischigualasto Formation. Today, the landscape is desert, but the rocks were laid down under the wetter conditions of the Triassic. Many animals and plants are found in this formation, but dinosaur bones are relatively rare. Paleontologists located the skull of a basal carnivorous theropod named *Daemonosaurus* in a siltstone member of the Chinle Formation in 2011 (figure 10.2).[11]

In the oceans, life rebounded from the extinction event at the end of the Permian. Paleontologists once thought that the marine reptiles came back more slowly than other marine dwellers, but recent fossil finds from China indicate that their recovery might have been swifter.[12] An animal named *Sclerocormus parviceps*, identified as an early type of *Ichthyosaurus*, was located in the Anhui Province of China. This animal was more massive than other contemporaneous marine reptiles and of a similar ancient *Ichthyosaurus* type to another species recently discovered. These specimens indicate the *Ichthyosaurus*-type

(a)

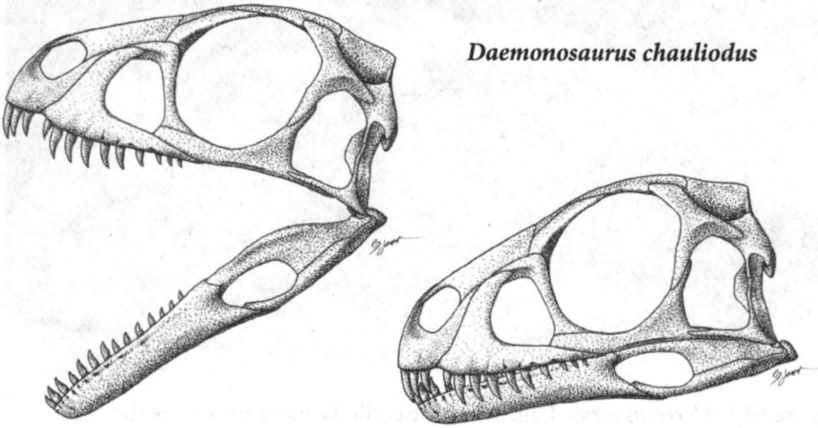

(b)

Figure 10.2 (a) Illustration of *Daemonosaurus* skull. (Sues, Nesbitt, Berman, and Henrici) (b) Life reconstruction of *Daemonosaurus chauliodus*. (Drawing by FunkMonk, Michael B. H., 2018)

reptiles spread out through different environments and diversified more quickly than paleontologists previously thought. Proto-dinosaurs developed in the late Triassic, among them *Plateosaurus*, an early sauropod. Types of land vegetation during the Triassic underwent a significant change from lycopsids and sphenopsids, which had emerged in the Carboniferous, to gymnosperms, seed ferns, cycads, and conifers.

The extinction event in the late Triassic is the fourth major extinction of the geologic time scale, particularly devastating to the marine organisms, killing about half of the genera in the oceans. It was the end of the conodonts that were essential biomarkers. Paleontologists had used them extensively for biostratigraphy of earlier periods, along with the ammonites and the radiolarians.[13] Geologists continue to research this extinction event for possible causes, such as an asteroid impact or continuing volcanism from the breakup of supercontinent Pangea. However, an analysis of biota and carbon isotopes in British Columbia, found that isotopic trends for the end of the Triassic were steady, with a few excursions indicating no climatic change. No isotopic reversal was recorded (unlike the carbon isotope pattern at the Cretaceous–Tertiary boundary when an asteroid hit the Earth).[14] Nevertheless, as the proto–Atlantic Ocean opened up, volcanoes erupted, sending carbon dioxide, other gases, and ash into the air, and may have caused the extinction.

Jurassic Period, 201.3–145.0 Million Years Ago

The Jurassic Period followed on the heels of the fourth mass extinction in the Triassic. It has three epochs—the Lower, Middle, and Upper—and eleven stages, and it lasted 56.3 million years. Ammonites that survived the boundary extinction event were used to create "standard zones," such as the Mariae Zone. Between 1849 and 1852, French naturalist Alcide d'Orbigny designated ten stages based on 3,717 species[15] of ammonites.[16] Between 1856 and 1858, German paleontologist Albert Oppel further refined his stages into thirty-three zones, twenty-two of which were Jurassic ammonite zones.[17] After that, stratigraphers have correlated the standard zones to formalize biostratigraphy based on ammonites, as discussed in chapter 3.[18]

Dinosaurs survived the end-Triassic extinction and flourished in the Jurassic Period. Theropod dinosaurs started to diversify during the early Jurassic. Their bones, however, beyond fragments, have been challenging to locate, and paleontologists have found sparse evidence. Researchers did find a nearly complete skeleton of a young juvenile theropod dinosaur near Lavernock Point in Cardiff, Wales, when the fossil weathered out of the marine cliffs of the Blue Liassic, interbedded clay, and limestone.[19] Localities for Jurassic dinosaurs also encompass the Liassic beds of Mary Anning on the Dorset coast of England, as well as a famous site in Argentina. Anning found *Plesiosaurs* and *Ichthyosaurs* (figure 10.3) in the Blue Liassic cliffs near Lyme Regis. William Conybeare identified some of the *Ichthyosaurus* skeletons and made stunning illustrations of the fossils.[20] Anning also discovered fossilized coprolites, dinosaur feces.

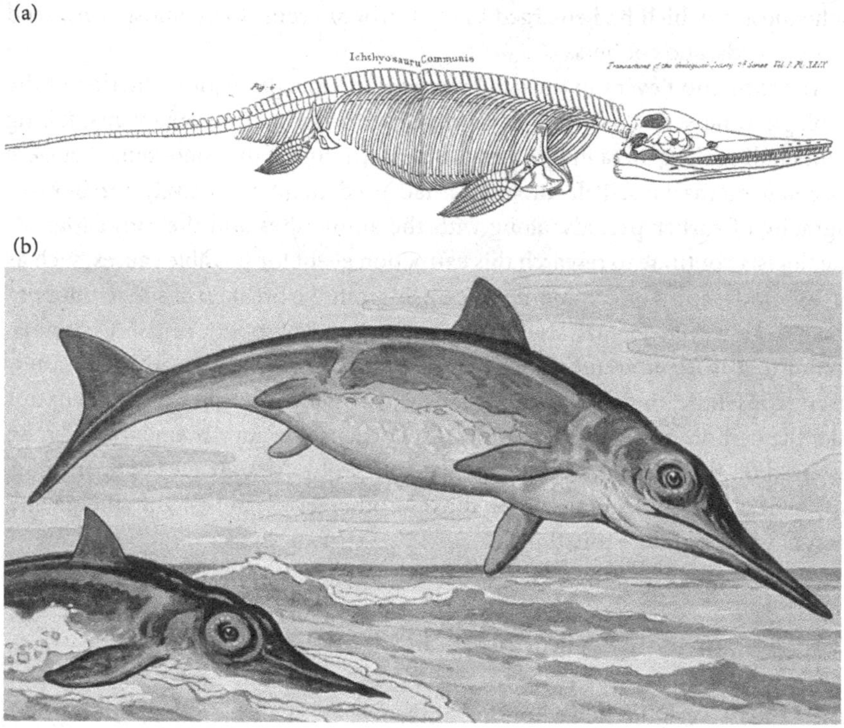

Figure 10.3 (a) Diagram of the skeletal anatomy of *Ichthyosaur communis*. (Conybeare, 1822) (b) *Ichthyosaurs* life drawing. (Henrich Harder, n.d.)

Another significant Jurassic fossil location is the Solnhofen Quarry in southern Germany on the Bavarian border, with its famous limestone. This rock was deposited in a shallow marine lagoon, which was cut off from the ocean and, as a result, was low in oxygen, forming a fine-limey mud. Bodies of animals washed into the basin during monsoonal flooding and became trapped in this mud. The low energy of the environment created excellent conditions for the preservation of the organisms. Like the Burgess Shale, it is a *Lagerstątten* deposit, and one of the rare fossil localities where soft parts of the creatures are preserved. The limestone breaks into sheets along natural planes of cleavage, revealing the fossils trapped in the rock. Paleontologists have identified more than 750 different species from the Solnhofen Limestone, including crinoids, ammonites, fish, crustaceans (figure 10.4), and the famous *Archaeopteryx*.[21] The first birdlike dinosaur found in the Solnhofen Quarry limestone, *Archaeopteryx* is the link between dinosaurs and birds. It had feathers, indicating it could fly, but other features that were more reptilian, such as a long, bony tail.

Some of the more massive sauropod dinosaurs—*Apatosaurus* and *Diplodocus*—died out at the end of the Jurassic Period from a lesser extinction

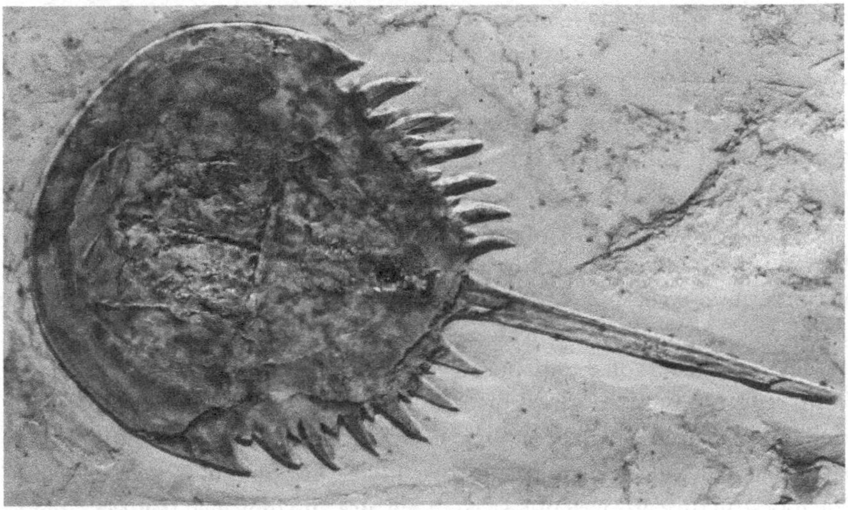

Figure 10.4 Fossil of a horseshoe crab, *Mesolimulus walchi*, Solnhofen Quarry. (Museum of Natural History, Berlin; photo by Anagoria, 2013)

pulse that included greenhouse conditions, warming, higher sea-level stands, and the ocean becoming anoxic.

Small mammals evolved in the late Jurassic from the cynodonts such as *Cynognathus*, which had made it through the Triassic extinction. One such group was the multituberculates, which were similar to but distinct from rodents, named for the many (*multi-*) cusps (*tubercula*) on their teeth and now extinct. They have the most extensive fossil record of any mammal group, more than 100 million years in span, but disappeared and became extinct in the late Eocene Epoch of the next era, the Cenozoic.[22] They are the only major mammal branch that has no living representatives in the modern world. Paleontologists think that rodents outcompeted them. The multituberculates lived in a variety of habitats, some in the ground and others in trees. It appears that they were proto-marsupials and bore their young alive, at a very early stage of growth, and the pups finished developing in their parent's pouch. Adults ranged from modern mice-sized to beaver-sized.

Cretaceous Period, 145.0–66.0 Million Years Ago

The Cretaceous Period is the final and longest-lasting of the Mesozoic Era, extending 79.0 million years, and it is the only period that has no clear lower boundary according to modern stratigraphic methods.[23] The border between the Jurassic and the Cretaceous is not clearly marked, because no significant shifts in life forms or geologic history occurred. This period consists of two epochs, the Lower and Upper, with twelve series based on marine fauna. Calcium-rich nanofossils (fossils of minute organisms, mainly plankton), foraminifera, and ammonites defined those epochs. Ammonites are the principal fossils for correlation in the Cretaceous. On the land, dinosaurs reigned, but their fossils were not able to be used for biostratigraphy. Although paleontologists developed a land-vertebrate age scale, it correlates poorly with marine stratigraphy. This lack of correspondence is because the dinosaur record is incomplete. Early angiosperms—flowering plants—emerged and diversified,[24] and along with them came the development of insects to pollinate them.

This period continued until the great age of the dinosaurs, in all their myriad forms and types, from the large carnivores like *Tyrannosaurus rex*—apex predators that dominated the landscape—to the small, non-avian dinosaurs such as *Microraptor*. *Microraptor* was discovered in China and described in 2000.[25] The foot structure was well preserved in the rock and indicated that *Microraptor* lived in trees. Many grazing dinosaurs such as *Iguanodons* and others roamed in herds, prey for the carnivorous dinosaurs. Dinosaurs also roamed the Western Interior Seaway and left their footprints in floodplain and lake deposits that bordered the seaway (figure 10.5) during the Jurassic and continuing into the next

Figure 10.5 Dinosaur trackways in the Dakota Sandstone, Dinosaur Ridge, Colorado. (Photo by James St. John, 2015)

period, the Cretaceous. Dinosaurs diversified into many niches and environments (color insert 10.1).

Not only did *Archaeopteryx* take to the air, as early as the Jurassic, but the Cretaceous skies were dominated by pterosaurs—flying reptiles—that used their wings to walk on the ground and soar high (figure 10.6). Pterosaurs varied from small-bodied to the super-sized *Quetzalcoatlus northropi*, discovered in Texas in the 1970s,[26] with a wingspan of nearly 11 meters (35 feet).

A bone of another huge pterosaur was discovered in the 1940s, though it was almost lost forever and took paleontologists decades to figure out just what they were dealing with. The story goes that workers along the Amman–Damascus rail line in Jordan found a bone measuring 0.6 meters (2 feet). Three years later, a local phosphate quarry owner purchased the bone and brought it to the attention of a British archaeologist. The bone was sent to France in 1953, where vertebrate paleontologist Camille Arambourg identified it as the wing bone of a giant pterosaur he later named *Titanopteryx philadelphiae*. After making a plaster cast of the unique bone, he returned it to the quarry owner in Jordan. The actual bone was forgotten and presumed lost. The story continues in 1975, when American geologist and paleontologist Douglas Lawson, who found the *Quetzalcoatlus* in Texas, noted from the cast that it was a cervical (neck) bone, not a bone from the wing. Another paleontologist, Lev Nesov of Russia, renamed the genus to *Arambourgiania* after he found out from an entomologist that the name *Titanopteryx* was already in use for a fly. Enter paleontologists David Martill and Eberhard Frey in the mid-1990s, who ventured to Jordan to get to the bottom of the matter and try to determine the particular pterosaur to which the bone belonged. They found some smaller pterosaur bones in a cabinet of the original

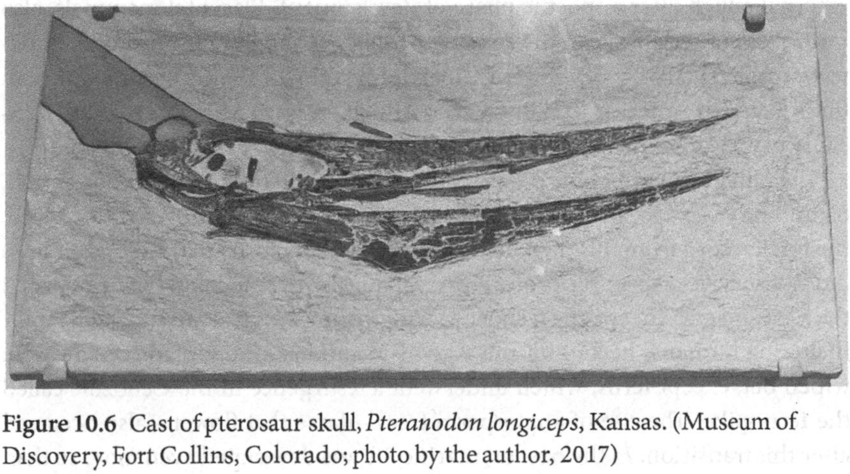

Figure 10.6 Cast of pterosaur skull, *Pteranodon longiceps*, Kansas. (Museum of Discovery, Fort Collins, Colorado; photo by the author, 2017)

phosphate-mining company, but never the famous neck bone, going home empty-handed. After they left, an engineer at the mine discovered the bone had been purchased by a geologist who donated it to the University of Jordan in 1973. Finally, in 1996, Martill and Frey were able to view the wayward bone. They identified it as a cervical bone from an *Arambourgiania philadelphiae*, which has an extremely long neck and a wingspan of 12–13 meters (39–43 feet).[27] In 2016, another cervical vertebra of *Arambourgiania philadelphiae* was found in Tennessee in the Coon Creek Formation, a near-shore marine deposit, extending the pterosaur's range to North America for the first time.[28]

These creatures were not related to birds or bats, and their ability to fly is an example of convergent evolution—when similar traits arise independently by different means in unrelated species. Paleobiologists and paleontologists have debated the origin of flight, whether flight evolving from tree-dwelling creatures who glided for their food or whether it developed from ground-based animals that hopped in the air to catch prey. Feathers became further specialized during this time, along with thermoregulation—the ability to regulate internal temperature—in the dinosaur lines.

Early mammals persisted into the Cretaceous Period but remained relatively small in size and did not evolve into a variety of taxa until after the extinction of the non-avian dinosaurs after the end of the Mesozoic Era. A recent find by paleontologists of an early mammal in Madagascar, named *Adalatherium hui*, is thought to be one of the oldest and largest mammals to date.[29] Paleobiologists speculate that Cretaceous mammals of this period were insectivores, similar to today's shrews and hedgehogs. Both primitive marsupials and mammals lived alongside many other species, including dinosaurs in settings like that of the Hell Creek Formation of North Dakota and Montana.

Marsupials, like today's kangaroos, are mammals that carry their young in an external pouch after a specific phase of development. Placental mammals, also known as eutherians, keep their young within their bodies for most of their development. The rocks of the Hell Creek form some of the earliest mammalian finds in North America. The Bug Creek Anthill fauna in Montana is another locality of early mammals on the continent.[30]

The fifth great extinction event occurred at the end of the Cretaceous Period, 66 million years ago, and was so dramatic that it ended the Mesozoic Era. Indeed, the fossil record from this time demonstrates that a majority of the dinosaur lines did not survive, except for proto-birds. Mammals persisted and have populated the next and current era, the Cenozoic, with all their species, varieties, and kinds, including humans, becoming the Age of Mammals. Many plant genera were wiped out, except ferns, which underwent a resurgence in the Cenozoic called the fern spike. The rise of angiosperm trees—those that flower—also occurred after this transition. In the oceans, marine life was devastated even more so than on land, with losses of numerous of the foraminifera, mollusks, diapsids, and echinoderms (marine animals of Phylum Echinodermata).

The majority of geologists attribute the cause of this extraordinary extinction event to an asteroid 11 kilometers (7 miles) wide that hit the Earth 66 million years ago at the Cretaceous–Paleogene, the K-Pg boundary (formerly called the K-T, Cretaceous–Tertiary boundary). In 1977, American physicist Luis Alvarez and his geologist son, Walter, initially discovered an unusual element—iridium—in Cretaceous clays from Italy and proposed an asteroid impact as the cause. Iridium is rarely found on the surface of the Earth. It is a dense metal, second only to platinum, which causes it to be present in the Earth's core but not in the crust. It is, however, a common element in asteroids. They wrote up their findings in 1980,[31] but there was no evidence of such a massive crater from that time. Confirmation, however, came ten years later. Alan Hildebrand and colleagues at the University of Arizona department of planetary sciences, discovered a 180- to 200-kilometer-diameter impact crater buried 20 to 30 kilometers deep (color insert 10.2), the Chicxulub Asteroid crater, near Mexico's Yucatán Peninsula, through remote sensing techniques.[32]

Research in 2016 by Jason Sanford and his colleagues at the University of Texas consisted of interpreting the Gulf of Mexico drilling logs and cores from forty deep wells, more than 300 meters (984 feet) in depth, and fifty-one shallow wells, less than 300 meters in depth.[33] They also examined 317 onshore and shallow wells, along with seismic surveys and sediment cores, to further understand the asteroid and deposits associated with it. The evidence took twenty years to obtain because of the proprietary nature of the data. When they finally received access to the core data and geophysics, the information revealed startling new details as to the extent and magnitude of the asteroid impact. When this asteroid hit the Earth, it released the energy of 100 teratons of TNT, equivalent to 1 billion times the energy released by the Hiroshima nuclear bomb, or 4–12 $\times 10^{23}$ joules. The result of this immense energy surge moved a sediment volume of more than 1.98 $\times 10^5$ cubic kilometers (48,000 cubic miles) over the entire gulf.[34]

The Chicxulub crater is distinctive not only for its size and far-flung effects but also because it has a rare inner circle of mountains called a "peak ring," the only one ever found on the Earth. These inner craters within the outer crater are rarely located on the Earth; most exist on other planets. The peak ring of Chicxulub is now covered in marine sediments. A recent study of cores drilled from two locations in the area of the peak ring, one in the ocean and the other on the edge of the Yucatán Peninsula, was undertaken by Expedition 364, the Chicxulub K-Pg Impact Crater Expedition organized and funded by the European Consortium for Ocean Research Drilling.[35]

When the asteroid hit, the force was so great it penetrated 32 kilometers (20 miles) into the Earth's crust, reaching a layer of basement crystalline rock made of granite. The granite, ordinarily hard and crystalline because it cools slowly underground from magma, became liquefied from the impact and flowed up and out.

Modeling of the asteroid and its effects has shown that it did not smash into the Earth directly but at an angle of approximately 60 degrees, causing more material to be released as high as 25 kilometers (15.5 miles) into the atmosphere. The asteroid first shattered surficial carbonate rocks, sending carbon dioxide and sulfur into the air; it then propelled vaporized rock airborne from the deeper granites. These particles cooled as they rose through the atmosphere, and the ones that did enter the stratosphere fell back to the Earth, heating up as they descended and creating a superheated cloud of soot and ash so hot that they started fires as the particles came in contact with the surface.

Materials released into the stratosphere caused a rapid cooling of the global climate of 26 degrees C (47 degrees F) for at least three years, as solar radiation was blocked.[36] The sulfur in the air combined with hydrogen to create sulfuric acid that rained back down on the land and oceans. A further consequence of the impact 66 million years ago was acidification of the oceans.

The amount of material released when the asteroid impacted vastly exceeds prior estimates by as much as twofold. This single event represents the most significant volume of material ever moved and redeposited at one time. Debris flows and turbidity currents were the primary agents of sediment transfer originating from mega-tsunami waves and earthquakes created by the impact-generated seismic energy.

A layer of clay containing platinum-group elements, mainly iridium, was laid down across extensive areas of the Earth, marking the event. Intense heat, pressure, temperature, and gas created changes in minerals, causing shocked quartz particles. Quartz made up of silica dioxide is customarily found on the Earth's surface and persists because it is hard and resistant to weathering. Impact-affected quartz has parallel bands called shock lamellae, which is a form of metamorphism. These are found not just near the impact crater but all over the world at the boundary layers.

Shocked quartz is not the only indicator of asteroid impact recorded in minerals. Glass spherules often accompany the deformed quartz grains, as do tektites—beads of glass formed as rock melted and cooled quickly in the air. Additionally, tsunami deposits from this time are found well inland from the Gulf Coast in Texas.

Geologists and paleontologists estimate that from 75 to 80 percent of all species living at the time were killed. Most life forms, including dinosaurs, were killed initially in the hot cloud of ash and smoke and resulting conflagration. Others perished later as their food supply dwindled because photosynthesis could not continue. In the oceans, increased acidity killed off the main predators at the time, the mosasaurs (large, swimming marine reptiles), as the lack of photosynthesis depleted the ocean food web from the bottom up. Smaller insectivore mammals survived the catastrophe, as did some dinosaurs—the lines that were the progenitors of modern birds. Plants of the Mesozoic were not left out of the destruction. The plants of the next era, the Cenozoic, were entirely different. Interestingly, plankton recolonized the crater area within 30,000 years.

Recent studies show that the specific location of the impact was one reason it killed so many. The asteroid hit where there were present not only abundant sources of sulfur and carbon but also hydrocarbon-rich, oil- and gas-containing sediments that were atomized to create soot. The chance of such a significant extinction event was calculated at just 13 percent if the asteroid had hit in an area not enriched in hydrocarbons.[37] If there had not been the mass extinction, which mainly was deadly to the dinosaurs, now-familiar ecological niches might not have opened up, allowing mammals to fill those spaces, and the Earth probably would look very different today.

Some scientists propose other, or additional, causes of the K-Pg extinction event. A second asteroid may have hit the Earth around the time of the bolide that created the Chicxulub crater.[38] The age of the interior of the Chicxulub crater—up to 300,000 years earlier than the end of the Cretaceous—may point to a different or additional cause of the K-Pg extinction event.[39]

Another theory is that the extinction resulted from Deccan volcanic traps, which opened up in the late Cretaceous. Eruptions from the Deccan traps have been age-dated starting 250 million years ago, before the boundary. Lasting 750,000 years, these eruptions spewed out ash, dust, gas, and 1.1 million cubic kilometers of lava and likely contributed to the mass extinction.[40]

Still, in 2010, forty-one international reviewers examined the accumulated data, logs, geophysics, findings, and literature and concluded that the asteroid event creating the Chicxulub crater was the predominant cause of the K-Pg extinction.[41]

Alternative theories for the mass extinction do not explain the ejecta distribution, the composition of the material, timing of the event, or its scale. Evidence for the impact being the leading cause is piling up as scientists examine additional data. The asteroid produced a shock wave and heat pulse as it hit, with massive tsunamis, global dust, debris, and gases, causing cooling of the climate, loss of light, and death of phytoplankton and algae, the primary food producers in the oceans. The recent data on the amount of sediment moved and the indisputably broad reach of the asteroid's influence confirm that the scale from this event was so large it is most likely the dominant factor in the K-Pg extinction. It marks the end of the Mesozoic Era and the beginning of the next, the Cenozoic.

11
Biography of the Earth
Cenozoic Era

The Cenozoic ("new life") Era, Quaternary Period, Holocene Epoch, Meghalayan Age is the time in which we, and all other life, currently exist. The Cenozoic extends from 66 million years ago to the present and has three epochs, the Paleogene, Neogene, and Quaternary. An extensive debate has been under way to delineate the divisions within this era. Stratigraphers in some countries proposed that older terms, Tertiary and Quaternary, be changed to Paleogene and Neogene. The arbiter of how systems, series, and stages are determined is the International Commission on Stratigraphy (ICS), the oldest and largest section of the International Union of Geological Sciences. The ICS ultimately divided the Cenozoic into (from older to younger) the Paleogene, Neogene, and Quaternary.[1] The USGS retains Tertiary and Quaternary as systems/subsystems but suggests that authors and researchers use the international divisions for the Cenozoic Era.[2] The Geological Society of America specifies Paleogene and Neogene for the Cenozoic.[3] In this book, the ICS geologic time scale is employed, with references back to original and historical names.

The final breakup of Pangea continued in the Cenozoic to the current configuration of the continents. The distance from the continental shelf around Cape Hatteras and the shelf around Morocco in western Africa is approximately 4,550 kilometers (14,290 feet) presently. At the beginning of the Cenozoic Era, 66 million years ago, the space between the two points was about 3,140 kilometers (10,300 feet), two-thirds of the width between them at present. To this day, the Atlantic Ocean widens by nearly as much as your fingernail grows in a year, 2 centimeters (0.8 inch) per year. On the other side of the world, the Pacific Ocean is shrinking, with its many convergent margins and the volcanoes of the Ring of Fire. During the Cenozoic Era, new niches opened up in the seas as North America was pushed farther away from Africa and Europe. The arrangement of the continents and ocean basins also had a bearing on climate during this era.

From 60 million to 50 million years ago, several volcanic island arcs smashed into the Asian plate, and 40 million years ago, not long in terms of geologic time, India collided with Asia. These two events have produced the most extensive mountain range on the continents, the Alpine–Himalayan chain. On the west coast of North America, the convergent margin continued to develop and

Song of the Earth. Elisabeth Ervin-Blankenheim, Oxford University Press. © Oxford University Press 2021.
DOI: 10.1093/oso/9780197502464.003.0012

change as the entire continent was pushed west. At one time, a trench could be found along the western coast of the United States, but what is now Southern California stopped its subduction, and a transform, a strike-slip fault zone—the San Andreas Fault System—developed in its place. The Rocky Mountains continued to rise, far from the nearest plate edge, possibly as a piece of the lithospheric plate flattened out like a spatula and created uplift. Along the western side of South America, a deep trench was established that is still an active and dangerous area of tectonics today. South of the northwestern U.S. Cascade Range, there were less compressive forces, and extension started to overtake the interior, creating the Basin and Range region. The track of the hotspot on the Snake River Plain records the movement of the North American plate through the Cenozoic, demonstrating the emergence of the hotspot beneath Yellowstone as a significant, and at times explosive, element of the volcanic activity of the West.

The climate in the Cenozoic was markedly colder than in the previous Cretaceous. The warming conditions of the prior era produced forests, swamps, and wetlands, but as the next era dawned, the regions that were formerly tropical dried out. Savannas replaced forests, and grasslands took root. The Cenozoic is also known as the Age of the Mammals. During this time, mammals expanded into the ecological niches that previously had been filled with dinosaurs.

Paleogene Period, 66.0–23.03 Million Years Ago

The Paleogene is composed of the Paleocene, Eocene, and Oligocene Epochs, along with nine stages. Historically, stratigraphers placed the Paleogene in the lower portion of the Tertiary Period. As mentioned, the USGS recognizes the Tertiary, although it suggests the use of the term "Paleocene."[4] The name "Tertiary" still appears in the literature, as it does in older references. Stratigraphers based Paleogene divisions on changes in foraminifera and nanoplankton (tiny, unicellular plankton).

The Paleogene Period is the beginning of the Age of Mammals. Small insectivore mammals survived the end-Cretaceous (K-Pg) extinction event and started to fill slots vacated by dinosaur species. Flowering plants, trees, and bushes with berries and fruits had taken hold in the late Cretaceous and flourished in the early Cenozoic. Not only did mammals expand into the various ecosystems that the dinosaurs had vacated, but they also exploited additional sources of food. Mammals increased rapidly, with 130 genera in the early Paleogene and more than 4,000 species. Not only did mammals live on land, but they took to the air (bats), tree limbs and forests (early primates), and the water. Ocean-dwelling mammals are of particular interest.

Finding fossil sites that record the period just after the Chicxulub Asteroid impact is vital to understanding how life bounced back and how long it took to do so, but few locations record the recovery time. However, a place in Colorado called Corral Bluffs, where researchers from the Denver Museum of Nature and Science (DMNS) have been studying mammal fossils, is providing answers to questions about the beginning of the Cenozoic and how mammals adapted.[5]

The Corral Bluffs site is an example of a chance occurrence turning into a significant find. Back in the 1940s, paleontologists noted that the area contained intriguing but sparse Paleocene fossils, with only fragments emerging from the sedimentary rocks.[6] Decades later, beginning in 2008, geologist Ken Weissenburger and citizen scientist Sharon Milito made a fantastic find. Both were volunteers with the DMNS, tasked to assess the potential impact of proposed projects (a motorcycle trail and reservoir, neither of which came to be) on paleontological resources. Milito found a round rock called a concretion (formed when sediments surround organic material), with a piece of jawbone sticking out. She gave the fossil to the DMNS, where it sat in a specimen drawer until curator Tylor Lyson noticed it years later, in 2016.[7] Lyson had a hunch and went back to Corral Bluffs with his team members to check out the concretions at the site. To his surprise, when he cracked one of the concretions open, he found a complete mammal skull; it was a "eureka" moment. Rarely do paleontologists pay much attention to concretions. Nevertheless, this unlocked a chapter in the story of how quickly and widely mammals recovered from the K-Pg extinction event.

Most rocks that record life after the Chicxulub extinction, early in the Cenozoic, do not represent a complete record of that time because of gaps and unconformities. Corral Bluffs is unique, with strata that contain nearly all of the 1 million years captured in their sediments. The researchers relied on age-dating techniques consisting of high-resolution magnetostratigraphy and palynology (age-dating by changes in pollen). Confirming by isotopic dating the ages of both fossil animals and plants, they pinned down a very detailed timeline of Corral Bluffs.[8] Based on this new information, geologists and paleontologists now speculate that life rebounded relatively swiftly after the asteroid impact and ensuing mass extinction. Indeed, mammals and plants coevolved in a stepwise fashion over the million-year interval.

The creatures surviving the post-apocalyptic time were small, omnivorous, rat-sized mammals, weighing only about 0.5 kilogram (1.1 pounds). They subsisted on what little food and vegetation were left after the impact, such as ferns.[9] One hundred thousand years into the new era, there were twice as many mammalian species, and they were about raccoon-sized, 6 kilograms (13 pounds). Plants diversified from ferns to palms, providing more food. After about another 200,000 years elapsed, a mammal called *Carsioptychus* became prevalent, a pig-like creature weighing 30 kilograms (66 pounds), with harder teeth to strip vegetation and crack nuts. Paleontologists now speculate that these are the ancestors of all hoofed animals. Meanwhile, the vegetation took a turn toward walnut-like plants. Nearly three-quarters of a million years after the mass extinction, a relatively short 720,000 years ago, mammals expanded in size, as represented by the *Taeniolabis taoensis*, named for their "ribbon-like" thin lips, weighing 34–56 kilograms (75–123 pounds). During this same period, the vegetation evolved into pea-pod-type plants, the oldest legumes ever found in North America. These legumes would have been packed with protein, needed by the mammals as they expanded into the niches the non-avian dinosaurs had vacated. Paleontologists will continue to study the unique fossil record of the Corral Bluffs site, which demonstrates, in terms of geologic time, a rapid diversification after the end-Mesozoic biotic collapse.

Mammals proliferated into every environment, even into the oceans. The cetaceans, made up of today's whales, dolphins, porpoises, beluga, and narwhals, returned to live in the sea to exploit niches not occupied by other animals. Just like other mammals, cetaceans are placental animals. How this transition occurred has been a point of debate. The fossil record in the Paleocene is lacking, but by the Eocene, whale ancestors surely were swimming in the oceans.

The first fossil whale ever found was that of *Basilosaurus*, by Philadelphia naturalist Richard Harlan in 1832, in Tertiary beds in Louisiana.[10] He named the creature "king lizard" but later discovered it was not a lizard at all. For many years afterward, scientists thought that the ancestor of whales was an insectivore-type mammal. However, research in Pakistan, Afghanistan, and Africa proved otherwise.[11] The *Basilosaurus* name, however, has been kept.

A proto-whale ancestor was found in Pakistan in 1977. Paleontologists determined this fossil to be a very early cetacean named *Pakicetus*, based on the ear structure located in the skull of the creature.[12] Whales and their predecessors have heavily boned small ear holes unique to cetaceans, to allow specialized hearing underwater. Ears of modern toothed whales have no external ear membrane, and the ear canal is not filled with air; neither are the bones of the ear attached to the skull. Sinuses particular to each ear direct vibrations so whales can discern sounds underwater. They also have specific limbs—fins—for paddling. In 1985, Vincent Sarich of Berkeley studied whale and hippopotamus blood proteins and discovered that the closest living relative to modern whales is the hippopotamus.[13] DNA later confirmed this finding. In 2001, additional fossil ancestor whales from Pakistan yielded joint bones with a "double-pulley" system that allowed attachment and movement of the tibia and foot.[14] This ankle bone (astralagus) is diagnostic of even-toed, hoofed mammals (Artiodactyla), proving a common ancestor with grazing land animals with double-pulley joints, such as the antelope, cow, sheep, deer, and hippopotamus. These results align with the earlier molecular biology and DNA findings.

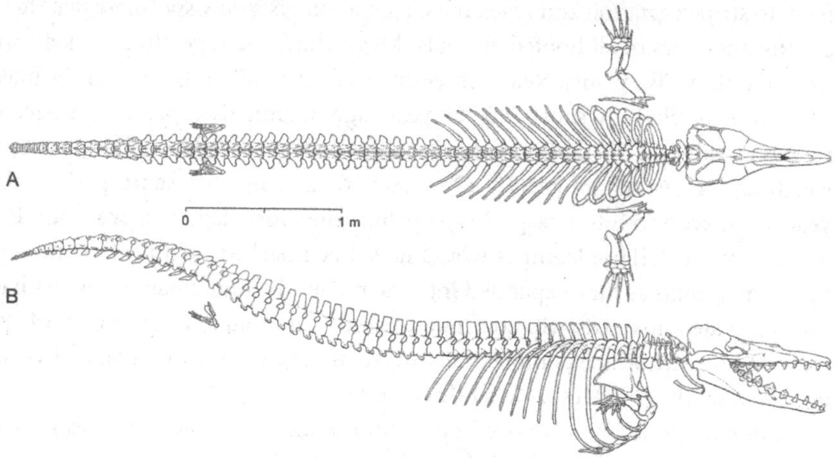

Figure 11.1 Skeletons of the Eocene Archaeocete whale *Dorudon atrox*: (A) view from top, (B) view from side (5.0 meters; 36.5 Ma). (Gingerich et al., 2009, figure 1, A and B)

Geologists identified the common ancestors of whales and hippos as anthracotheres, based on the fossil finds in Africa.[15] These were land-based animals first described from Tertiary lignite (low-rank coal) beds in Europe. Stratigraphers classify them in a supergroup, the even-toed ungulates. They base this classification on data indicating that the whale line evolved from several ancestors. One, *Pakicetus*, was semiaquatic. Others, such as *Maiacetus*, were more aquatic, and the *Dorudon atrox* (figure 11.1), a type of basilosaurid that was entirely aquatic, had vestigial hind limbs and a flattened or fluted tail based on the spinal vertebrae found in specimens from Afghanistan and Egypt in 1983 and 2004, respectively.[16]

Besides mammals, insects and plants adapted to life after most of the dinosaurs died out at the end of the Mesozoic. The Florissant fossil beds represent a treasure trove of well-preserved fauna and flora from the Upper Eocene Epoch. Geologists and paleontologists have recovered more than 40,000 specimens from the lakebed shales from this site in Colorado, representing 1,500 species of arthropods—insects and spiders.[17] Two lakes were created, one after the other, by volcanic lahars (a fast debris-type of flow the consistency of wet concrete, made of a mixture of volcanic ash and ice and snow from stratovolcanoes) that dammed up rivers to create the lakes. These arthropods lived near or in the lake. It was likely a site of abundant life. Paleontologists have uncovered more fossil butterflies in the Florissant Formation than anywhere else in the world, and 140 plant species have been found, including flowers (figure 11.2) and pollen.

Figure 11.2 Fossil flower, collected in August 2010 from the Commercial Clare Family Florissant Fossil Quarry, Florissant, Colorado. This fossil dates back to between the late Eocene Epoch, 35 million years ago, and the Neogene, 23.03–2.58 million years ago. (Photo by Slade Winstone, 2010)

The late Paleocene transition to the early Eocene, the time of the Florissant beds, 56 million years ago, represents the warmest period in the Cenozoic Era besides present times on the Earth—called the Paleocene–Eocene Thermal Maximum. This interval underwent immense carbon dioxide emissions, from unknown sources, possibly volcanic eruptions, as well as methane emissions released from the seabed, creating massive greenhouse gas increases in the air. Again, temperatures warmed—this time by 5–8 degrees C (41–46 degrees F), and the oceans became acidic. Benthic foraminifera died off, but this time, a mass extinction was avoided. Animals, including mammals, seem to have adapted by migrating to cooler northern latitudes where possible and undergoing rapid evolution, getting smaller in size when resources and food were scarce. It took another 150,000 years for the carbon and methane cycles to recover their natural equilibrium.[18]

After this event, there was considerable global cooling related to shifts in the continents from the tectonic movement of the plates. As Antarctica separated from Australia, a deep channel formed, allowing the circum-Antarctic current to develop between the two continents. This current changed the heat-transport mechanism in all the oceans and cooled the global climate. Stratigraphers locate the base of the Oligocene at this cooling event, but a specific age is still being pinned down by the ICS. Related to the cooling of the Oligocene in the oceans, marine life migrated toward the warmer waters near the equator. Numbers of plankton, supplying the base of the food chain in the seas, died off. On land, mammals dominated. Grasslands opened up, and herbivores grew in size.

Neogene Period, 23.03–2.58 Million Years Ago

The Neogene consists of the Miocene and Pliocene Epochs (Upper Tertiary Period), lasting 20.44 million years. Within the Miocene, 23.03–5.333 million years ago, were six epochs based on further foraminifera and nanoplankton changes. The Pliocene, 5.333–2.58 million years ago, is made up of two epochs.

The Miocene was a time of warming, after cooling in the Oligocene. Grasslands further developed, and there were many arid areas of the continents. In the oceans, forests of kelp, a vital source of food for marine mammals, grew for the first time on a significant scale. An increasing number of mammal species inhabited North America.

Mountain-building episodes in the Miocene created the Andes and uplift of the Sierra Nevada and Cascade Mountains, exotic terranes, as the Pacific plate dove under the South and North American plates. Weather and global air and rainfall circulation patterns were impacted by the height and extent of these ranges, creating a rain shadow effect on the leeward side of the mountains, high deserts, and plains.

The middle part of the Miocene recorded an increase in temperatures, up to 5 degrees C (41 degrees F), known as the Miocene Climatic Optimum (MCO). The MCO lasted from 17 million to 15 million years ago, before the onset of cooling and the ice ages later in the Cenozoic Era. Paleoclimatologists think warming initially to be the result of mountain uplift, with changes to weather and air circulation patterns and warmer ocean current patterns resulting in a reduction in the extent of the Antarctic ice sheet. Paleoclimatologists have researched temperature and carbon dioxide information from boron-based oxygen isotope data. These data are from bottom-dwelling marine foraminifera that lived during the Miocene, and they show a strong correlation with increased atmospheric carbon dioxide levels. The climate warmed and ice sheets melted in response to increases in carbon dioxide and methane.[19] Foraminifera, for example, are indicative of the extent of ice and the temperature near the bottom of the ocean. Climatologists think the Antarctic ice sheet responded to moderate carbon dioxide levels

(350–400 parts per million) more than models predicted. This melting may have been more likely because part of the ice sheet in the Aurora subglacial basin was below sea level. Additionally, northern hemisphere ice sheets may have contributed by building and retreating more swiftly.

The MCO impacted land vertebrates as well, although they were not as dependent on their particular niches as non-vertebrates, because the vertebrates can migrate to find better conditions. Vertebrates of myriad species adapted to the warmer conditions by migrating to other areas during this time, from 18 million to 16.5 million years ago, when average temperatures were 22 degrees C (71.6 degrees F). A die-off of some of the fauna occurred when the climate shifted abruptly some 2 million years later, around 14 million years ago, when the average temperature dropped by more than 11 degrees C (51.8 degrees F).[20]

The Mediterranean Sea today spans 2,500,115 square kilometers (965,300 square miles). It seems incomprehensible that it would ever dry up and become a waterless basin, but at the very end of the Miocene, an event the magnitude of which scientists have not seen in the last 20 million years in the history of the oceans occurred. From 5.59 million to 5.33 million years ago, the beginning of the Pliocene, the Mediterranean was entirely or nearly cut off from the replenishing waters of the Atlantic Ocean, and it dried up. This period, starting 5.93 million years ago, is known as the Messinian salinity crisis (named for the Messinian Epoch of the Miocene). Salt and evaporite deposits formed as the waters became more concentrated in salts and through evaporation without fresh ocean water coming into the basin. The salt and evaporite deposits consist of an upper layer hundreds of meters thick and a lower layer thousands of meters thick. These layers show up in cores drilled by the *Glomar Challenger* and the Deep Sea Drilling Project in 1970. The composition of the strata indicates that they formed in a relatively short span in geologic terms, less than 1 million years.[21] The volume of evaporite deposits and salt layers is so vast that it represents 5 percent of the salt in all the world's oceans, bound up in these strata measuring more than 1 million cubic kilometers.[22]

Once the Mediterranean seabed was relatively dry, erosion occurred at a rapid rate as terrestrial deposits formed. Researchers think the floor of the Mediterranean was lowered by hundreds of meters based on how deeply streams and rivers incised, including the Rhône and the Nile. Ground-water flow changed in response to the lowering of the base level, carving, and eroding karst and limestone regions in central Europe. Animals from Africa most likely migrated across the dry ocean basin.[23]

Researchers speculated, initially, that the cause of the Messinian salt crisis was related to changes in the level of the land from glacial ice. Still, in 1999, geologists determined that the Mediterranean was cut off as a result of shifts in positions of the tectonic plates, with the possible ancillary climatic variations related to periodic changes in the Earth's orbit and wobble.[24] Whatever the cause, there were several stages of partial closing and then full closure of the strait to the open ocean.

The straits of Gibraltar opened in a massive flood event 5.333 million years ago, at the boundary of the Miocene and Pliocene Epochs. The flood, nearly unimaginable from the knowledge of today's floods, is called the Zanclean flood. The event allowed fresh seawater back into the Mediterranean basin, rapidly ending the Messinian salinity crisis within several years. Through borehole drilling and seismic data, geologists discovered a 200-meter (656-foot) channel cut by the floodwaters, indicating that the water rushed back in with an estimated velocity as high as 300 kilometers per hour (186 miles per hour).[25] A layer of benthic ooze produced by foraminifera indicates a return to average salinity and temperature conditions in the Mediterranean.

The relatively recent Pliocene Epoch, the upper part of the Neogene, marks a gradual cooling in the climate of the Earth with the spread of grasslands and savannas. Forests and closed plant communities decreased. Grazers on land migrated more widely, encouraged by new forage, and ranged over emerging land bridges. The Panamanian land bridge connecting North and South America was created when the Caribbean plate shifted east.

In response to the cooling conditions, the mammals of the Neogene began to increase in size. Paleontologists have studied this evolutionary trend of body size increase through examining thousands of living species of birds and mammals during the Cenozoic and have run model predictions of body size in response to climatic pressures. They find that evolution occurs more rapidly during cold spells, and such conditions favor a larger body size.[26] Researchers have also linked an increase in body size to a colder climate. German biologist Carl Bergmann in the 1800s formulated Bergmann's rule, which states that the body mass of endothermic animals (those generating their internal heat) have larger body sizes, as a species, at higher latitudes and altitudes.[27] In other words, the rule correlates larger body size to colder climes, indicating that such sizes may have the adaptive advantage of better thermoregulation in the cold. Critics of the Bergmann's rule argue that it does not apply in all cases. There is support for it among those who study Cenozoic mammals, although they disagree about whether the rule applies to thermal regulation.[28]

Carbon dioxide levels decreased to approximately 100 parts per million,[29] much lower than in the preceding epoch, encouraging ice sheets to grow at both poles. Ocean temperatures were similar to those of today, but the change of temperature with depth in the ocean was less than it is currently. Factors that may be responsible include how ocean currents mix and less warming due to an increase in clouds.

Australopithecus, an early hominid, diverged from other primates around 6 million to 5 million years ago on the African continent, around the time of the Messinian salt crisis.[30] For many years, scientists had looked for the ancestors of humans in places such as Europe and Asia. Africa was not considered a candidate, in part because of racial bias and prejudice. In 1974, an American archeologist, Donald Johanson, along with graduate student Tom Gray, found fifty-two bones of a fossilized skeleton near the Hadan of Ethiopia. The bones were identified later as the ancient hominid, *Australopithecus afarensis*, and was given the name Lucy by the investigators.[31] The Hadan is in the northern portion of the East African Rift Valley, in the Afar Triangle—a tectonically active triple junction (where three plates meet) undergoing dynamic rifting and spreading. The site sits in Ethiopia north of Addis Ababa toward the border with Djibouti, along the lower valley of the Awash River. The researchers determined that the Lucy skeleton is 40–50 percent complete, one of the most complete *Australopithecus* skeletons ever found (figure 11.3). This individual stood about 1 meter (3.3 feet) tall and walked upright on two feet, as evidenced by her pelvic structure and leg bones. Based on isotopic dating, the bones are 3.18 million years old. This fossil find was the first significant breakthrough in the understanding of the early hominin lines, and the site was given UNESCO World Heritage status.

The Leakeys, Kenyan-born Louis and British-born Mary, both anthropologists and archeologists, had been working for decades in Olduvai Gorge, also within the East African Rift Valley to the south of the Hadan site. There they found bones of other branches of the hominin family, but one of the most vital discoveries was that in 1978 of fifty nine footprints of two footed, upright walking tracks preserved in volcanic ash beds (figure 11.4).[32] These prints were located mainly in two beds, along with raindrop imprints, bird and hare tracks, and a deeply carved game pathway cutting across the hominin tracks. An examination of the geology shows that the volcanic tuffs came from showers of ash and dust emanating from nearby Mount Sadiman and occurred over a short time. The upper volcanic unit containing footprints is age-dated at 3.6 million years old. The environmental reconstruction shows that the prints were made and covered up with ash that solidified into volcanic tuff in a single rainy season, thus preserving them. Anthropologists think several different *Australopithecus* made these trace fossil footprints.

Quaternary Period, 2.58 Million Years Ago to Present

The epochs of the Quaternary are the Pleistocene and the Holocene, which brings life on the Earth to the present time. Stratigraphers are debating whether

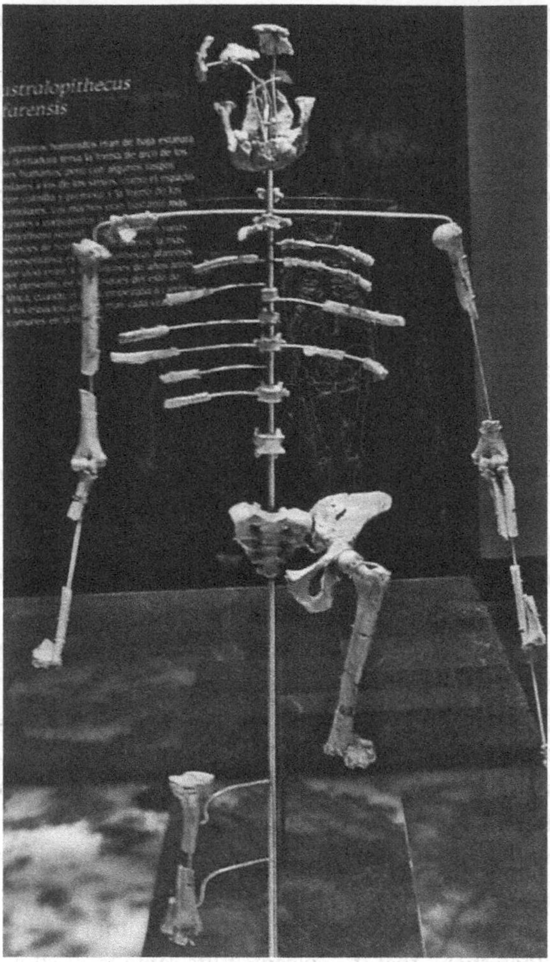

Figure 11.3 A full replica of Lucy's skeleton (*Australopithecus afarensis*). (Museo Nacional de Antropología, Mexico City; photo by Daniel Acosta, 2006)

to end the Holocene and begin a new epoch, the Anthropocene, recognizing the impact, for better or worse, that humans have had on the Earth and in the geologic record.

The Pleistocene lasted from 2.58 million years ago to 11,700 years ago. Climate during this time cooled dramatically as ice sheets covered continents and sea level fell as much as 130 meters (426 feet) at the glacial maxima 20,000 years ago. Land bridges emerged with the lowering of ocean water. Mammals filled more niches and grew in size—this was the time of the megafauna. They radiated out and claimed all of the ecosystems that dinosaurs had once dominated with a large

Figure 11.4 Casts of the Laetoli footprints. (Natural History Museum, Vienna; photo by Wolfgang Sauber, 2013)

variety of types. There were camels and many other animals that became extinct as the ice ages came to a close. The first scientist to describe the undisputed horse ancestors from deposits in the American east was paleontologist Joseph Leidy of the Academy of Natural Sciences in 1847. The same year, he also identified camels, *Poebrotherium wilsoni*, from western North America. Megafauna of the Pleistocene also includes mammoths, mastodons, giant ground sloths, long-horned bison, dire wolves, and saber-toothed cats, among others, existing on most of the continents.

One of the most stunning Pleistocene fossil sites is located on the west coast of America. Los Angeles is a sprawling, bustling, busy city, the largest in California and home to much of the movie and television industry. However, few visitors know that beneath their feet is an iconic site of Pleistocene terrestrial life and megafauna. To many geologists, the best part of Los Angeles is the Rancho La Brea Tar Pits. These pits of naturally occurring asphalt were rediscovered in 1769 by Spanish explorer Gaspar de Portola. However, Native Americans had known about and used the asphalt for hundreds of years as a resource for tar to seal canoes. The first scientific study of the site was made by in 1875, when the fossils were noticed during an evaluation of the area for oil deposits.[33] In 1901, petroleum geologist William W. Orcutt (1869–1942) began excavating fossils from La Brea, building a substantial collection. He uncovered the complete skull of a saber-toothed cat in 1906.[34] The site is between 23,000 and 40,000 years old based on radiocarbon dating of leafcutter bee nests found in one of the pits.[35]

There are many oil seeps in the region, most located in inaccessible canyons. The La Brea Tar Pits are in a more open area uplifted by tectonic forces on which Pleistocene rivers and streams eroded and deposited sediments. The asphalt is formed when the oil bubbles to the surface and its lighter elements fraction and evaporate, leaving sticky tar behind. Tar is an excellent preservative of the fossils and has conserved the thin, hollow bones of birds, which usually would be crushed or destroyed. The delicate shells of insects, such as beetles, have been saved in the asphalt, too. The La Brea Tar Pits represent an entire ecosystem, complete with fossilized mammals, birds, fish, reptiles, arthropods, insects, plants, and microbes.

Paleontologists have recovered 59 species of mammals from the pits and 135 bird species; it is one of the best sites for fossilized birds in the world. Of the animals found at the La Brea Tar Pits, a high number are predators, 90 percent of the mammals. The most common carnivore is *Canis dirus*, the dire wolf, followed by saber-toothed cats (*Smilodon californicus*, figure 11.5) and bears. Most of the birds discovered were songbirds, but there are a significant number of scavengers and predators—vultures, condors, eagles, and the extinct *Teratornis*, a short-legged, rounded-bodied bird with a 3.2-meter (10-foot) wingspan that preyed on smaller animals. The predators were drawn to the tar pits when other animals became entrapped after seeking out the water supplies nearby and could not extricate themselves. Many of the carnivores became victims as well.

The late Pleistocene was marked by a general warming trend that continued well into the Holocene Epoch. However, a period called the Younger Dryas, named for a European flower that thrives in cold conditions, began around 14,500 years ago, following on the heels of the melting ice sheets and glaciers, and lasted until abruptly ending 11,500 years ago. Temperature swings caused a rapid cooling episode with increases in snowfall, possibly caused by changes in ocean circulation patterns from glacial meltwaters, impacting mostly Europe, Greenland, and the northern Atlantic.[36] More recent data have been found in the southern hemisphere for these climate perturbations.[37] In 2007, researchers proposed that an impact from an asteroid exploding above the Earth 12,900 years ago was responsible for the cooling of the Younger Dryas.[38] This theory has been in dispute, but in 2019, scientists found an impact crater dating to 12,800 years ago that may be responsible.[39]

During this time, the extinction of 70 percent of the megafauna in North America and other areas, including Europe, occurred. The cause of the event is not well known. Still, researchers have theorized that it resulted from ecosystem changes because the climate shifted, or the animals may have been hunted to the point of no return by humans migrating across the Bering Strait to inhabit North America. Some proto-horse lines in Europe and Asia fared better than their North American counterparts. Research shows that these early horses were

(a)

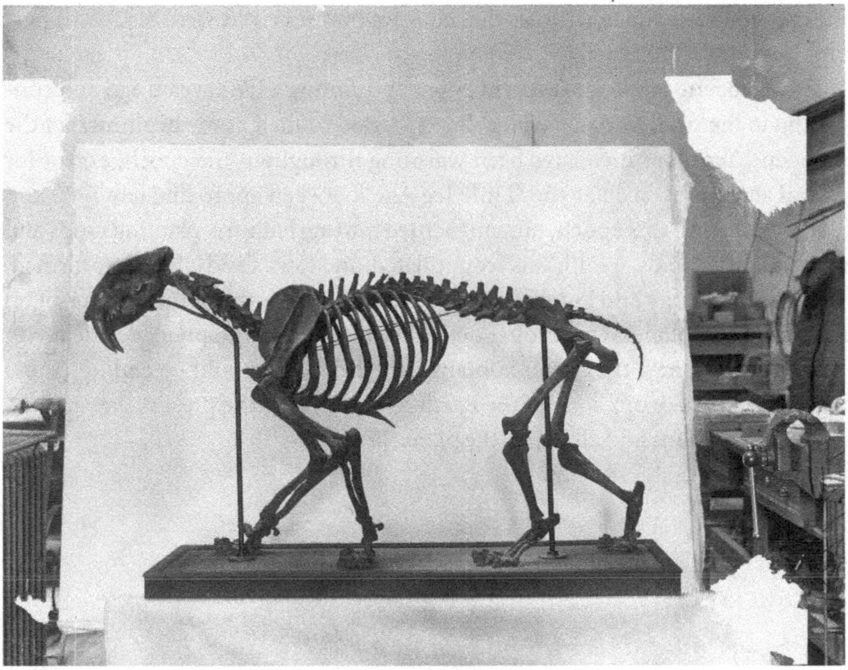

(b)

Figure 11.5 (a) *Smilodon californicus* from the La Brea Tar Pits, Los Angeles. (Field Museum Library, Chicago, 1917, via Getty Images; photo by Charles Carpenter) (b) *Smilodon fatalis* life drawing. (Dantheman9758, 2008)

more adaptable to environmental change and went back into the forest to survive on leaves. DNA evidence shows that equines that survived were black and perhaps better able to hide.[40]

The Holocene is the most recent epoch, beginning 11,700 years ago and continuing to the present day. The ice sheets started to melt at the beginning of the Holocene. Temperatures have been warming throughout the epoch, except for several anomalies, such as the "Little Ice Age" between approximately 1300 and 1850 CE. During this epoch, humans settled into agriculture, planting crops and husbanding animals, wealth was accumulated, and towns and cities were formed. Human technology and knowledge expanded, as well as the human footprint on the face of the planet. Depending on what the ICS stratigraphers decide about the Anthropocene—the age of humans—the Holocene could be ending. These developments take our song of the Earth to its outro, in the present day, and the challenges humans and all life on the planet are facing.

12

The Earth's Impact on Life and Life's Impact on the Earth

Our journey through geologic time has considered the history of geology and how geologists know what they know, the basic tenets that frame geology (geologic time, plate tectonics and evolution), and a biography of the Earth from the Precambrian Supereon to the present. Now we examine geology and life from a perspective of mutual dependency—each influences the other, and both have implications for humanity and the Earth's future.

Geologic processes that shape the face of the planet operate slowly and steadily, as detailed by the principle of uniformitarianism, "the present is the key to the past," and they also operate catastrophically, as when a tipping point is exceeded or by events like an asteroid hitting the Earth and ending the age of the dinosaurs. Indeed, both gradual and rapid forces have been seminal in the formation of the Earth through the span of geologic history.

Looking with a view toward the future of the Earth and life on it is of particular relevance given the challenges of overpopulation, sustainability of resources, and human impact. In this light, we may restate the principle of uniformitarianism as "the past is the key to the present and the future." In 1748, Scottish philosopher David Hume formulated the basis for the concept that geologic processes operating throughout the Earth's history may predict events yet to come: "For all inferences from experience suppose, as their foundation, that the future will resemble the past."[1] Key findings in geology can guide future decisions and actions because of the record of past events. It is the Earth's message translated into human terms, much like what we can learn from the lessons of history as stated by Spanish philosopher George Santayana, who cautioned that "those who cannot remember the past are condemned to repeat it."[2]

From the very beginning of life on the Earth, starting in the Precambrian Supereon, organisms have had a bearing on the direction taken by the planet, and vice versa. Life affected the makeup of the planet in the Great Oxygenation Event during the Archaeon Eon (see chapter 8), in which oxygen was released into the Earth's seas and atmosphere, kicking off the planet's third atmosphere. Early microorganisms, cyanobacteria, influenced the carbon cycle of the Earth nearly 4,000 million years ago by changing the carbon-rich atmosphere into one with oxygen. Another example of organisms intersecting in the processes

Song of the Earth. Elisabeth Ervin-Blankenheim, Oxford University Press. © Oxford University Press 2021.
DOI: 10.1093/oso/9780197502464.003.0013

of the planet is the explosion of life in the Cambrian Period, 541 million years ago, which brought about further changes in the atmosphere and oceans and the rapid diversification of animals and plants.

The Earth: Interconnected Systems

The Earth is a system (color insert 12.1), and within it are five spheres (see chapter 8): the magnetosphere, the geosphere, the atmosphere, the hydrosphere, and the biosphere. The spheres are simply constructs, or a way to see and identify that all matter, life, air, and water are interconnected parts of a whole.

Each sphere has specific forces acting on it and cycles acting within it. The hydrosphere is dominated by the hydrologic cycle: water is evaporated from the salty oceans to become freshwater in clouds, which move over landmasses, creating precipitation and rain. Some of this excess runs off into streams that ultimately carry it back into the oceans. Part of it percolates into the ground to form ground water; water in some of these forms evaporates again, some freezing into ice and, if enough ice accumulates, glaciers. Ultimately, surface water flows back to the ocean, and the process repeats.

In the geosphere, the rock cycle dominates. All rocks—igneous, metamorphic, and sedimentary—are subject to this cycle in which rocks are created, destroyed, and recycled; one type of rock can be transformed into any of the other types.

The spheres of the Earth naturally interact and are dependent upon one another. Mass and energy move among the four areas. The interface between the spheres, where they touch and overlap, is where the intensity of dynamic processes increases. An example of a material traversing the intersections is soil. A plant (biosphere) grows in soils derived from rocks (geosphere) that have been weathered, eroded, and transported from mountains and deposited onto the edge of a floodplain by a stream (hydrosphere), to be nourished by the sun (atmosphere) and rain (hydrosphere), only to return nutrients from its cells (biosphere) back to the soils (geosphere). A change to one part of the system will impact the dependent spheres, a tenet of basic systems theory, which applies nowhere more than to the Earth.

The Earth is a closed system, for the most part, which means it receives energy from outside but usually not influxes of mass or matter. The sun radiates charged particles, some of which are deflected by the magnetic field and others that reach the surface as solar radiation. The energy of the sun drives the hydrologic cycle and photosynthesis in the biological realm, as well as wind and weather that influence the climate in the atmosphere. Circulation in the mantle causes geospheric processes, through plate tectonics and the magnetosphere. Once in a while, asteroids and meteorites hit the planet, and then it becomes an open system, such as when the asteroid hit the Earth at the end of the Cretaceous Period, causing the Cretaceous–Paleogene extinction event and the end of the Mesozoic Era.

Within the spheres of the Earth, matter and energy move back and forth between the various systems; these are measured as fluxes in units of mass per year. Residence time is how long a form of mass stays in one part of the system, from days to years or millions of years. Thus, time varies widely depending on the portion of the system in which the mass resides. Water is evaporated from the ocean into clouds with a residence time of ten days. River water has a long residence time, on the order of a couple of weeks to six months. Once water percolates down to the ground water, the residence time can be hundreds of years for shallow ground water and tens of thousands of years for deep ground water. These lengths of time influence the cleanup of pollutants. If something toxic enters a river system, the river is refreshed by new water much more quickly than is ground water, in which a contaminant can persist for millennia.

Sometimes matter gets caught up in a particular part of the system, referred to as a "sink," where it lingers for a very long time, almost like storage. Banded iron layers, which now are often mined as economic deposits, are examples of a sink for iron that formed when free oxygen mixed with dissolved iron in the oceans of the early (Precambrian) Earth. Other sinks, for instance, those associated with the element carbon, are in limestone and coal, which bind up carbon for millions of years. These geologic deposits trap carbon dioxide until some change occurs, and the material is released back to the environment through dissolution, for example, in the case of burning coal for heat. An example of a non-economic iron sink is the Permian red beds so common around the world, especially in South Africa and west Texas. The oxygen still locked in these rocks changed the level of oxygen in the atmosphere and impacted the evolution of breathing systems in animals, particularly dinosaurs and birds.

Relatively small changes of state (or condition) can lead to substantial changes elsewhere in the system. Indeed, some researchers think a "tipping point" from which the climate cannot quickly recover, or which could even be irreversible, might occur in the not-too-distant future.[3] The tipping-point theory, as applied to climate studies and mainstream discussions, emerged from epidemiology but was popularly refined in sociology by Canadian author and *New Yorker* writer Malcolm Gladwell. He defined a tipping point as "the moment of critical mass, the threshold, the boiling point."[4] Scientists describe a tipping point as a critical threshold beyond which the state or the condition or the future parameters of a system are altered measurably by small inputs to the system (figure 12.1). Small changes in initial condition that magnify to cause larger perturbations have been identified as the "butterfly effect," the consequences of which can lead to

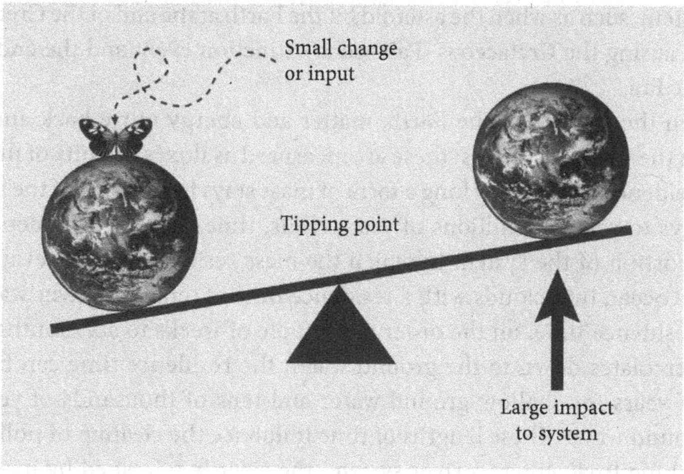

Figure 12.1 Tipping point and butterfly effect. (Concept and initial diagram by the author, 2020; illustration by R. Gary Raham, 2020)

unexpected, nonlinear results. American meteorologist Edward N. Lorenz first discussed the butterfly effect in a talk he gave at the 193rd meeting of the AAAS in 1972.[5]

A variety of tipping points may be reached if carbon dioxide continues to climb in the atmosphere. Such events could result in drastic environmental shifts. The International Panel on Climate Change (IPCC) details numerous tipping points at different temperature-increase scenarios, 1.5 degrees C and 2 degrees C. These thresholds include wide-scale events, such as the collapse and irreversible loss of the Greenland and Antarctic ice sheets, along with changes in ocean circulation patterns. Regional tipping points are more localized, resulting in challenges to food production, rainforest dieback (Amazon Rainforest), and increases in heatwaves.[6]

Population Matters

The world's human population growth is another factor to consider in assessing the impact of one system on another. The population of the Earth is projected to reach 11.2 billion by 2100.[7] It exploded from 0.5 billion in 1650 to 7.5 billion in 2017, showing exponential growth, a nonlinear trend.[8] A population's doubling time is the amount of time it takes for that population to increase twofold in size at a constant growth rate. The world's population increased from 1 billion to 2 billion in seventy-five years, from 1850 to 1925, but it took just forty years for the number of people on the Earth to increase from 3 billion to 6 billion, from 1960 to 2000.

The human population was around 2.6 million until 12,000 years ago. After the melting of the Pleistocene ice sheets, as climatic conditions stabilized and warmed, the population increased to 5 million. Numbers grew as societies transitioned to agriculture and living in one location, but hunter-gatherer populations also increased during this time.[9] Researchers correlate the population growth at the beginning of the Holocene, 10,000 years ago, to the development of agriculture. Stepping forward in time, since the 1960s, the rise in the number of people on the planet is a result of the green revolution in agriculture. During the early part of this revolution, 1966 to 1985, more food was produced on the same or less acreage of land by the introduction of higher-yielding crops and more robust plants. New farming methods, advances in machine technology, and the application of nitrates and other fertilizers to the soil created more plant growth and yield.[10]

When the number of births is higher than the number of deaths, there is population growth, resulting in an exceedance of what's known as replacement capacity, but these increases are not equal among countries. In more developed nations, such as the European countries, the number of people has stabilized or declined and is replenished in some cases by immigration. In emerging countries, population numbers are increasing because births exceed deaths, as in India.

Such growth in overall numbers places stress on the Earth systems and resources within those systems: food, shelter, land usage, water supplies, and the ecosystems of other species. Additionally, the number of people on the Earth impacts the atmosphere through carbon dioxide emissions released through industrial processes and the use of other resources, but this relationship is not direct. Affluence and related lifestyle create more demand for resources, fossil fuels, and water, to name several. Five factors can affect the relationship of population, wealth, and impact on emissions: how many people live in a region, their per capita gross domestic product (GDP), how much energy they use to produce a dollar of GDP, the carbon content of their source of that energy, and whether or not they use methods to remove any of the carbon produced from the air.[11]

The human footprint—the impact of humans on the extent of resources, land, water, air, and minerals needed to sustain life on Earth—was measured through satellite imagery from 1993 to 2009.[12] Variables measured include population density, amount of cropland, percentages of developed or constructed environments, amount of pastureland, nighttime light density, and access to railroads, roads, and waterways. During this sixteen-year study, the human population increased by 23 percent, and economic growth increased by 153 percent. The impact on the Earth by humans, the human footprint, by comparison, rose

9 percent, overall less than expected in developed countries. However, overall, 75 percent of the planet has undergone significant alteration by humankind. The study demonstrated that in 2009, there still was time to conserve ecosystems and the environment.

Systems theory predicts that each sphere will be dependent on what occurs in the other areas; what occurs in one sphere will likely impact all other spheres.

Geosphere–Human Interaction

The geosphere, the solid-earth part of our planet, is what we usually think of as geology. Minerals and rocks make up this sphere. Elements compose minerals, which make up the components of rocks. Rocks seem permanent and unchanging, but, as we have learned, they transform from one type to another through the rock cycle (figure 12.2). Rocks that arise from cooling hot magma are igneous rocks. With weathering and erosion, these rocks can transform into sediments that, with time and pressure, become lithified to create sedimentary rocks. And any rock can become a metamorphic rock with increased temperature and pressure, up to the point of melting. Even metamorphic rocks may be subjected to repeated metamorphic processes. In fact, any type of rock can become any other type of rock, including a repeat of its own type. For example, Jurassic sandstone is today becoming the sand of the Coral Pink Sand Dunes of Utah, which will most likely become a future generation of sandstone. Plate tectonics drives the rock cycle and allows geologists to predict the locations of particular rocks.

Economic resources of minerals, oil, and natural gas are the literal fuel of modern life. Yet many do not realize that geosphere resources are nonrenewable. These deposits took a long time, sometimes millions or tens of millions of years, to form, but society is exhausting them at a much faster rate—by orders of magnitude—than they form. This truth does not mean, necessarily, that the world will run out of vital minerals, rocks, or fossil fuels, but these may be more challenging to locate and increasingly costly to extract.

Fossil fuels arise from deposits of ancient organic matter over time, subjected to temperature and pressure changes. Fossil fuels include oil, natural gas, and coal. The first oil well (the Drake Well) was drilled in Pennsylvania in 1858. Before that time, nineteenth-century civilization used whale oil to light lamps in many parts of Europe and America. A *Vanity Fair* cartoon from 1861 shows sperm whales celebrating the find of oil with a grand ball (figure 12.3). Prior to this, oil had been known from natural seeps but was not available in quantities adequate for commercial use. In fact, the development of a technique to extract fossil fuels for use spurred, in large part, the second Industrial Revolution.

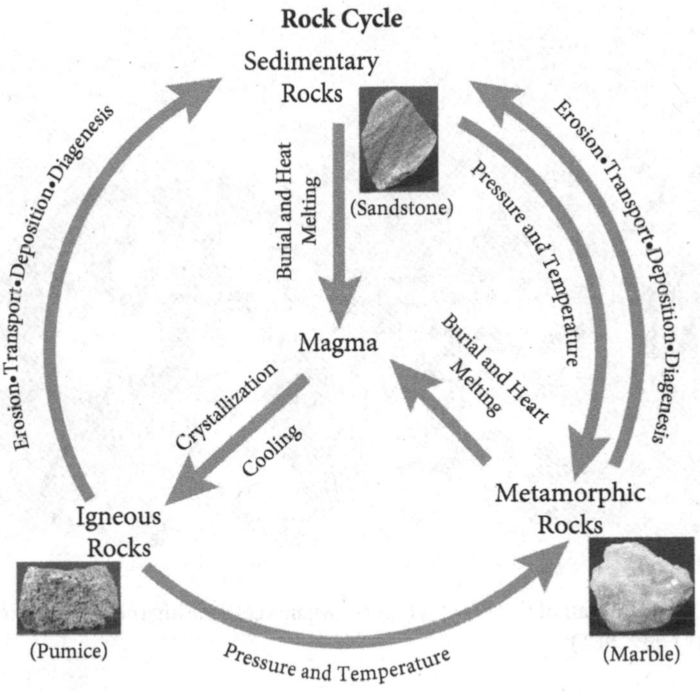

Sedimentary rocks include sandstones, shales, limestones, dolomites, and coal. Erosional material is deposited in layers in water, then compressed at low temperatures.

Metamorphic rocks include slates, phyllite, schist, gneiss, and marble. They form when other rocks (usually sedimentary) become partially melted under high temperatures.

Igneous rocks solidify from a molten state. **Volcanic** rocks include basalts, obsidians (volcanic glass), and rhyolites. Granite, diorite, gabbro, periodite, or pyroxenite form deep in the Earth and are referred to as **Plutonic.**

Figure 12.2 Simplified rock cycle. (Minnesota Geological Survey, University of Minnesota; illustration by R. Gary Raham, 2020, based on illustration by Andrew Wickert, 2009)

People had long known about and burned coal for fuel in small amounts before it became a driver of the economy and industry in the first Industrial Revolution. Beginning in the mid-eighteenth century, its burning was used to power steam engines.

Coal is an organic sedimentary rock formed mainly during the Carboniferous Period in fresh to brackish water in swampy areas or in near-shore environments where vast amounts of plant matter existed. The coalification process begins when a significant amount of vegetative material accumulates as the plants die, fall into

Figure 12.3 Grand Ball of the Whales (sperm whales celebrating the discovery of oil). (*Vanity Fair*, 1861)

the water, and start to decay. Over geologic time, the organic material accumulates in layers and becomes compressed as the pressure squeezes the water out. Peat then begins to form in the oxygen-poor environment. The rise and fall of sea levels build new layers on top of the peat, creating further compressive forces. The burial of this peat, together with temperature and time, forms lignite, low-ranked coal. Over yet more time, this material becomes denser and drier, forming sub-bituminous, then bituminous coal, and finally, under high temperature and pressure, as occurred during the Appalachian mountain building, creating a metamorphic, organic rock, anthracite. With increasing coal rank from lignite through anthracite, the coal is older, water content is reduced, specific gravity increases, and the material becomes denser and more enriched in carbon.

As noted, one of the most significant coal-forming times was the Carboniferous Period in the Paleozoic Era, when seed-bearing plants, such as *Calamites*, *Lepidodendron*, and *Sphenophyllum*, and swamps dominated. No coal exists that is younger than 2 million years old.[13]

Miners extract coal through surface (strip) mining if the beds are close enough to the Earth's surface or by other methods, such as room and pillar mining, if coal is deep. These methods can have environmental consequences. Not only are large areas of land carved up in strip mining, but acid mine drainage can degrade water bodies, streams, and soils.

Whether coal is surface-mined or brought up through subsurface extraction processes, a mineral called pyrite (fool's gold), containing iron and sulfur, comes along with it; pyrite forms in the same environment as coal. Coal and associated pyrite are crushed, as part of processing the ore to make a more usable product and for shipment, which increases the surface area of both materials. Pyrite forms in a low-oxygen environment and is in equilibrium there, but at the Earth's surface, it is no longer stable. The pyrite starts to weather (degrade), and the additional surface area of the crushed mineral speeds up the weathering process. The degradation of pyrite causes acid mine drainage. The presence of bacteria speeds up the process, resulting in runaway acidification. Other elements and metals in surrounding rock, or in mine tunnels in which increasingly acidic water is stagnant, leach out because of the substantial acidity. Thus, sometimes streams in coal-mining areas run red.

Beyond economic and mineral resources, the geosphere and the biosphere have been instrumental in each other's creation through geologic time. This interaction is first evident in the Precambrian banded iron formations that were built up by the early life forms, the stromatolites. These form some of the largest deposits of iron mined today. Weathering of continental rocks allowed vital minerals and nutrients to flow to the seas and ocean. Such minerals were present in the hard calcite shells of the first creatures in the Cambrian explosion of life in the early Paleozoic, enabling them to protect themselves from the first predators seen on the Earth. Life would not have developed without rocks and the elements they contain, and many rocks would not exist but for their predecessor biological forms.

Hydrosphere–Human Interaction

The hydrosphere's hydrologic cycle connects water everywhere on the Earth. Seventy percent of the surface of the Earth is composed of water, most of it in oceans.

Water is so much a part of everyday life that people do not often think about it. If not for water, and its unique attributes, life on the planet would not exist. Water on the Earth exists in three states—liquid, solid, and gaseous—unlike on most of the other bodies in the solar system. When water turns to ice, it expands, is less dense than liquid water, and has a crystalline lattice. In this form, it meets the definition of a mineral. Because of its lower density, ice forms on the surface of the water rather than at the bottom of a lake or other surface water body. This property allows life to continue during cold spells. When the ice melts, the water shrinks in volume and loses its crystalline structure. This makes the Earth the only planet in the solar system that is in the habitable zone, though some moons in the solar system are thought to harbor liquid water along with ice and thus possibly fit to live on.

Water has been called the universal solvent, since, over time, it will dissolve almost anything with which it comes in contact. The ability of water to dissolve things, along with its capacity to expand when frozen, is a driving factor in the weathering of rocks. Even granites, which are hard crystalline rocks, will start to break apart as water seeps into joints and creates crevices on the rock surface. The water freezes and widens the crack, allowing more water to seep in; then lichen and plants gain a foothold. Once the original rock starts to crumble, water acts as the primary mode of transportation of weathered materials via streams, rivers, and glaciers.

The only water on the Earth during the Hadean Eon was in the form of steam, if the theory of a hot-early-Earth is assumed. Oceans consisted of hot magma, and no landmasses had emerged—it was indeed hell-like. Oceans of liquid water arose once the Earth had cooled during the Archean. Where that water came from is a subject of debate among geologists. Meteorites or asteroids, many of which have icy components, may have supplied water; alternatively, the mantle could have emitted water through early tectonics and volcanic eruptions and condensation of steam in the atmosphere. Whatever the source of water on the planet, once it accumulated in the oceans, evaporation and condensation occurred, and it may have rained for millennia.

The natural hydrologic cycle (figure 12.4) describes the way water circulates on the Earth.[14] From the oceans, freshwater is evaporated and condensed to

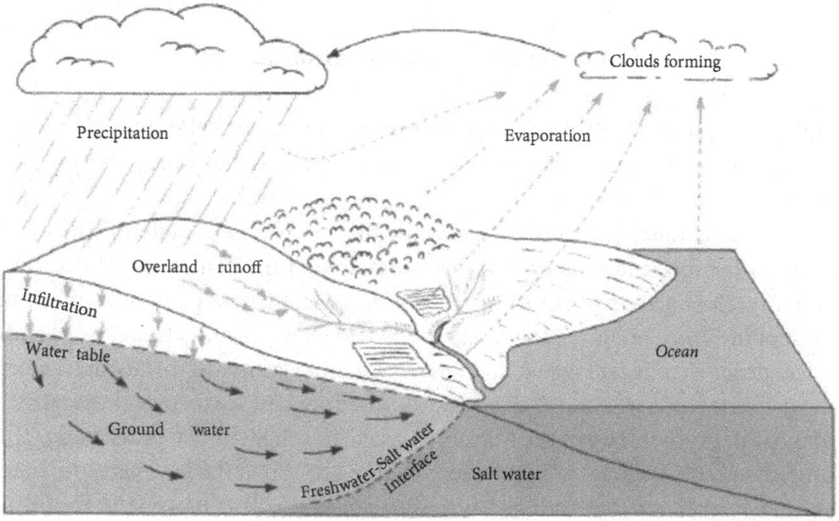

Figure 12.4 The natural water cycle. (Heath, 1998)

form clouds that move over the land and release the precipitate as rain. Some water from the precipitation is re-evaporated; some is bound up in glaciers and ice sheets; some makes it overland to replenish streams, lakes, and rivers; and some of it percolates into the soil and below to become ground water. A portion of the water in streams and rivers flows back to the ocean, carrying freshwater and minerals dissolved from continental weathering.

Most of the Earth's water exists in the oceans, representing 96.5 percent of all the water on the planet. Ocean water is salty, with total dissolved solids around 35,000 parts per million, or 35 percent salinity. Though it contains many dissolved ions, ocean water has achieved a relative balance of materials weathered from the continents flowing in from rivers and other inputs along with the evaporative outputs. In 1715, English astronomer Edmond Halley (1656–1743), who discovered the comet named after him, postulated that scientists could determine the age of the Earth through measuring the relative amount of salinity over time, in essence using a salinity clock. However, he was unable to calculate an actual age because there was no measure of the initial saltiness of the oceans. Other scientists in the late 1800s came up with salinity rates based on "chemical denudation" of the continents via rivers and arrived at 25 million years as the age of the Earth, but that number did not correlate with other observations and data. In addition, researchers found several problems invalidating the use of such a salinity clock. For one, the oceans are not closed systems, mainly as a result of plate tectonics. Large volumes of water are subducted at ocean trenches and carried into the mantle. Shifting plates can cut off parts of an ocean from the open seas, so that the isolated water evaporates, creating layers and layers of salt—evaporite deposits. Ice sheets can generate lower salinity, just as higher evaporation rates near the equator increase salinity. Researchers also discovered that the rates of erosion from continents, amount of rainfall, and runoff vary too much overtime to calculate the age of the Earth. Scientists ultimately realized that a salt clock was not a viable way to determine the Earth's age.

In 2011, NASA sent an instrument called Aquarius onboard an international satellite to measure worldwide concentrations of salt in the oceans. The Atlantic Ocean is known to have higher salinity than the Pacific. More rain falls over the Pacific than the Atlantic, and trade winds carry water vapor and dissolved solutes across Central America to the Atlantic Ocean basin, making it saltier.

Salinity in the oceans also varies as glaciers and ice sheets advance and retreat. The ice sheets bind up millions of gallons of freshwater. Sea levels, too, are determined by how much ice is around. During the maximum extent of the Pleistocene glaciation 20 million years ago, sea levels fell 120 meters (394 feet), exposing much of the continental shelf and land bridges by which animals and humans migrated. The North American ice sheets melted 14,000 years ago and left behind the Great Lakes, an area of commerce and recreation as well as a freshwater source.

Freshwater makes up only a little more than 3 percent of all the water on the planet, but most of that exists in ice sheets and glaciers (nearly 69 percent). Of the remainder, 30 percent exists as ground water and only a little more than 1 percent as streams and surface water bodies, swamps, soil moisture, and life forms. Ground water is an important drinking-water resource and accounts for about 30 percent of all freshwater on the Earth, with surface water—lakes, streams, and rivers—just 2 percent of those resources.

The natural hydrologic cycle is impacted by myriad human activities such as diversions for dams, tapping of surface water and ground water for drinking-water supply, and overpumping ground water along coasts inducing saltwater intrusion. Other human impacts are those of irrigation (lowering the water table), deforestation (increasing runoff and flooding), drainage of wetlands (impacting ground water and increases storm damage), and urbanization (increasing runoff), to mention some major effects.[15] The result of anthropogenic water use reduces the amount of water flowing back to the oceans and has a lowering effect (small compared to melting glaciers) on sea level. Freshwater that is suitable for drinking and human use does not have an unlimited supply. Much of the world's population, up to 2.1 billion people, do not have access to these resources, according to the United Nations.[16]

Not only is the amount of drinking water in question for future use, but the quality of that water can be jeopardized by human activities and even natural sources of contamination.[17] Ground water is especially prone to contamination by pollutants because it flows much more slowly than surface water.[18]

Water is the most precious of all the Earth's resources, but the scarcity of freshwater is a growing concern.[19] Ground water, which is a vital source, is essentially being mined and is a nonrenewable resource. It takes much longer for ground water to replenish compared to the rate at which it is being used up.[20] Some predict that water issues will be the leading cause of geopolitical strife in the coming decades and centuries as changes to the climate create instability, droughts, and saltwater intrusion. The problem with rising sea levels is not just the absolute amount of the increase but that the saltwater penetrates under coasts and links to ground water (figure 12.5). Saltwater is 2.5 percent denser than freshwater. Thus, it sinks relative to fresh ground water and creates a saltwater wedge under coasts and also invades coastal rivers.[21]

Many of the planet's rice-growing areas, a staple in many cultures, are in deltas. Deltas occur where rivers meet the ocean and drop their sediment load, creating low, flat areas with estuaries, the delta structure itself, and the inter-fingering of fresh- and saltwater. These areas are productive for life, and not only have people used them for agriculture, but they also built cities on them, such as New Orleans and Venice. Rising sea levels threaten such cities but especially the river deltas of

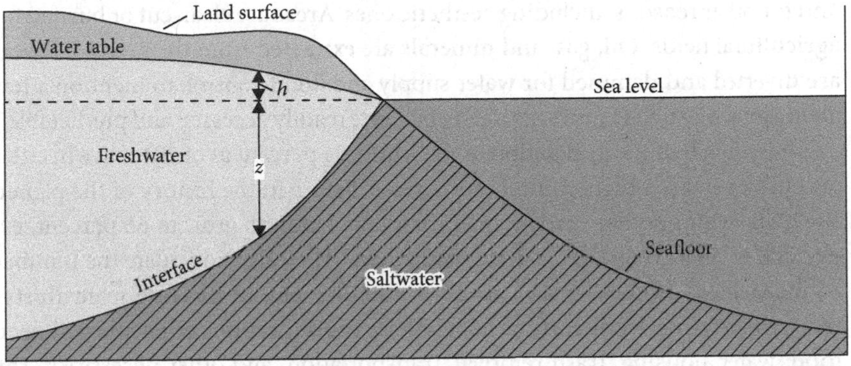

Figure 12.5 Saltwater intrusion in a coastal aquifer. (Barlow, 2003)

the Yangtze, Nile, Indus, Ganges, Mississippi, Congo, Niger, and Zambezi, all of which constitute the most significant rice-farming areas of the world. Saltwater intrusion invades deltas and poisons crops. Deltas slowly sink over time, as the sediments compress, making them even more susceptible to sea-level rise and coastal storms.

Today, the melting of ice sheets and glaciers is occurring in the hydrosphere at a much-increased rate. Many researchers have documented the loss of ice, but American photographer and geomorphologist James Balog's Extreme Ice Survey has captured the retreating glaciers in stunning time-lapse photos, logging the death throes of some of these massive features. Balog and his team, starting in 2007, placed cameras aimed at the glaciers in bedrock with circuitry designed to withstand the world's harshest conditions. Their photos recorded changes in the ice sheets and glaciers in Greenland, Alaska, Canada, the Rocky Mountains, the Andes, Iceland, and the Alps, providing visible proof of climate change.[22] Balog calls glaciers the "canary in the coal mine," because their loss is directly related to warmer temperatures and variations in precipitation patterns linked to global warming.

Many, if not most, of the Earth's glaciers, outside of Antarctica, are in retreat.[23] The rate of ice loss is not steady but is accelerating as human activities pump more carbon dioxide and methane into the air. Since ice sheets and glaciers make up 69 percent of the freshwater on the planet, these observations are cause for concern.

Biosphere–Human Interaction

Over thousands of years, people have shaped, prodded, and poked the environment to change it and rearrange it to provide food, water, shelter, and energy and for other reasons, including aesthetic ones. Areas are clear-cut or burned for agricultural fields. Oil, gas, and minerals are extracted from the ground. Rivers are diverted and dammed for water supply and flood control, to mention a few alterations, all so life's necessities may be more readily accessed and predictable.

More than half the population of the world, 54 percent as of 2014, live in cities and urban areas, a higher number than ever before in the history of the planet. By 2050, demographers expect the urban population to grow to 66 percent, especially in regions of India, China, and Africa. They also anticipate the number of megacities, those with at least 10 million people, to increase from thirty-three to forty-one by as early as 2030. These concentrations of people will need food, water, housing, trash removal, transportation, and other necessities. The International Geographical Union Megacity Task Force says of these population centers:

> Megacities are major global risk areas. Due to highest concentration of people and extreme dynamics, they are particularly prone to supply crises, social disorganization, political conflicts and natural disasters. Their vulnerability can be high.[24]

Coastal megacities are prone to storm surge and sea level rise.

Animal species are disappearing at an alarming rate, in what some call a sixth major extinction event.[25] Biodiversity is being challenged on a level no one in recorded history has ever experienced.

Reefs are cities of animals in the oceans. Their history goes back to the earliest times of geologic history. Today reefs of coral populate the seas, but it was not always so. In the past, reefs looked very different. No matter what type they are, they are nurseries for life itself. The history of reefs is a story that informs us about the present and possibly even where life will head in the future.

Stromatolites and cyanobacteria made up the first reefs known and were some of the most extensive reefs on the Earth during Precambrian times. These organisms adapted to live cooperatively, to gather together and function as community-building mounds and long linear reefs. They are the first forms of life to take carbon dioxide from the air, use it, and return it as oxygen. This development started to shape conditions on the planet, as the oxygen dissolved into the ocean water and ultimately into the air. Reefs evolved in complexity in the Phanerozoic Era in the early Paleozoic times. The first intricate animals are associated with conditions on the reef. These structures grow and expand upward like human cities where concentrations of life forms occur.

Imagine the world going from full color to fading color and then no color at all, just black and white. This is what is happening to the coral reefs. Not only are the reefs being lost, but opportunities we never imagined and resources, along with the food and livelihoods people depend on, are lost as well.

Atmosphere and Climate–Human Interaction

The atmosphere is the most dynamic of the Earth's spheres. The gases that make up the current atmosphere—Earth's third—are dominantly nitrogen and oxygen. Atmospheric changes result in both weather over the short time frame and climate over a longer time span. As discussed in chapter 8, the atmosphere, which varies in thickness, is held to the planet by gravity and protected by the magnetic field. Temperature defines the layers with no outer boundary, as it grades into space. The atoms and molecules of gases that make up air—carbon dioxide, for example—get farther and farther apart with increasing height.

Reconstructing temperatures back in time involves analyzing multiple data sets and normalizing the data to provide for the same areal coverage and baseline. The National Oceanographic and Atmospheric Administration has developed a system of buoys and other ways to record air-temperature data over the oceans to come up with global temperature readings.

Temperature data are difficult to correlate, because some records are localized and other data sets cover entire countries, using various measurement techniques. Temperature readings became more reliable around the mid-1800s as scientists developed more accurate thermometers. Other variations in recorded temperatures occurred when weather stations changed from manual measurements to electronic measurements because of increased consistency. Many temperature measurements on land are taken at stations checked by volunteers and collected when convenient, rather than being recorded at a standard time. These values are biased toward warmer temperatures as a result of insulation over the land, which requires the data to be adjusted.[26] And siting differences based on the location of the temperature measurement can produce another type of glitch in the data.

The earliest-known system of temperature measurements was the Medici Network, recording observations from 1654 to 1670. The network consisted of eleven stations—seven in Italy and the others in Paris, Warsaw, Innsbruck, and Osnabruck.[27] Each station was operated by monks who all used the same equipment and techniques.

Climatologists today have other measures to extend temperature readings back in time, using paleoclimate indicators, for example. Paleoclimate data are recorded by tree rings, ice cores, ocean and lake sediments, corals, and pollen. Trees that grow in temperate climates have annual growth rings, which scientists use to decipher past climates. Rings are thicker when the climate is hotter and wetter, thinner when the climate is colder and drier. Air trapped in ice cores is critical in studying and understanding paleoclimates from particular times. Climatologists use mass spectrometry to analyze the ancient air to calculate oxygen isotope ratios and to determine the content of gases, such as carbon dioxide. Climatologists can age-date ocean and lake sediments utilizing radioisotope techniques and fossils that indicate particular climatic conditions. Corals are useful indicators of water temperature and contain oxygen isotopes useful for age-dating. Pollens are well preserved in lakes and other bodies of water as part of the sedimentary record. Particular species of pollen are produced by plants that live in particular environments and provide data on the climate under which the plants lived. These indirect paleoclimate measures are sometimes called proxy data.

Michael Mann, climatologist and geophysicist at Pennsylvania State University, first published paleoclimate and recent temperature data in 1998.[28] His initial graph went back to 1400 CE and showed temperature anomalies from a baseline of 0 degrees C (32 degrees F). He used some reconstructed data because direct temperature records were available only from 1902 to 1995. The pre-1902 temperature data came from historical temperature readings and proxy data from tree rings, ice cores, and isotopes from corals.

Mann later extended his graph to 1000 CE (color insert 12.2) in 2001.[29] In 2003, he and an English colleague, Philip D. Jones, a professor from the University of East Anglia, using another method, extended the curve up to 2000 CE using data from ice cores from the Ohio State University Byrd Polar and Climate Research Center and atmospheric carbon dioxide readings from the Mauna Loa Observatory in Hawaii. Other independent research data sets have confirmed Mann's data.

Mann's graph shows temperatures starting to increase sharply around 1900, presumably from the impact of the second Industrial Revolution and the burning of fossil fuels. The gray zone around the proxy data points illustrates, statistically, the possible high and low range of each data point. The iconic graph is better known as the "hockey-stick" temperature graph because its curve is relatively flat over an extended period and then increases dramatically, resembling a hockey stick turned on its side, blade up. The publication of the graph brought attention to temperatures over time and the human influence on the climate, with both acclaim and notoriety. Temperature charts published by Mann,[30] along with his colleagues' data and graphs, point to a strong influence of anthropogenic impact on the climate.[31]

Interfaces among All Spheres

All of the Earth's spheres—the geosphere, the hydrosphere, the biosphere, and the atmosphere—intersect and overlap to varying amounts through time. More changes and transitions occur depending on where and how the spheres interact within a particular sphere. Two factors of particular note are soils and the carbon cycle.

Soils and Their Significance

Soils, a part of the geosphere, are essential to life; they are the Earth's permeable skin and act as an interface between all four spheres of the planet. They are part of the "critical zone," which extends from the vegetation above to the uppermost ground water below, and demonstrate the health of the interrelated systems. Soil contains air in pores from the atmosphere; water from the hydrosphere infiltrates and develops soil structure, layers, volume, and thickness; animals such as insects and arthropods, part of the biosphere, inhabit soils and impact soil structure and aeration; and bedrock (the geosphere) is the parent material of soil as it weathers. Not only do soils exist at the intersection of the spheres of the Earth, but they also represent an interdisciplinary field that brings together geologists, biologists, chemists, physicists, engineers, agriculturalists, soil scientists, and others.

Time, rainfall, and humidity are needed for soils to develop from bedrock. Their development involves weathering of the original rock which occurs as water invades, the rock breaks down, and minerals are released. Soil layers are thicker where precipitation and moisture in the form of humidity are higher. Thus, thick, well-developed soils occur in the eastern part of the United States and the Pacific Northwest, as well as in the mid-portion of the country where farmers produce much of the wheat and corn for the nation in fertile prairie soils. Each of these areas has soils unique to it, but the soils have similar development of layers and variations in chemistry and acidity. In the arid regions, soils are thinner and may consist of little or no organic matter. They are alkaline as well (higher than 7 on the pH scale). Even in very humid areas, soils can take thousands of years to form. The thickest soils are in the tropics, but there is so much precipitation that they often are leached and not particularly useful for agriculture.

Soils that have developed under the right conditions and time, on the order of hundreds to thousands of years, have distinct layers. The dark-colored top horizon is mostly organic matter, consisting of undecomposed plant litter. The next-lower layer is the topsoil, made up of decayed organic material (humus), clay, sand, and silt. The lower layer after that is called the subsoil. Prairie soils have a very well-developed topsoil zone. The subsoil is usually redder in color and contains clays and other minerals that have migrated down from above. It is in the subsoil that clay may form an impervious layer that traps water so it may pool and not soak into the subsequent layers. Below the subsoil is weathered rock, which is on its way to becoming soil. The lowest layer in a soil column is the original parent of the soil: the bedrock unit. Soil and its viability have implications for myriad subjects and applications, including biodiversity, the biogeochemical cycle, hydrology, human health, and climate.

Biologists measure biodiversity in the soil in different manners and at a variety of scales, from macro to micro, for example, from arthropods to microbes, including burrowing organisms such as amphibians, reptiles, birds, and mammals. These life forms provide essential functions and improvements to this zone when they create organic matter by transferring nutrients through soil layers, via aeration and the creation of pore spaces and tunnels in the soil, and through their remains when they die. Also, the microbial health of the soil provides a balance to the system.

Biogeochemical cycling occurs as matter is transferred through substrata via natural processes, by air and water, and artificially through the use of nitrogen-rich fertilizers for farming. Soils have become reservoirs of the artificially introduced nitrogen during the past few decades and emit it back into the atmosphere in the form of nitrous oxide, a greenhouse gas. Phosphorus is another nutrient applied to and stored in soils by animal manure of various methods, including slurries. The release of phosphorus stored in the soil layer has caused surface water bodies to have increased nutrient content that stimulates algal growth followed by bacteria that feast on the dead algae. This results in eutrophication that can negatively impact surface and ground-water supplies through algal blooms and hypoxic (low-oxygen) conditions. Soils are the most extensive reservoir of carbon on the Earth other than the deep oceans, containing more carbon than exists in the plants and the atmosphere combined; they are a significant contributor to carbon in the carbon cycle.

Water from precipitation percolates through the soil. Some enters the zone above the ground water (the vadose zone), and some makes it down to the water table (the phreatic zone) and to lower aquifers. Water, in all its locations, and soils are intimately connected. Drought and unusually dry weather patterns, along with some farming practices, initiated what became the "Dust Bowl" of the Midwest in the 1930s. In the early twentieth century, settlers tilled 5.2 million acres of thick prairie soils to plant crops in what is known as the "Great Plow-Up." Weather conditions at the time favored rain, and the area was very productive. However, all of that changed in the decade that followed, when warmer and drier conditions prevailed. Four periods of drought occurred beginning in 1931. Crops failed, and massive windstorms created clouds of dust that engulfed areas as far away as Chicago and New York.

The migrating dust clouds contributed to the loss of millions of acres of topsoil, and recognition of the causes resulted in the 1935 formation of the Soil Conservation Service (now the Natural Resources Conservation Service, a federal agency within the U.S. Department of Agriculture) to prevent such disasters. The Dust Bowl is considered by many to be the most massive natural disaster of the twentieth century and by several, including American ecologist and hunger expert George Borgstrom, to be one of the top three ecological disasters in human history.[32] The Dust Bowl made the Great Depression of the 1930s all the worse, and historians estimate that nearly 3 million people abandoned the southern Great Plains states, many moving to California. The amount of topsoil lost during 1935 alone is thought to be 850 million tons—some eight tons of dirt for every American then alive—and was deposited thousands of miles away, some even reaching the Greenland ice sheet.[33]

The "critical zone" includes soils and is the Earth's permeable, breathing layer that supplies food, water, and, in large part, oxygen. Unfortunately, today, changes in climate are having an impact on the health of the critical zone. Across the United States, there are nine Critical Zone Observatories, originated by the National Science Foundation in 2007, focusing on studies of this region by interdisciplinary teams of scientists.

Toward the polar latitudes, some soils are frozen, forming permafrost, which remains in a deep freeze throughout the year. The formal definition of permafrost is ground that has been at or below 0 degrees C (32 degrees F) for more than two years, according to the International Permafrost Association. Only the top few centimeters of the active zone may thaw, but the rest of the soil stays frozen, representing time capsules from the glacial times. These permafrost areas are located at the margins of glaciers from the last ice age and preserve relict ground ice. They also preserve glacial features formed as the ice sheets rode over them and retreated. Gases, residual pollutants, and living matter may also be trapped within glaciers. Mammoths, mastodons, bison, and seeds are examples of organisms frozen in Pleistocene glaciers. With the recent warming of the oceans and the atmosphere, permafrost has begun to melt more quickly than predicted, releasing stored carbon dioxide and methane, microorganisms, and sometimes heavy metals such as mercury. As the frozen soils thaw, some of the slope materials that had been stable become landslides and slumps of wet, gooey material called thaw slumps.

Temperature, Weather, and Climate

Weather, as noted, is the atmospheric changes that humans experience from day to day and is related to climate, which represents longer trends in the atmosphere. The weather and climate of today are not the same weather and climate known for more than a generation.[34] Data based on actual instrument readings show that warming of 1 degree C since the 1950s has led to increasing temperatures of up to 4 degrees C in some areas and more highs and lows in temperatures in general. Longer growing seasons measured by the length of frost-free days are occurring, and there are longer and more intense fire seasons. Warming through the year is not spread out evenly; December and January have been warming the most rapidly, according to one study in Canada, and that country is undergoing double the rate of warming compared to the rest of the world because of the loss of ice sheets and changes to the snowpack.[35]

Temperatures are higher on average and have been breaking records every subsequent year (color insert 12.3).[36] According to NASA, 2019 was the second-warmest year since 1880, when temperature records were first kept.[37] The potential impacts on the hydrosphere and water supplies over the entire planet will be critical going forward. Consider this: it is possible to replace hydrocarbon fuels, such as oil, natural gas, and propane, with renewable sources, such as wind, hydropower, and solar, but it is not possible to replace water. Human impact on the hydrosphere is just one example of the challenges future generations will face.

The Dynamic Carbon Cycle

Organic molecules contain carbon bonded to other elements, such as oxygen, hydrogen, and nitrogen. Carbon has many different forms because it can bond with other atoms. It also is an example of an element that cycles among all parts of the Earth in a biogeochemical cycle. Carbon is taken in by plant tissue via photosynthesis and transferred by roots to the soil, where it is stored. Carbon also is fixed by photosynthesis through marine plants and corals and diffuses from the atmosphere into the oceans, where it is stored in cold deep waters or transformed into marine sediments or bound up in reservoirs of limestone made by the shells of invertebrates. Cold water can hold more dissolved carbon than warm water can. Other sinks of carbon are coal and petroleum. Carbon is released back into the atmosphere through respiration by animals, decomposition after plants and animals die, deforestation and burning of trees, and the use of fossil fuels. Over the oceans, carbon diffuses back into the atmosphere, where surface and deep ocean water mix depending on wind, currents, ocean water temperatures, and global weather and climate patterns.

USGS geologists estimate that volcanoes today produce an average of 140–440 million tons of carbon dioxide each year. Still, though significant, these numbers pale in comparison to anthropogenic emissions of carbon dioxide, a scale of billions of tons per year. On the surface of the Earth, these gases interact with water, rocks, and organisms and can be influenced significantly by them.

At the same time that volcanoes project volatiles into the air, material in other locations, mainly along trenches and subduction zones, is drawn back down into the asthenosphere. Enormous amounts of water and carbonate (the mineral form of carbon) are consumed at deep-sea trenches and other convergent boundaries. Some of the water and carbonate minerals are then released through new volcanoes, and some are recycled into the mantle. However, no matter what is happening on the surface of the Earth—even during the frozen conditions during Snowball Earth—volcanic processes continue to discharge carbon and other volatiles despite the ice and snow, with a heating effect that results ultimately in the melting of even Snowball conditions.

Scientists can ascertain that the increase of carbon in the atmosphere is related to the burning of fossil fuels based on the isotopic makeup of carbon dioxide, which acts as a fingerprint. The isotopic composition of carbon, first discussed in chapter 4, consists of three varieties: carbon-14, which is heaviest and slightly radioactive, carbon-13, and carbon-12. The amount of carbon-14 decreases naturally over time as it decays to form carbon-12. Plants preferentially take up lighter isotopes of carbon—carbon-13 and carbon-12—in their tissues. Coal is made from plants, and so here carbon-13 dominates over carbon-14; it is not radioactive. As coal burns, it releases stored carbon. Since the atmosphere is enriched in carbon-14, there is a dilution effect as coal's carbon-13 spews into the air, resulting in an overall atmospheric decrease in carbon-14. This is a by-product of the human need for energy through fossil fuels, and in this case, the signature points to the impact of human activity: scientists tracking the amount of carbon-14 in the atmosphere have noticed a marked decrease over time since the onset of the Industrial Revolution.

Changes to the air and the atmosphere through increased levels of carbon dioxide (a greenhouse gas) have been at the forefront of scientific findings and the news and media. The IPCC Synthesis Report documents that climate is changing through the human use of fossil fuels.[38] Oil, natural gas, and coal deposits are carbon reservoirs that took millions of years to form but whose carbon is released back into the atmosphere over a much shorter period. Methane (CH_4) is another exceptionally potent greenhouse gas produced by agricultural practices, in particular raising livestock. Human activity has changed the background state, or condition, of the atmosphere, both through the burning of fossil fuels and agriculture—modifying the character of the very air that all organisms, including humans, need.

Carbon dioxide and temperature are correlative and vary together; as the atmospheric content of carbon dioxide gas increases, so temperature follows, and vice versa. The record of carbon dioxide and temperature reaching back 800,000 years (color insert 12.4) is based on carbon dioxide and temperature data determined from ice cores drilled in Antarctica. The color figure illustrates cycles of highs and lows in both measures. But the maximum carbon dioxide level rarely exceeded 300 parts per million (ppm), with only one instance over this period. The graph records glacial periods, times of great cold when ice abounds and carbon dioxide and temperature are low, and inter-glacial episodes when the two variables soar. Planetary variables such as the shape of the Earth's orbit, the amount of axis tilt, and the wobble at the pole (precession) impact how much solar radiation reaches the Earth and thus determine cold and warm periods. These factors vary through time, each with its own periodicity.[39] The cycle is called the Milankovitch cycle, named for the Serbian mathematician Milutin Milankovitch (1879–1958), who in 1930 related cyclical changes in the Earth's axis tilt (obliquity), orbital shape (eccentricity), and wobble at the pole (precession) to glacial and interglacial periods in the geologic record. He did this through pen-and-paper calculations, since computers were not around at the time of his work.[40] These cycles lead to variations in solar radiation by as much as 10 percent from the average, causing warming and cooling. Until the last several hundred years, the ice core and other proxy data indicate that this was the dominant carbon dioxide and temperature trend. The Milankovitch cycles do not account for, and are not the cause of, the recent warming of the Earth's atmosphere, which has occurred since pre-industrial times and is the result of human activities.[41]

Since the advent of agriculture and the phases of the Industrial Revolution, the average concentration of carbon dioxide in the Earth's atmosphere has soared nearly 49 percent from pre–Industrial Revolution values of approximately 280 ppm. In May 2019, carbon dioxide measured by NOAA at the Mauna Loa Observatory was 414.7 ppm, the highest peak on record at that time.[42] NOAA recently released the carbon dioxide concentration at the Mauna Loa Observatory for April 2020 of 416.21 ppm,[43] a sobering number that, at a level 25 percent greater than the historic high over the past 800,000 years, portends dire consequences.

Interestingly, during the COVID-19 pandemic, the amount of greenhouse gas emissions decreased globally by 17 percent during initial "stay at home" orders in the two-month period from mid-March to mid-May 2020, for the United States and the rest of the world.[44] Accordingly, NOAA undertook a mission to track the impact of reduced carbon dioxide and other pollutants to the environment since the pandemic began.[45] Initial results show better air quality as people fly and drive less. These societal effects of the pandemic, along with a reduction in industrial output, account for this most likely temporary decrease in emissions and cleaner air. Even though carbon emissions had declined because of less car and air travel, atmospheric carbon dioxide levels, as measured at the Mauna Loa Observatory, simultaneously reached 417.2 ppm, 2.4 ppm higher than the previous peak in 2019.[46] Carbon dioxide generation would need to decrease by 10 percent worldwide for at least a year to see an impact on the "Keeley Curve," the chart that graphs daily carbon dioxide over time. Nothing in the sixty-two-year history of the Keeley Curve, including the 2008 economic depression or even the collapse of the Soviet Union in the 1980s, has yet to cause a dent in these ever-increasing carbon dioxide levels recorded by the graph.[47]

Although it does not seem possible that warming of a degree or two Celsius would be significant or that humans could impact such a massive system as the Earth, evidence reveals just what has occurred through people's activities. Changes are happening in many of the most extensive ice sheets and glaciers on the Earth. A 2015 study of 5,200 glaciers shows "unprecedented loss of ice," especially from those for which there are one hundred years of data.[48] The magnitude of these ice sheets and glaciers makes them essential to climate and weather on the Earth. They reflect sunlight and are part of what keeps the planet cool. As more ice melts, there is additional warming by the sun's rays because of the lowered reflection surface of white ice sheets.

The implications of greenhouse gas emissions and climatic warming go beyond the atmosphere itself. Perhaps more sobering is the impact on the other spheres of the Earth. Climate warming affects many factors, including sea level rise, changes in weather and precipitation patterns, challenges to freshwater resources, reduction in marine biodiversity, ocean acidification, issues of food security and production (especially cereal crops), human health impact, heatwaves, population dislocations, and conflicts over resources.[49] The IPCC estimates that the poorest countries will be impacted the most, at least initially. These outcomes define the range and tenor of the human bearing on the planet—the Anthropocene.

Among these many problems generated by rising carbon dioxide levels and global warming, changes in long-term weather patterns, including where and when rain falls and the intensity and duration of that precipitation, is of major concern.[50] Climatic zones are shifting from south to north as well as west to east, as certain areas dry out and other regions receive more rain. Climate researchers predict droughts and floods to be more severe in the next century. Already, some of these changes are occurring today.

Munich RE, the largest reinsurer in the world, keeps an extensive database of more than forty thousand natural disasters that have occurred since 1980. Part of its mission is to determine if natural disasters are increasing in number and strength, in order to predict future natural catastrophes and prevent loss of life and property damage. Data from 1980 to 2016 show an increase in climatological (droughts, extreme temperatures, and forest fires), meteorological (cyclones and storms), and hydrological (floods and landslides) events, from fewer than three hundred in 1980 to nearly eight hundred in 2016, more than a twofold increase. Natural disasters resulting from geophysical sources—earthquakes, tsunamis, and volcanoes—have, in contrast, remained relatively steady.

With an upsurge in atmospheric and hydospheric natural disasters has come increased severity of storms and extreme weather events.[51] Why these are occurring and why their intensity has grown are issues that Munich RE and others attribute to changes to the atmosphere, based on data from climate models. Climatologists run complex computer models multiple times and with different variables. They then aggregate the results, akin to running an experiment over and over to average out random effects. Models without the input of greenhouse gases (including carbon dioxide), aerosols, and changes in land use do not demonstrate the increase in extreme events, but when these factors are input, the models match what observed conditions indicate is happening.

In 2013, intense rainfall occurred over two weeks along the northern Front Range of Colorado, leading to floods not seen since the historic flash flood of 1976 that killed 144 people in Big Thompson Canyon and caused millions of dollars in damage. The 2013 flood resulting from the rainfall impacted a larger area than the 1976 flood, extending from the Poudre drainage, Fort Collins, Greeley, Severance, and Big Thompson Canyon in the north to the St. Vrain River to the south. The rainfall also inundated Boulder and towns on the plains to the east. As much as 46 centimeters (18 inches) of rain fell within ten days, with most of the precipitation occurring in a span of thirty-six hours.[52] On some reaches of the rivers in the impacted area, hydrologists estimated the flows to be larger than a five-hundred-year flood event.[53] The one-hundred-year flood is the standard flood height for Federal Emergency Management Agency maps based on past flood events and data, having the probability of occurring once every hundred years. Probabilities being what they are, a one-hundred-year flood theoretically could happen three years in a row.

The 2013 flooding was worsened by 2012 wildfires that destroyed much of the vegetation and changed the structure of the soils along the Front Range drainage basins. As a result of the subsequent major rainfall, runoff increased in the severe burn areas, leading to greater flood heights; slopes in the fire-impacted regions became unstable and failed.[54] Alteration of the climate has led to the wildfire season extending for longer periods during the year. In the United States, the number of western wildfires has increased fourfold from 1970 to 1986.[55] Not only are the fire seasons longer, but wildfires are burning more extensive areas. Moreover, climatologists predict that wildfires will become even hotter as areas undergo droughts and lack of rainfall. Increased erosion of fire-impacted soils, slope failure, and more runoff leading to additional flooding illustrate the real-life impact of one sphere, the atmosphere, on other parts of the system.

Earth's equilibrium is sensitive to small changes in carbon concentration. In the geologic past, high levels of carbon in the atmosphere have triggered greenhouse conditions, leading to a dramatic impact on life. Consequences of carbon cycling include dictating the course of global climate and storage of carbon in various reservoirs. Changes in carbon concentration, in turn, change the development of organisms by extreme measures in the environment, including the atmosphere. Again, life and all of the Earth's processes are integral to one another and bound together in an elegant but ever-changing dance.

Song of the Earth and Implications

People ignore or possibly forget about the environment and the planet, and they do so at our mutual peril. Perhaps most do not understand or take time to reflect on the source of their water, food, land, and air. But neglecting or discounting this relationship with the Earth threatens human survival and the existence of the biosphere.

Life has altered and changed the Earth—the very substrate from which it springs. The Earth has co-created life, and life has co-created the Earth. The song of the Earth is an ancient tune, one that has many harmonies and melodies. It has ebbed and flowed through geologic time with the voices of organisms past—stromatolites, trilobites, dinosaurs, and humans—all of which have blended into the symphonies of today.

The Earth's extensive history and future scenarios are revealed through the lens of geology. Geology takes the long view of the biography of the planet, shows how life and the Earth are intertwined and how all systems work together to shape the world. Geologists examine the Earth at different scales, from atoms to minerals to rocks to the immense lithospheric plates and global processes. The entire past of the Earth, starting from its emergence 4,600 million years ago, ought to be considered when making decisions about both short-term activities and future choices. To repeat the quote from Santayana near the beginning of this chapter, "those who cannot remember the past are condemned to repeat it." This concept applies to the biography of the planet as well.

What can people glean from the treasure trove of geology to help solve international, national, and local problems? Reading the Earth through geology, we know why earthquakes happen where they do; we understand that particular volcanoes pose increased risks because of their explosive type and location of human populations. We also realize that when ice sheets melt, sea levels will rise, and not only are there incursions landward, but the seawater will overtake usable groundwater supplies, and vital coastal agricultural areas and fields could be ruined. It is an existential danger to ignore the messages of geology or leave consideration of the Earth out of our business and economic decisions, favoring short-term goals over long-term survival. It is not that those consequences will never be realized; it is just that most prefer not to name, plan, or pay for them upfront. When disaster comes, whether in the form of a pandemic or economic, political, or environmental collapse, it costs even more than it would have had there been preparation. Understanding geology and its messages are one way forward in addressing the issues of climate change and other imbalances at the human–Earth interface.

Through its long history, life on the planet has shaped and been shaped by the forces of geology, but often blindly. Life is resilient, particular species and genera not so much. Most of the life forms that have ever lived are now extinct. Lessons from the Burgess Shale deposit show that life is more like an ever-branching vine than a tree-like structure. Extinction wiped out parts of the vine, and those creatures appear to be gone forever. Predation and competition between species have engendered new behaviors and mechanisms of survival, leading to complexity and differentiation among species.

Systems of the planet are especially sensitive to small changes in carbon dioxide and other greenhouse gases. Variations in the carbon cycle account for four of the five mass extinctions and most of the smaller extinction events. Tipping points have happened throughout geologic time, hurtling the world from one state of climate to another—from icebox and Snowball Earth to greenhouse conditions. Many life forms do not survive such sudden devastation of their environment.

Lessons can be learned from past greenhouse conditions, when temperatures warmed, to predict where the climate may go in the future. For instance, in studying the Eocene Epoch in the early Cenozoic from 55.8 million to 34 million years ago, there was corresponding warming both in waters of equatorial and in those of the higher-latitude polar regions. Carbon dioxide levels exceeded 560 ppm. Scientists have known that seawater temperatures in the polar region are more sensitive to increases in carbon dioxide than the areas near the equator, an effect called polar amplification, but there have been few data to detail the relationship. In a study conducted in 2018, researchers examined the shells of Eocene foraminifera—microscopic ocean dwellers—to reconstruct ocean temperature and chemistry from that time.[56] Examining the foraminifera data, they found, based on projections, that tropical waters were 6 degrees C (10 degrees F) warmer than today, and polar waters were 20 degrees C (38 degrees F) warmer. Yet current models used by the IPCC underestimate this difference by more than 50 percent and are significantly minimizing the impact of carbon dioxide on the polar latitudes. The new temperature data from the foraminifera will improve upcoming climate models so climatologists can make better predictions going forward.

Looking Backward to Look Forward

Educator, Astrid Steele, Nipissing University, wrote about advice her mentor once gave her when she feared she was lost on a Canadian Lake a distance from shore as dusk fell. He told her to "Always look backward to see where you are coming from. You won't know where you are unless you look backwards [*sic*] once in a while."[57] This wisdom applies to the role of geology as humans navigate increasingly complex issues related to how to live with and on a changing planet. Geology provides context for life; a framework and guidebook to move forward by examining past events in the long biography of the Earth.

Early geologists and scientists thought "outside of the box" to glean the concepts that make up the field of geology today. The three foundational ideas in geology—geologic time, plate tectonics, and evolution—in harmony create the biographical narrative of the Earth's long past. Each precept is instructive and provides lessons to be learned. Taken together, they uncover the trajectories of an ever-changing, evolving planet, as well as events that occur over and over in the geologic record. They additionally reveal as false the illusion that life is separate from that of the planet.

Stephen J. Gould ruminated on the trajectory of time—time's arrow—and the repeating patterns of time—time's cycle—as critical concepts for understanding geologic time. To put it another way, perhaps deep time can be experienced as a vessel that not only encloses life like an amphora that stores precious liquid (a) and simultaneously moves life forward through uncharted waters like the schooner *Rembrandt* navigating the Arctic Ocean (b) (color insert 12.5). Indeed, the comprehension of deep time and geology need not elicit only warnings but also certain comforts in realizing that the Earth and life on it have persisted through phases of monumental destruction and healing for billions of years. Beyond the lessons from geology and the biography of the Earth, knowledge and appreciation of geology can enrich and enliven people's lives. The human role on the planet is placed in a different scale and perspective when examining the history of the Earth and life on it. Geology provides a certain substrate and grounding (pun intended) for anyone who enjoys nature; wherever you go, geology is there for anyone to see with their own eyes and enjoy.

From John Hutton and Charles Lyell came the wisdom and understanding that "the present is the key to the past." Turning the thought around, the past is the key to the present and even beyond, philosophers and academics speculate, as Hume said, that the past will govern the future. The question is what future will the Earth and its inhabitants have?

Humans are at a unique juncture, standing at a crossroads, where several paths lie ahead. Never before has a wealth of scientific knowledge and data become available to light the way for influencing possible outcomes for the planet's future. Geology and understanding how the Earth sings to humanity can be part of the solution.

The hour is ripe not to be paralyzed by desolation but to take positive steps toward healing the human relationship with the natural world. The Earth can bounce back more quickly than many realize; natural waters, habitats, and environments can recover successfully and often more quickly than predicted. But there is not a minute to lose. The time to listen to this song of the Earth is now.

Glossary

Term	Definition
Absolute age-dating	See numerical age-dating.
Ammonite	An extinct mollusk group, belonging to the Sub-class Ammonoidea, Class Cephalopoda, which had a spiral shell. It is the main index fossil of the Mesozoic Era.
Ammonoidea	A sub-class of cephalopod mollusks including the ammonites.
Amniotic egg	Eggs laid by tetrapods that can survive dry periods because of an outer shell or leathery covering; also refers to animals that lay eggs, or amniotes.
Anaerobic	Oxygen-poor; environments that favor preservation of organic remains that can become fossilized.
Angular unconformity	A specific type of unconformity whereby the rock strata are at an angle to one another.
Anomalocaris	An arthropod living in the middle Cambrian Period, also called the strange shrimp.
Anthropogenic	Made or created by humans.
Anticline	A folded set of beds that look like an upside-down U.
Archaeopteris	An extinct tree-like plant, common in the Carboniferous Period, and contributed to coal formation.
Archean Eon	An eon in deep time named for its ancient nature, from 4 billion years to 2.5 billion years ago.
Asthenosphere	Lies below the lithosphere and consists of a ductile, plastic zone in the upper mantle of the Earth, with convection cells moving through heat.
Atmosphere	One of the five systems of the planet, the air envelope around the Earth, its composition and climatic impacts, past and present.
Basalt	A dark-colored, fine-grained extrusive igneous rock produced by volcanic eruptions, makes up the rocks of the ocean basins.

Term	Definition
Basilosaurid	A type of extinct cetacean (whale ancestor).
Belemnite	A marine cephalopod, now extinct, Order Belemnitida, resembled squid with a straight shell, often utilized as an index fossil.
Benthic	The seafloor or bottom of the ocean including the most shallow sediments, representing an ecosystem.
Biodiversity	A measure of the variability among life on the Earth.
Bioherm	Mounds of algal mats, sometimes captured as fossils in the rock record.
Biosphere	One of the five systems of the planet, comprises all life on the Earth, past and present.
Biostratigraphy	The field of geology in which past life forms and fossils are related to geologic strata and ordered according to age.
Brachiopod	A marine bivalve with a hard shell; fossils of these creatures are used to correlate geologic strata.
Calcareous	Made of calcium carbonate ($CaCO_3$); one example is limestone.
Carbonaceous shale	A shale containing a high percentage of carbon.
Cetaceans	Marine mammals, Order Cetacea, such as whales, dolphins, and porpoises.
Chalk	A soft, fine-grained limestone, made from the shells of tiny marine organisms.
Chemical fossils	Created as an organism decomposes and leaves behind molecules that indicate indirect evidence of life, its isotopic signature; oil is an example: the tiny marine animals that were the source of the oil are long gone but leave behind a chemical imprint.
Chemically formed rocks	Rocks that have been created through chemical means, such as precipitation of chemicals from water, such as chert
Chevron fold	Folds that have a repeated pattern of V shapes or sharp angles between the limbs of the folding, indicating intense pressures and compressional forces.
Chordate	An animal with a central nerve running down its back (dorsal nerve cord), supported by cartilage (notochord), a series of openings that connect the inside of the throat to the outside of the neck (pharyngeal slits), and an extension of the body past the anal opening (post-anal tail); Chordata is an animal phylum that includes vertebrates.

Term	Definition
Chronometric dating	Age-dating by the use of numerical age-dating methods; *see* chronostratigraphy.
Chronostratigraphy	The field of geology that covers numerical age-dating methods, including radioactive decay of elements and magnetostratigraphy, used to find ages and boundaries between rock units.
Clade	A group of organisms arising from a common ancestor.
Cladistics	The study of how closely groups of animals and plants are related based on measurable characteristics.
Climatostratigraphy	The field of geology in which past climatic conditions are related to geologic strata and ordered according to age.
Coal measures	An older term referring to coal deposits of the upper Carboniferous Period, which were mined beginning in the 1800s for coal, beginning in England and extending to the Americas.
Concretions	Hard, rounded structures formed around a particle or grain, found in sedimentary rocks and formed before the sediments undergo lithification. Shapes and sizes can vary. Fossils have been found at the center of some concretions.
Conodont	An extinct marine chordate, of Class Conodonta, thought to be eel-like, whose fossils mainly consist of toothlike structures; these fossils are used to correlate geologic strata.
Continental shelf	The area adjacent to a continent and extending from it, currently under the ocean but at relatively shallow depths (60 meters, or 200 feet).
Coprolites	Fossilized feces, also known as bezoar stones.
Craton	The most stable and oldest parts of the continents, usually made up of igneous rocks.
Crinoids	Also known as sea lilies, related to starfish, brittle star, and sea urchins, these animals once dominated the Paleozoic seas; their fossils are used to correlate rock units.
Crystalline	Rocks that have crystallized, or cooled, from a magma and whose crystals are interlocking because of the cooling process; *see* igneous rocks.
Cyanobacteria	Blue-green algae that were photosynthesizing bacteria, one of the earliest life forms on the Earth.
Cynognathus	An extinct genus of therapsids, protomammals of the late Permian Period.

Term	Definition
Dendrochronology	A numerical age-dating method using tree rings, of both modern and fossilized trees.
Dentition	The way in which teeth are formed and arranged.
Detrital	Weathered grains or particles of material that often make up new sedimentary rocks, such as sandstone and conglomerate.
Diagenesis	Physical and chemical processes that occur when sediments are made into rocks.
Diapsids	A group of amniote tetrapods, with two holes in their skulls, Class Reptiles.
Dike	A vertical intrusion of magma that cools forming a rock layer into surrounding rock.
Dimetrodon	A large, extinct reptile, part of the informal group of Pelycosaurs.
Dinosaur/Dinosauria	Translated as "terrible lizard," a superorder of reptiles, which dominated during the Mesozoic Era. Originally two orders (clades), the saurischia (lizard-hipped) and the ornithischia (bird-hipped) and later the Therapoda were added. In 2017, the ornithischians and the theropods reclassified into the "Ornithoscelida" clade. Non-avian dinosaurs became extinct at the end of the Cretaceous. Avian dinosaurs are not extinct and are our modern birds; all are technically and anatomically the same as other avian dinosaurs.
Dipole	A molecule with two poles of different charges, separated in space.
Dolomite	A sedimentary rock made up of calcium, magnesium, and carbonate, usually altered from limestones in ancient reefs when a magnesium ion substitutes for one of the calcium ions in the limestone; may contain fossils.
Domains of life	The three domains of life are archaea, bacteria, and eukarya.
Eccentricity	The variation in the shape of the earth's orbit around the sun, from more circular to more elliptical, which occurs over a 100,000-year cycle.
Echinoderms	Marine animals of Phylum Echinodermata, having a radial symmetry.

Term	Definition
Electron spin resonance	A numerical age-dating method that measures the number of free radicals in a sample that increase over time from cosmic radiation.
Enalio-Saurian	An extinct group of fossil saurian dinosaurs, including *Plesiosaur* and *Ichthyosaur*.
Epicontinental seas	Shallow seas that extend inland over continental interiors when sea levels are high, also called epeiric seas.
Epigenetics	Variations in organisms caused not by a change in DNA but by modification of gene expression.
Erratics	Rocks and boulders that have been entrained into a glacier and dragged to a place far from their origin, sometimes creating grooves in the bedrock as they are dragged along.
Eukaryotes	Organisms whose cells are organized with a nucleus.
Evaporite deposits	Layers of salts created when oceans or enclosed basins dry up and evaporate, leaving behind the salts.
Exotic terranes	Pieces of other continents and ocean floor smashed onto continental cratons by tectonic forces.
Extremophiles	Microorganisms that can live under extreme conditions, such as high temperature, very acidic or basic conditions of pH, or chemical environments not usually suited to life.
Fauna	Animal life represented by modern animals or animal fossils from a particular region in geologic time.
Faunochrons	A division of geologic time based on classification of fauna of time, as found in fossils from rocks of the particular time.
Fieldwork	Work by geologists out in nature to see rocks, fossils, or other features where they were formed, aka "going out into the field."
Fishapod	An intermediate form between fish and tetrapods, with a reptile-like head, a fish-like body, and a tetrapod-like pelvic girdle.
Flora	Plant life represented by modern plants or plant fossils from a particular region in geologic time.
Foot wall	Term comes from the mining field and means the block of rock below a fault, as if a tunnel were drilled parallel to the fault; the block on which the miner would be standing.

Term	Definition
Foraminifera	The phylum or class of single-celled, tiny, planktonic marine animal, also known as forams.
Forams	The shortened name for foraminifera.
Fossil assemblages	Groups of animals that lived together and whose fossils appear together in the rock record.
Fusulinids	Extinct, tiny, grain-sized foraminifera that are an important index fossil.
Genera	A taxonomic rank, the plural of genus, for groups of related species.
Genetic drift	The change in the frequency of an existing gene in a population resulting from certain individuals passing on their genes more readily than others, by random chance, as opposed to an adaptation.
Geognosy	An archaic term for the study of rocks, their formations, and how they relate in time.
Geologic time	The long span of the biography of the Earth, recorded from the birth of the planet 4,600 million years ago to the present.
Geomorphology	The field of geology in which the land forms and features on the surface of the Earth are studied.
Geosphere	One of the five systems of the planet, all solid material, the structure of solid Earth, plate tectonics, and rocks on the Earth, past and present.
Geothermal gradient	The increase of temperature with depth in the Earth's interior.
Glauconite	A green silicate mineral with iron and potassium, often associated with sandstones and clays.
Glossopetrae	Triangular stones, originally called "tongue stones," found by N. Steno to be fossilized shark's teeth.
Glossopteridales	An extinct order of seed plants, including the genus *Glossopteris*.
Glossopteris	The genus designation of an ancient seed fern, whose distribution was pivotal in indicating continental drift; one of the members of the extinct order of *Glossopteridales*.

Term	Definition
Graphite	A mineral composed of pure carbon, classified among the Native Elements in the Chemical Classification of Minerals; a dark-gray, soft mineral used as pencil "lead"; some deposits are linked to ancient life.
Graptolites	An extinct marine animal, Subclass Graptolithina, Class Pterobranchia, which lived mainly during the Paleozoic Era; their fossils are used to correlate rocks.
Greensand formation	Marine silts and clays colored green by the mineral glauconite.
Guild	Pertaining to ecology, a group of species that utilize similar resources or resources in similar ways, even though they may live in different ecosystems.
Gymnosperms	"Naked seed" plants that arose and dominated in the Permian to the Cretaceous Periods, the source of much coal; a famous example is *Glossopteris*.
Hadean Eon	An eon in deep time named originally for its theorized hell-like conditions, from the birth of the Earth, 4.55 billion to 4 billion years ago.
Half life	The amount of time it takes for half of the parent isotope to decay to the daughter isotope, used in chronometric dating.
Hanging wall	Term comes from the mining field and means the block of rock above a fault, as if a tunnel were drilled parallel to the fault; the block above the miner.
Hotspot	A feature in the Earth's crust indicating a plume of magma from the mantle, which rises up and creates volcanoes or other features.
Hydrosphere	One of the five systems of the planet, all water on the Earth, both ice and liquid water, past and present.
Hylonomus lyelli	The first land-dwelling reptile, found by Charles Lyell and William Dawson at the Joggins Fossil Cliffs, Nova Scotia.
Ichthyosaur	Swimming reptile of considerable size that swam in an undulating manner.
Ichthyosaurus trigonus	A large marine carnivorous reptile, now extinct.

Term	Definition
Igneous rocks	Rocks formed from magma.
Inclusions	Any matter (another mineral, rock fragment, fossil, gas, or even oil) that has been trapped within a mineral at the time of its formation; for example, diamonds contain inclusions of ancient mantle material.
Index fossil	Particular fossils of animals or plants that lived over a wide geographic area, are distinct from other subspecies or species, and are abundant, allowing them to be identified easily.
Invertebrates	Animals with external shells and soft body parts.
Isotope	A chemical element that has the same number of protons but a different number of neutrons in its nucleus; an example is carbon-12, which has six protons and six neutrons, the most common form of carbon, and carbon-14, which has six protons and eight neutrons and is unstable.
Kerogen deposits	Sedimentary deposits which contain a high concentration of organic material, such as oil shales.
Lagerstätten	Fossil preservation that captures soft parts of organisms along with the exoskeletons, German for "mother lode."
Lahar	A fast debris-type flow the consistency of wet concrete, made of a mixture of volcanic ash and ice and snow from stratovolcanoes.
Laramide orogeny	The tectonic, mountain-building event responsible for the current Rocky Mountains, which occurred from 70 to 40 millions of years ago.
Lepidodendron	An extinct vascular plant, also called the "scale tree," related to the lycopsids and prevalent during the Carboniferous, important for the formation of coal.
Lignite	A low-ranked coal.
Limestone	A chemical sedimentary rock made of calcium carbonate.
Lipids	A class of organic compounds made of fatty acids.
Lithification	The process by which loose sediments become rock through compression, compaction, and cementation.
Lithoherms	A domal type of stromatolite.
Lithosphere	The crust and upper mantle of the Earth, which has rigid properties and in which the tectonic plates are located.

Term	Definition
Lithostratigraphy	The field in geology in which strata are correlated by looking at characteristics of the rocks—grain size, cement type, the environment of deposition, presence of unconformities, and chemistry.
Lycopsids	Ancient seedless, vascular plants that grew prolifically in the Carboniferous, were around in the Silurian, and were the most common coal-forming plant; modern descendants are the club mosses.
Lystrosaurus	An extinct genus of therapsids, Clade Therapida, that lived from the late Permian Period to the Early Triassic Period.
Magma	Melted, molten rock, which when cooled forms igneous rocks.
Magnetosphere	One of the five systems of the planet, the protective magnetic shield around the Earth, past and present.
Magnetostratigraphy	A numerical age-dating method using the magnetic signature captured in rocks from the Earth's magnetic field as they cool, based on magnetic reversals in the Earth's core.
Mathematical models	Complex representations of a system, often three-dimensional, based on computer simulations, using input data and boundary conditions and employed by scientists to look deep into the past or into the future.
Marl	A sedimentary rock made of limestone and clay.
Marsupial mammals	One of the three subdivisions of the class of mammals, pouched mammals that carry their young in an external pouch after a certain phase of development.
Megalosaurus	An extinct, land-dwelling, sizable carnivorous theropod dinosaur.
Mesosaurus	An extinct marine reptile, Family Mesosauridae, Order Mesosauria, that lived during the early Permian Period, sometimes referred to as a mesosaur.
Metamorphic rocks	Rocks altered by heat and pressure up to the point of melting; any rock can be the parent or precursor and undergo metamorphism to create a metamorphic rock.
Migmatite	A type of metamorphic rock that is nearly melted from the forces of shearing.

Term	Definition
Mineralogy	The field of geology in which minerals are studied.
Minerals	The building blocks that make up rocks, composed of various atoms; for example, calcium carbonate, $CaCO_3$ is composed of the atoms of calcium, oxygen, and carbon.
Molecular fossils	Fossils that contain DNA, RNA, or other organic molecules.
Mollusca	The second-largest phylum of invertebrates, also referred to as mollusks.
Moment magnitude scale	A measure of the strength and size of earthquakes, based on the energy released by the earthquake.
Mosasaur	A large, swimming, extinct marine reptile.
Multituberculates	An extinct taxon of mammals, named for the configuration of their teeth with "many tubercles" (cusps), now completely extinct with no living descendants.
Myomeres	Chordate blocks of skeletal muscle.
Nanofossils	Fossils of minute organisms, usually plankton.
Nanoplankton	Tiny, unicellular plankton.
Neptunist school	An early theory of how rocks are created, named for Neptune, the Roman god of the sea, which held that rocks had formed by sedimentation in water.
Neuropteris	An extinct seed fern, common during the Carboniferous Period.
Normal fault	A type of fault resulting from extensional forces, where the hanging wall moves down relative to the foot wall.
Notochord	A skeletal rod made of cartilage, which supports the body in all embryonic and some adult chordate animals.
Numerical age-dating	A way to get very precise age dates of geologic materials using chemistry and the rate of radioactive decay of elements. Also includes dendrochronology, magnetostratigraphy, and other absolute age-dating methods.
Obliquity	The variation in the tilt of the earth's axis, which occurs over a 41,000-year cycle.
Offlap	A time of falling sea level, also called a regression.

Term	Definition
Onlap	A time of rising sea level, also called a transgression.
Ooids	Rounded sphericals of calcium carbonate, whose round shape is formed by rolling back and forth in waves and which make up some limestones.
Oolites	Calcium-carbonate-rich rocks related to stromatolites, also made up of carbonate particles called ooids
Ornithischians	An extinct clade of herbivorous, birdlike dinosaurs.
Ornithoscelida	A proposed clade including the ornithischians and the theropods.
Orogeny	The process of mountain building through folding of the Earth's crust and other processes, such as uplift, from tectonic forces.
Outcrop	An exposure at the surface of the Earth of a rock unit, often visited by geologists "in the field" to study the rocks and the environment in which they formed.
Palaeotherium	An extinct family of herbivorous mammals, related to horses but not a horse.
Paleobotanists	Those who study fossil plants and their ancient environments.
Paleomagnetism	The past record of the Earth's magnetic field, including polar reversals, as recorded in rocks.
Paleontology	The field of geology in which past life forms and fossils are studied.
Palynology	Age-dating by changes in pollen.
Pelycosaur	An informal group of large, extinct reptiles, whose ancestors were the synapsids; among them was *Dimetrodon.*
Peroxide	Any chemical molecule in which two oxygens are bonded together; molecules of peroxides are negatively charged and act as a weak acid.
Peroxidated	An old term used to describe deep ocean sediments that have been buried and undergo chemical changes, including crystallization of the hardest minerals and metals being reduced.
Phanerozoic Eon	The geologic eon of "visible life" from 542 million years ago to the present.

Term	Definition
Photosynthesis	The process by which plants respirate, taking in carbon dioxide and releasing oxygen into the atmosphere.
Physiographic province	Geographic areas grouped by similar surface features, based on topography, physiology, and geology.
Pikaia	An early Paleozoic marine creature with a notochord, possibly the ancestor of all vertebrates.
Placental mammals	One of the three subdivisions of the class of mammals—those that carry their young for most of their development with their bodies, bear and nurse their young with milk, breathe air, and are warm-blooded (internally regulating their temperature).
Plectrodus	A jawed fish first found in the Silurian Period.
Plesiosaur	An extinct genus of marine reptiles with a long neck, a rounded body, and flipper-like appendages.
Plutonic rocks	Rocks originating from magma, those that cool slowly underground, causing them to have a visible crystalline texture to the naked eye; an example is granite.
Plutonist school	An early theory of how rocks are created, named for Pluto, the god of the underworld and heat, which held that igneous rocks were created in the Earth's fiery magma.
Precession	The variation in the earth's wobble at the poles, which occurs over a 26,000-year cycle.
Primitive rocks	An archaic term for igneous rocks, thought to be the oldest rocks at the time, which made up the lowest level of the early geologic time scale.
Produced water	Water existing at depth in geologic formations, which is drawn up during the pumping of oil or gas, also known as formation fluids which can have high concentrations of solids, ammonia, and salts.
Prokaryotes	Organisms whose cells are not organized with a nucleus.
Proterozoic Eon	The geologic eon beginning 2.5 billion years ago and extending 542 million years to the beginning of the Phanerozoic Eon.
Pterosaur	An extinct flying reptile that lived during the Jurassic and Cretaceous Periods in the Mesozoic Era and had a lengthened fourth finger that supported the wing.

Term	Definition
Relative age-dating	A method used to place geologic materials or events in relationship and in order relative to each other, usually fossils, to obtain a qualitative idea of the age of something.
Reverse fault	A type of fault resulting from compressional forces, related to convergent plate boundaries; low-angle reverse faults (less than 35 degrees) are called thrust faults.
Richter scale	A measure of the size of an earthquake, developed by Charles Richter in 1935 for Southern California earthquakes
Rock cycle	The processes by which one rock is transformed into another, such as when sedimentary rocks get buried and melted through intense heat and/or pressure, making magma and ultimately igneous rocks.
Rostrum	The hard shell portion of belemnites that are typically found in the fossil record.
Sandstone	A sedimentary rock made of grains of sand, usually silica dioxide.
Sauropsids	Animals whose skulls have two temporal holes (compared to the synapsids which have one), more akin to reptiles and leading to the dinosaurs and bird lines.
Saurischians	One of the two main divisions of dinosaurs, the other being the ornithischians, having a different hip structure; some were carnivorous and some plant eaters, both of which grew to large sizes.
Schistus	An archaic term for a rock made of shale and limestone.
Sedimentary rocks	Rocks formed from sediments.
Seismometer	An instrument, consisting of a drum and a pen, used to record the amplitude of earthquake waves.
Shale	A sedimentary rock created from clay and/or clay and silt-sized particles.
Shelf seas	Shallow seas that exist on top of the continental shelf areas and provide rich environs for life.
Silicate minerals	The most common mineral class in the Chemical Mineral Classification System, containing the element silica. Also termed 'Silicates.'

Term	Definition
Sauropsida	A clade made up of the Reptilia and the extinct Parareptilia.
Sauropsids	Animals whose skulls had two temporal holes; the sauropsids, Clade Sauropsida, were the ancestor of the diapsids.
Spagodus	A jawless fish first found in the Silurian Period.
Species	A group of individuals that make up a gene pool or are capable of breeding and passing on their genes to offspring.
Sphenacodontia	A clade emerging from Pelycosaur ancestors that resulted in the therapsids.
Sphenodiscus	An ammonite species used as an index fossil for relative dating of rocks.
Squaloraja	A genus of fossil fishes that was the link between sharks and rays.
Stem group concept	A technique of sorting out phylogenetic relationships by tracing relationships of the ancestor and all its descendants through related groups and graphing them.
Stratigraphic order	Age relationships of one strata to another.
Stratigraphic section	A pictorial way of portraying rock units in a vertical sense arranged by oldest at the bottom to youngest at the top, as seen in the field, similar to a cross section.
Stratigraphy	The field of geology that relates one rock unit or stratum to another and places them in geologic time order.
Strike-slip fault	A type of fault resulting from lateral forces, where the fault is vertical or nearly so, also called transform faults.
Stromatolites	Algal mounds made up of calcareous cyanobacteria, they formed some of the earliest fossils on the planet and still are in existence.
Subduction	The process by which a tectonic plate dives down under another, lighter-density plate, creating a trench.
Supercritical fluid	A substance, based on extremely high temperature and pressure beyond its critical thresholds, that exists as neither a liquid nor a gas and has properties of both.
Suture zone	An area where tectonic plates have been stitched together by shearing and compressive forces.

Term	Definition
Swash zone	The area of a beach where the ocean water rushes up and back, creating turbulent flow.
Synapsids	Animals whose skulls had one temporal (near the eye socket) opening, thought to be ancestral to mammals, and an older term for the vertebrate group of mammal-like reptiles.
Taphonomy	The study of past life, through how that life is buried and fossilized, a branch of paleontology.
Terrane	A tectonic term for a piece of a plate that has broken off from a larger plate and often smashed into another plate, sometimes called suspect or accreted terranes.
Tetrapod	A general term for four-limbed animals.
Thelodus parvidens	An extinct jawless fish that lived during the Silurian and Devonian Periods of the Paleozoic Era.
Therapsids	A major order of reptile-like animals, in the synapsid group, ancestors of mammals.
Thermoluminescence	A numerical age-dating method measuring the accumulated dose of radiation in an object since it was last heated.
Theropod	A sub-order of dinosaurs with hollow bones and three-toed feet.
Thrust fault	A low-angle reverse fault, created by compressive forces related to convergent plate boundaries.
Trace fossils	Fossils created by the actions of an animal or plant, such as impressions of burrows and feeding tubes or footprints.
Transform faults	A type of fault resulting from lateral forces, where the fault is vertical or nearly so, also called strike-slip faults; these faults are often found in the ocean perpendicular to mid-ocean ridges.
Transitional fossils	Fossils that exhibit characteristics of the ancestral form of the species and also newer traits.
Trilobites	An extinct marine arthropod, Class Trilobita, that lived mainly during the Paleozoic Era, Cambrian Period, the main index fossil of the Paleozoic Era.
Triple junction	A type of boundary in plate tectonics where three plates meet, creating a "Y"-shaped boundary.

Term	Definition
Type fossils	Representative specimens utilized to define a species.
Unconformities	A break in the geologic record, leading to the absence of expected rock units caused by erosion or destruction of the intervening layers.
Vertebrates	Animals with internal skeletons.
Viscosity	A measure of the "stickiness" of a magma (melted rock); higher viscosity indicates more silicate minerals and explosive tendencies, while lower viscosity indicates less silica and magma that will flow rather than erupt.
Volatiles	Chemical compounds with low boiling points, such as nitrogen, water, carbon dioxide, ammonia, hydrogen, methane, and sulfur dioxide, generally found in the crust of a planet, on the moon, or in the atmosphere.
Volcanic rocks	Rocks originating from magma, which cool quickly, under the air or water, causing them to not have a crystal texture visible to the naked eye; an example is basalt.
Volcanic sill	A volcanic deposit injected horizontally into surrounding rock, like the sill of a door.
Volcanic traps	Fissures in the Earth from tectonic rifting forces, from which flow voluminous amounts of basaltic lava, creating layers upon layers of lava deposits spreading over a wide area, called large igneous provinces.
Weathering	The process of breaking down a substance, rock, or mineral through freeze-thaw, water, temperature, biological, or chemical means.
Zircon	A mineral composed of zirconium and a silicate, classified among the silicate minerals in the Chemical Classification of Minerals, one of the oldest minerals found so far on the Earth.
Zooids	Individuals living within a colonial organism.

Notes

Introduction

1. The term "dinosaur," when discussed in the text, refers to "non-avian" dinosaurs, those that became extinct at the end of the Mesozoic Era. Avian dinosaurs, the birds, are referred to as such.
2. Dodick, J., and Argamon, S., 2006, Rediscovering the historical methodology of the earth sciences by analyzing scientific communication styles, *in* Manduca, C.A., and Mogk, D.W., eds., Earth and Mind: How Geologists Think and Learn about the Earth: Geological Society of America, Boulder, Colorado, Special Paper 413, p. 105.
3. Intergovernmental Panel on Climate Change, 2013, Climate change 2013: The physical science basis, *in* Stocker, T.F., Qin, D., Plattner, G.K., Tignor, M., Allen, S.K., Boschung, J., Nauels, A., Xia, Y., Bex, V., and Midgley, P.M., eds., Contribution of working group I to the fifth assessment report of the Intergovernmental Panel on Climate Change: Cambridge, UK and New York, NY: Cambridge University Press, .
4. United Nations, Department of Economic and Social Affairs, Population Division, 2017, World population prospects: The 2017 revision, key findings and advance tables, Working Paper ESA/P/WP/248: New York, United Nations.
5. Archer, M.O., Hietala, H., Hartinger, M.D., Plaschke, F., and Angelopoulos, V., 2019, Direct observations of a surface eigenmode of the dayside magnetopause: Nature Communications, v. 10, no. 1, p. 1–11.
6. Schumann, W.O., 1952, Über die strahlungslosen Eigenschwingungen einer leitenden Kugel, die von einer Luftschicht und einer Ionosphärenhülle umgeben ist: Zeitschrift für Naturforschung A, v. 7, no. 2, p. 149–154.
7. Soriano, A., Navarro, E.A., Paul, D.L., Portí, J.A., Morente, J.A., and Craddock, I.J., 2005, Finite difference time domain simulation of the Earth-ionosphere resonant cavity: Schumann resonances: IEEE Transactions on Antennas and Propagation, v. 53, no. 4, p. 1535–1541.
8. First, D., 2003, The music of the sphere: An investigation into asymptotic harmonics, brainwave entrainment and the Earth as a giant bell: Leonardo Music Journal, v. 13, p. 31–37.
9. Sagan, C., 1980, Cosmos: New York, Random House, p. 151.

Chapter 1

1. Leonardo da Vinci, 1888, The Notebooks of Leonardo da Vinci (translated by Richter, J.P.): Milan, n.p.

2. Hansen, J.M., 2009, On the origin of natural history: Steno's modern, but forgotten philosophy of science: Bulletin of the Geological Society of Denmark, v. 203, p. 15.
3. Steno, N., 1667, Elementorum Myologiae Specimen seu "Musculi Descriptio Geometrica" cui accedunt "Canis Carchariae Dissectum Caput" et "Dissectus Piscis ex Canum Genere": Florence, Stellae, 123 p.
4. Scott, M., 2004, Nicolaus Steno (1638–1686): The head of a shark, http://earthobservatory.nasa.gov/Features/Steno/steno3.php.
5. Steno, N., 1669, De solido intra solidum naturaliter contento dissertationis prodromus: Florence, n.p., 78 p.
6. Hansen, 2009, p. 21.
7. Hutton, J., 1785, Concerning the system of the Earth, its duration and stability, a paper read to the Royal Society of Edinburgh, on the 7th of March and 4th of April 1785: Edinburgh, Royal Society of Edinburgh.
8. Hutton, J., 1795, Theory of the Earth with Proofs and Illustrations (2 volumes): Edinburgh, William Creech.
9. Hutton, J., 1899, Theory of the Earth with Proofs and Illustrations (Vol. 3, edited by Sir Archibald Geikie): London, Geological Society.
10. See Steno, N., 1916 [1669], The prodromus of Nicolaus Steno's dissertation, concerning a solid body enclosed by process of nature within a solid (translated by John Garrett Winter, English version with an introduction and explanatory notes): London, Macmillan, p. 173.
11. Carozzi, A.V., 1976, Horace Benedict de Saussure: Geologist or educational reformer?: Journal of Geological Education, v. 24, no. 2, p. 48. Abraham Werner's thoughts on geology were influenced de Saussure, and he speculated a "Neptunian" water source for the origin of the layers, a hypothesis later disproved.
12. Hutton, 1795, Vol. 1, p. 369.
13. Dean, D.R., 1992, James Hutton and the History of Geology: Ithaca, Cornell University Press, p. 221.
14. Hutton, 1899.
15. Dean, D.R., 1975, James Hutton on religion and geology: The unpublished preface to his *Theory of the Earth* (1788): Annals of Science, v. 32, no. 2, p. 189.
16. Ibid., p. 187.
17. Royal Society of Edinburgh, 1826, Transactions, p. 452.
18. Playfair, J., 1802, Illustrations of the Huttonian theory of Earth: Edinburgh, Neill & Co. for Caddell and Davis, London, William Creech.
19. Playfair, J., 1805, Biographical account of the late Dr. James Hutton, F.R.S.: Transactions of the Royal Society of Edinburgh, v. V, part III, p. 39–99.
20. Hutton, 1788, p. 304.
21. Werner, A.G., 1971 [1786], A Short Classification and Description of the Various Rocks (translated with an introduction and notes by Ospovat, A.M.): New York, Hafner, 194 p.
22. Philosophical Magazine, 1817, Memoir of Abraham Gottlob Werner, late professor of mineralogy at Freiberg: Philosophical Magazine, series 1, v. 50, no. 233, p. 182.

23. Werner, A.G., 1805 [1774], A Treatise on the External Characters of Fossils (translated by Weaver, T.): Dublin, M. N. Mahon, 312 p.
24. Guntau, M., 2009, The rise of geology as a science in Germany around 1800, *in* Lewis, C.L.E., and Knell, S.J., eds., The Making of the Geological Society of London: Geological Society of London, Special Publications, v. 317, no. 1, p. 168 n. 12.
25. Schuchert, C., 1916, Correlation and chronology in geology on the basis of paleogeography: Geological Society of America Bulletin, September 1, p. 491.
26. Werner, A.G., 1787, Kurze Klassification und Beschreibung der verschiedenen Gebirgsarten: Dresden, Waltherischen Hofbuchhandlung, 28 p.
27. Greene, M.T., 1985, Geology in the Nineteenth Century: Changing Views of a Changing World (Cornell History of Science Series): Ithaca, Cornell University Press, p. 41.
28. Oldroyd, D.R., 1971, The Vulcanist–Neptunist dispute reconsidered: Journal of Geological Education, v. 19, no. 3, p. 124. The original work is Lehmann, J.G., 1756, Versuch einer Geschichte von Flötz-Gebürgen: Berlin, n.p.
29. Şengör, A.M.C., 2002, On Sir Charles Lyell's alleged distortion of Abraham Gottlob Werner in *Principles of Geology* and its implications for the nature of the scientific enterprise: Journal of Geology, v. 110, no. 3, p. 361.
30. Philosophical Magazine, 1817, p. 187.
31. Werner, 1787.
32. Oldroyd, 1971, p. 125.
33. Pickford, S., 2015, "I have no pleasure in collecting for myself alone": Social authorship, networks of knowledge and Etheldred Benett's Catalogue of the Organic Remains of the County of Wiltshire (1831): Journal of Literature and Science, v. 8, no. 1, p. 73.
34. Benett, E., 1831, A catalogue of the organic remains of the County of Wiltshire: Warminster, Vardy, 9 p.
35. Torrens, H.S., Benamy, E., Daescher, E.B., Spamer, E.E., and Bogan, A.E., 2000, Etheldred Benett of Wiltshire, England, the first lady geologist: Her fossil collection in the Academy of Natural Sciences of Philadelphia, and the rediscovery of lost specimens of Jurassic Trogoniidae (Mullusco: Bivalva) with their soft parts preserved: Proceedings of the Academy of Natural Sciences of Philadelphia, April 14, v. 150, p. 67.
36. Benett, E., Letter to Samuel Woodward, dated 12 April, 1836, Woodward MSS, Norwich Museum, vol. 10, p. 52
37. Torrens, Benamy, Daescher, Spamer, and Bogan, 2000, p. 60.
38. Spamer, E.E., Brogan, A.E., and Torrens, H.S., 1989, Recovery of the Etheldred Benett collection of fossils mostly from the Jurassic-Cretaceous strata of Wiltshire, England, analysis of the taxonomic nomenclature of Benett (1831), and notes and figures of type specimens contained in the collection: Proceedings of the Academy of Natural Sciences of Philadelphia, v. 141, p. 118.
39. Torrens, H.S., 2015, Rock stars: William "Strata" Smith: GSA Today, September, p. 38.
40. Smith's map—and the story of his life—have been documented in Simon Winchester's 2001 book The Map That Changed the World: New York, HarperCollins.

41. Smith, W., 1817, Stratigraphical system of organized fossils with reference to the specimens of the original Geologic Collection in the British Museum explaining their state of preservation and their use in identifying British strata: London, E. Williams, 118 p.
42. Woodward, H.B., 1908, The History of the Geological Society of London: London, Longmans, Green, p. xv.
43. Greenough, G.B., 1820, A geological map of England and Wales: London, Longmans, Hurst, Rees, Orme & Brown, scale 5 nautical miles to 1 inch, 4 sheets.
44. Torrens, H.S., 2001, Timeless order: William Smith (1769–1839) and the search for raw materials 1800–1820: Geological Society of London Special Publications, v. 190, no. 1, p. 78.
45. The Wollaston Medal is awarded by the Geological Society for "researches concerning the mineral structure of the Earth . . . or of the science of Geology in general" and to enable the Council of the Geological Society to reward "the researches of any individual or individuals, of any country, saving only that no member of the Council . . . shall be entitled to receive or partake of such aid or reward." "Wollaston Medal," Geological Society, https://www.geolsoc.org.uk/About/Awards-Grants-and-Bursaries/Society-Awards/Wollaston-Medal.
46. Buckland, W., 1824, Notice on the *Megalosaurus* or great fossil lizard of Stonesfield: Transactions of the Geological Society of London, series 2, no. 1, p. 390–396.
47. Turner, S., Burek, C.V., and Moody, R.T. J., 2010, Forgotten women in an extinct saurian (man's) world, *in* Moody, R.T.J., Buffetetaut, E., Naish, D., and Martill, D.M., eds., Dinosaurs and Other Extinct Saurians: A Historical Perspective: London, Geological Survey, p. 111–153.
48. Owen, R., 1841b, Report on British Fossil Reptiles, Part II, Report of the British Association for the Advancement of Science, 11th Meeting: London, Richard and John E. Taylor, p. 85.
49. Buckland, W., 1823, Reliquiae diluvianae; Or, Observations on the Organic Remains Contained in Caves, Fissures and Diluvial Gravel, and on Other Geological Phenomena, Attesting the Action of an Universal Deluge: London, John Murray, 279 p.
50. Boyer, P.J., 1984, William Buckland, 1784–1855: Scientific institutions, vertebrate palaeontology and Quaternary geology [Ph.D. thesis]: Leicester, University of Leicester, v. 1, p. 220.
51. Ibid., p. 220.
52. Lyell, C., 1830, Principles of Geology (Vol. 1): London: Murray, p. 40.
53. Wilson, L.G., 1998, Lyell, the man and his times, *in* Blundell, D.F., and Scott, A.C., eds., Lyell: The Past Is the Key to the Present: London, Geological Society, p. 21–37.
54. Ibid.
55. Dott, R.H., 1998, Charles Lyell's debt to North America: His lectures and travels from 1841 to 1853: Geological Society of London Special Publications, v. 143, no. 1, p. 53–69.
56. Dean, 1992.
57. Conybeare, W.D., 1830, Letter on Mr. Lyell's Principles of Geology: Philosophical Magazine, new series 8, p. 215–219.

58. Wilson, 1998.
59. Wool, D., 2001, Charles Lyell—"the father of geology"—as a forerunner of modern ecology: Oikos, v. 94, no. 3, p. 385–391.
60. Eiseley, L.C., 1958, Darwin's Century: Evolution and the Men Who Discovered It: Garden City, NY, Doubleday, p. 105; Coleman, W., 1962, Lyell and the "reality" of species, 1830–1833: Isis, p. 326; Bartholomew, M., 1973, Lyell and evolution: An account of Lyell's response to the prospect of an evolutionary ancestry of man: British Journal for the History of Science, v. 26, no. 3, p. 261.
61. Dean, 1992, p. 229.
62. Cope, J.C.W., 2016, Geology of the Dorset Coast (second edition, with contributions from Malcolm Butler; Geologists' Association Guide No. 22): London, Geological Society, 222 p.
63. Owen, R., 1840, A description of a specimen of the *Plesiosaurus macrocephalus, Conybeare*, in the collection of Viscount Cole, MP, DCL, FGS: Transactions of the Geological Society of London, v. 2, no. 3, p. 515–535.
64. Torrens, H.S., 1995, Mary Anning (1799–1847) of Lyme: "The greatest fossilist the world ever knew": British Journal for the History of Science, v. 28, p. 264.
65. Ibid. See Dorset County Council, 2000, Nomination for the Dorset and East Devon Coast for inclusion in the World Heritage List, UNESCO, p. 25–27.
66. Davis, L.E., 2009, Mary Anning of Lyme Regis: 19th century pioneer in British palaeontology: Headwaters, Faculty Journal of the College of Saint Benedict and Saint John's University, v. 26, p. 105–106.
67. Anning, M., 1839, Note on the supposed frontal spine in the genus *Hybodus*: Magazine of Natural History, v. 12, p. 605.
68. De la Beche, H.T., 1848, Obituary notes: Quarterly Journal of the Geological Society of London, p. xxi–cxx.
69. Agassiz, L., and Bettannier, J., 1840, Études sur les Glaciers (Studies on Glaciers): London, Dawsons of Pall Mall.
70. Scott, M., 2018, Louis Agassiz. https://www.strangescience.net/agassiz.htm
71. Taquet, P., 2009, Geology beyond the channel, *in* Lewis, C.L.E, and Knell, S.J., eds., The Making of the Geological Society of London: Geological Society of London Special Publications, v. 317, no. 1, p. 155–162.
72. Gupta, S., Collier, J.S., Palmer-Felgate, A., and Potter, G., 2007, Catastrophic flooding origins of the shelf valley system in the English Channel: Nature, v. 448, July 19, p. 343.
73. Cuvier, G., and Brongniart, A., 1810, Carte geognostique des environs de Paris, scale 1:200,000.
74. Cuvier, G., and Brongniart, A., 1811, Essai sur la géographie minéralogique des environs de Paris, avec une carte géognostique, et des coupes de terrain (2 volumes): Paris, Baudouin.
75. Cuvier, G., 1796, Memoire sur les fossiles des environs de Paris (talk): National Institute of France; Cuvier, G., 1799, Mémoire sur les espèces d'éléphans vivantes et fossiles: Mémoire de l'Academie des Sciences, v. 2, p. 1–32.
76. Lamarck, J.B., 1801. Système des animaux sans vertèbres . . .: Paris, Chez Deterville, 468 p.

77. Cuvier, G., 1812, Researches sur les ossemens fossiles (4 volumes): Paris, Chez Deterville.

Chapter 2

1. Halpern, J.M., 1951, Thomas Jefferson and the geological sciences: Rocks and Minerals, v. 74, p. 601.
2. Narendra, B.L., 1979, Benjamin Silliman and the Peabody Museum: Discovery, v. 14, no. 2, p. 1–29.
3. Frazer, 1888.
4. Agassiz, L., 1866, Geological Sketches (Vol. 2): Boston, Ticknor and Fields, 311 p.
5. Agassiz, L., 1860, [Review of] *On the origin of species*: American Journal of Science and Arts, series 2, July 30, p. 154.
6. Winsor, M.P., 1979, Louis Agassiz and the species question: Studies in the History of Biology, v. 3, p. 89–117.
7. Irmscher, C., 2013, Louis Agassiz: Creator of American Science: New York, Houghton Mifflin Harcourt, p. 4.
8. Menand, L., 2001, Morton, Agassiz, and the origins of scientific racism in the United States: Journal of Blacks in Higher Education, v. 34, p. 110–113.
9. Falcon-Lang, H.J., and Calder, J.H., 2005, Sir William Dawson (1820–1899): A very modern paleobotanist: Atlantic Geology, v. 41, p. 103–114.
10. McCulloch, A.W., 2010, Sir John William Dawson: A profile of a Nova Scotian scientist: Proceedings of the Nova Scotian Institute of Science, v. 45, no. 2, p. 3–4.
11. Eakins, P.R., and Eakins, J.S., 1900, Dawson, Sir John William, *in* Dictionary of Canadian Biography (Vol. 12): Toronto, University of Toronto, p. 1892.
12. Moodie, R.L., 1916, The coal measures Amphibia of North America (No. 238): Carnegie Institution of Washington, plate 9.
13. Davis, W.M., 1915, Biographical memoir of John Wesley Powell: National Academy of Sciences Biographical Memoirs, v. 83, p. 12.
14. Ibid., p. 14.
15. Rabbitt, M.C., McKee, E.D., Hunt, C.B., and Leopold, L.B., 1969, The Colorado River Region and John Wesley Powell (U.S. Geological Survey Professional Paper 669–A): Washington, D.C., U.S. Government Printing Office, p. 3.
16. Brewer, W.H., 1902, John Wesley Powell: American Journal of Science, v. 14, p. 381.
17. Aton, J.M., 1994, John Wesley Powell (Western Writers Series 114): Boise, Boise State University Printing and Graphic Services, p. 12–13.
18. Rabbitt, McKee, Hunt, and Leopold, 1969, p. 6.
19. Powell, J.W., 1875, Exploration of the Colorado River in the West and Its Tributaries; Explored in 1869, 1870, 1871 and 1872, under the Direction of the Secretary of the Smithsonian Institution (Monograph): Washington, D.C., U.S. Government Printing Office, 291 p., 2 plates.
20. Powell, J.W., 1876, Report on the Geology of the Eastern Portion of the Uinta Mountains and a Region of the Country Adjacent Thereto (Monograph): Washington, D.C., U.S. Government Printing Office, 218 p., 8 atlas sheets.

21. Powell, J.W., 1877, Report on the Geological and Geographical Survey of the Rocky Mountain Region (Monograph): Washington, D.C., U.S. Government Printing Office, 19 p., 1 map.
22. Powell, J.W., 1895b, The Exploration of the Colorado River and Its Canyons: New York, Dover, 458 p.
23. Powell, J.W., 1895a, Canyons of the Colorado River: Meadville, Flood & Vincent, 127 p.
24. Aton, 1994, p. 23.
25. Aalto, K.R., 2004, Rock stars: Clarence King (1842–1901): Pioneering geologist of the West: GSA Today, February, p. 18.
26. King, C., 1874, Mountaineering in the Sierra Nevada (fourth edition): New York, Charles Scribner's Sons, p. 131.
27. Six additional volumes written by King'scolleagues were Report of the Geological Exploration of the 40th Parallel, Vol. 2, Descriptive Geology (Hague and Emmons, 1877); Vol. 3, Mining Industry (Hague, 1870); Vol. 4, (Meek, Hall, Whitfield and Ridgway, 1877) Palaeontology Part I (Meek), Palaeontology Part II (Hall and Whitfield), Ornithology Part III (Ridgway); Vol. 5, Botany (Watson, 1871); Vol. 6, Microscopic Petrography (Zirkel, 1876); and Vol. 6, Odontornithes. (Marsh, 1880) .
28. King, C., 1877, Catastrophism and evolution: American Naturalist, v. 11, n. 8, p. 449–470.
29. Aalto, 2004, p. 19.
30. Sandweiss, M.A., 2009, Passing Strange: A Gilded Age Tale of Love and Deception across the Color Line: New York, Penguin, p. 359.
31. Arnold, L., 1999, Becoming a geologist: Florence Bascom in Wisconsin, 1874–1887: Earth Science History, v. 18, no. 2, p. 159–179.
32. Schneiderman, J.S., 1997, Rock stars: A life of firsts: Florence Bascom: GSA Today, July, p. 8.
33. Bascom, F., Clark, W.B., Darton, N.H., Knapp, G.N., Kuemmel, H.B., Miller, B.L., and Salisbury, R.D., 1909, Philadelphia folio, Norristown, Germantown, Chester and Philadelphia, Pennsylvania–New Jersey–Delaware: U.S. Geological Survey Folios of the Geologic Atlas 162.
34. Rosenberg, G.D., 2009, Introduction: The revolution in geology from the Renaissance to the Enlightenment, *in* Rosenberg, G.D., ed., The Revolution in Geology from the Renaissance to the Enlightenment: Geological Society of America Memoir 203, p. 2, doi: 10.1130/2009.1203(00).
35. Dean, 1975, p. 191.

Chapter 3

1. Cohen, K.M., Finney, S.C., Gibbard, P.L., and Fan, J.-X. 2013; updated 2020, The ICS International Chronostratigraphic Chart. Episodes 36: 199–204. http://www.stratigraphy.org/ICSchart/ChronostratChart2020-03.pdf.
2. Scott, G.R., and Cobban, W.A., 1965, Geologic and biostratigraphic map of the Pierre Shale between Jarre Creek and Loveland, Colorado: U.S. Geological Survey Miscellaneous Geologic Investigation Map I-439, scale 1:48,000.

3. Gould, 1989.
4. Werner, 1787.
5. Ospovat, A., 1969, Reflections on A.G. Werner's "Kurze Klassifikation," *in* Schneer, C.J., ed., Toward a History of Geology: Cambridge, Massachusetts, MIT Press, p. 251.
6. Ospovat, A., 1960, Abraham Gottlob Werner and his influence on mineralogy and geology [Ph.D. thesis]: Norman, University of Oklahoma Graduate College, p. 165.
7. Greene, 1985, p. 39.
8. Walsh, S.L., 2008. The Neogene: origin, adoption, evolution, and controversy: Earth-Science Reviews, v. 89, no. 1–2, p. 42–72.
9. Rudwick, M.J.S., 2005, Bursting the Limits of Time: The Reconstruction of Geohistory in the Age of Revolution: Chicago, University of Chicago Press, p. 93.
10. Salvador, A., ed., 2013, Internationalstratigraphic Guide (second edition): Boulder, International Union of Geological Sciences and Geological Society of America, ch. 10, p. 3.
11. Ibid.
12. Thackery, J.C., 1976, The Murchison–Sedgwick controversy: Journal of the Geological Society, v. 132, p. 367–372.
13. Berry, 1987, p. 86.
14. Lapworth, 1879.
15. Berry, 1987, p. 97.
16. Ibid, p. 99.
17. Ross, R.J., Jr., 1984, The Ordovician System, progress and problems: Annual Review of Earth and Planetary Science, v. 12, p. 309.
18. A note on the historical use of geologic time terms. The time scale of geology has changed and evolved since its inception; therefore, in a historical context the terms lower, middle, and upper (not capitalized) are used when discussing the development of the time scale.
19. Murchison, 1839, p. 11.
20. Ibid., p. 579.
21. The use of the term 'Upper'is capitalized because it is referring to a formal rock unit, see endnote 54.
22. Murchison, 1839, p. 605.
23. Sedgwick, A., and Murchison, R.I., 1839, On the classification of the older stratified rocks in Devonshire and Cornwall: London and Edinburgh Philosophical Magazine and Journal of Science, series 3, v. 14, no. 89, p. 241–260.
24. Bate, D.G., 2010. Sir Henry Thomas De la Beche and the founding of the British Geological Survey: Mercian Geologist, v. 17, no. 3, p. 162.
25. De la Beche, H., 1839, Report on the Geology of Cornwall, Devon, and West Somerset: London, Longman, Orme, Brown, Green, and Longmans, p. 40.
26. Rudwick, M.J.S., 1988, The Great Devonian Controversy: The Shaping of Scientific Knowledge among Gentlemanly Specialists: Chicago, University of Chicago Press, p. 280.
27. Sedgwick and Murchison, 1839, p. 254.

28. Sharpe, T., and McCartney, P.J., 1998, The papers of H.T. De la Beche (1796–1855), Geological Series 17 (short summary letter, not complete): National Museum of Wales, Cardiff, p. 69–70.
29. Barclay, W.J., 2005, Introduction to the Old Red Sandstone of Great Britain, *in* Barclay, W.J., Browne, M.A.E., McMillan, A.A., Pickett, E.A., Stone, P., and Wilby, P.R., eds., The Old Red Sandstone of Great Britain (Geological Conservation Review Series 31): Peterborough, Joint Nature Conservation Committee, illustrations, A4, p. 11.
30. Ibid., p. 13.
31. Conybeare and Phillips, 1822, p. 233.
32. Barclay, 2005, illustrations, A4, p. 11–13.
33. Conybeare and Phillips, 1822, p. 334.
34. Williams, H.S., 1891, Correlation Papers: Devonian and Carboniferous (U.S. Geological Survey Bulletin 80): Washington, D.C., U.S. Government Printing Office, p. 136.
35. Berry, 1987, p. 101–102.
36. Murchison, 1841.
37. Benton, M.J., Sennikov, A.G., and Newell, A.J., 2010, Murchison's first sighting of the Permian at Vzackniki in 1841: Proceedings of the Geologists' Association, v. 121, p. 317–318, .
38. Alberti, von, 1834.
39. Smith, 1816–1819.
40. Conybeare and Phillips, 1822.
41. Phillips, J., 1860, Life on Earth: Its Origin and Succession: Cambridge, Macmillan, p. 51.
42. Stuart, S., 1972, Biography Luis W. Alvarez, *in* Nobel Lectures, Physics, 1963–1970: Amsterdam, Elsevier, p. 291-292.
43. Alvarez, L.W., Alvarez, W., Asaro, F., and Michel, H.V., 1980, Extraterrestrial cause of the Cretaceous-Tertiary extinction: Science, v. 208, no. 4448, p. 1095–1108.
44. Vacarri, 2006.
45. Cuvier, G., and Brongniart, A., 1822, Description géologique des environs de Paris: Paris, Chez G. Dufour et E. D'Ocagne, 428 p.
46. Lyell, 1833.
47. Naumann, 1866, p. 8.
48. Hörnes, 1853, p. 808.
49. Desnoyers, 1829.
50. Reboul, H.P.I., 1833, Géologie de la période Quaternaire: Paris, F.G. Levrault, p. 1–2.
51. Aubry, M-P., Berggren, W.A., Van Couvering, J., McGowran, B., Pillans, B., and Hilgen, F., 2005, Quaternary: Status, rank, definition, and survival: Episodes, v. 28, no. 2, p. 118.
52. Schuchert, C., 1910, Paleogeography of North America: Geological Survey of America Bulletin, February 5, v. 20, p. 513.
53. Phillips, 1860, p. 51.
54. Ibid., p. 66, fig. 4.

55. “Peroxidated” is an old term used to describe deep ocean sediments that have been buried and undergo chemical changes including crystallization of the hardest minerals and metals being reduced. See Section E: Geology and Geography, 1885, Proceedings of the American Academy of Science, 33rd meeting, Philadelphia, September 1884: Salem, Massachusetts, Salem Press, p. 437.
56. A note on the modern use of geologic time terms. The following geochronological (time) terms are used to describe events in the geologic record in a period or other subdivision of time: early, middle, and late (all lower case). For discussion of geologic units of time for specific rocks, rock units, or stratigraphic layers where the periods or subdivisions are identified as such, according to the International Commission on Stratigraphy, the following chronostratigraphic (time-rock) terms are used to describe a period or other formal time division: Lower, Middle, and Upper. When capitalized, “Middle” refers to chronostratigraphic terms; when lower case “middle” refers to geochronological terms. See Haile, N.S., 1987, Time and age in geology: the use of upper/lower, late/early in stratigraphic nomenclature: Marine and Petroleum Geology, v. 4, no. 3, p. 255.

Chapter 4

1. Williams, H.S., 1893, Studies for students, the elements of the geologic time scale: Journal of Geology, v. 1, p. 294.
2. Valley, J.W., et al., 2014, Hadean age for a post-magma-ocean zircon confirmed by atom-probe tomography: Nature Geoscience, v. 7, p. 219–223.
3. O’Neill, J., Boyet, M, Carleson, R.W., and Paquette, J.L., 2013. Half a billion years of reworking the Hadean mafic crust to produce the Nuvvuagittuq Eoarchean felsic crust: Earth and Planetary Science Letters, v. 379, p. 13–25.
4. Lowrie, W., 2020, Fundamentals of Geophysics (third edition): London, Cambridge University Press, p. 211.
5. Becquerel, H.A., 1896, Sur les radiations émises par phosphorescence: Comptes Rendus des Séances de l’Académie des Sciences, v. 122, p. 420–421.
6. Friedlander, G., Kennedy, J.W., and Macias, E.S., 1981, Nuclear and Radiochemistry: New York, Wiley, p. 2.
7. Froman, N., 1996, Marie and Pierre Curie and the discovery of polonium and radium (originally delivered as a lecture at the Royal Swedish Academy of Sciences, Stockholm, February 28, translated by Marshall-Lundén. N.): https://www.nobelprize.org/nobel_prizes/themes/physics/curie/, p. 6.
8. Ibid, p. 10.
9. Soddy, F., 1923, The origins of the conception of isotopes: Scientific Monthly, v. 17, no. 4, p. 305–317.
10. The Selection Committee for the Nobel Prize in 1921 decided that none of the nominees met the criteria outlined in the will of Alfred Nobel, thus no prizes were awarded that year; however, Soddy did receive his Nobel the following year. https://www.nobelprize.org/prizes/chemistry/1921/summary/.

11. Rutherford, E., and Soddy, F., 1902, The cause and nature of radioactivity: London, Edinburgh, and Dublin Philosophical Magazine and Journal of Science, v.4, 6th series, p. 370–396 (Part I), and 569–585 (Part II).
12. Rutherford, E., 1904, Radio-activity, *in* Neville, F.H., and Whetham, W.C.D., eds., Cambridge Physical Series: London, Cambridge University Press, p. 4.
13. Sustainability of semi-arid hydrology and riparian areas (SAHRA), University of Arizona, http://web.sahra.arizona.edu/programs/isotopes/hydrogen.html (accessed June 15, 2017.
14. International Atomic Energy Agency, 1992, Statistical treatment of environmental isotope data in precipitation (revised edition), Technical Reports Series No. 331: Vienna, IAEA, p. 34.
15. Kresic, N., 2006, Hydrogeology and Groundwater Modeling (second edition): Boca Raton, CRC Press, p. 393.
16. Gould, S.J., 1989, Wonderful Life: The Burgess Shale and the Nature of History: New York, Norton, p. 54.
17. McPhee, J., 1980, Basin and Range: New York, Farrar, Strauss & Giroux, p. 127.
18. Alden, A., 2013, A New "Golden Spike" Monument in Colorado Marks Geologic Time, KQED, https://www.kqed.org/science/10292/a-new-golden-spike-monument-in-colorado-marks-geologic-time.
19. Crutzen, P.J., and Stoermer, E.F., 2000, The "Anthropocene": Global Change Newsletter, v. 41, May, p. 17–18, http://www.igbp.net/download/18.316f18321323470177580001401/1376383088452/NL41.pdf.
20. Chen, A., 2014, Rocks made from plastic found on Hawaiian beach, Science, http://www.sciencemag.org/news/2014/06/rocks-made-plastic-found-hawaiian-beach.
21. Subramanian, M., 2019, Anthropocene now: influential panel votes to recognize Earth's new epoch: Nature, v. 21, p. 2019.
22. Willis, B., 1912, Index to the stratigraphy of North America: U.S. Geological Survey Professional Paper No. 71, 894 p., with Map 1:5,000,000 by Willis, B., and Stose, G.
23. Frazer, P., 1890, The American Association for the Advancement of Science, of 1890: American Naturalist, Proceedings of Scientific Societies, v. 24, part 2, p. 987.
24. King, P.B., and Beikman, H.M., 1974, Explanatory text to accompany the geologic map of the United States: U.S. Geological Survey Professional Paper 901, Washington, D.C., U.S. Government Printing Office, p. 25–26.
25. Frazer, P., 1888, A short history of the origin and acts of the International Congress of Geologists, and of their American Committee delegation to it: American Geologist, January, v. 1, no. 1, p. 8.
26. Ibid., p. 99.
27. Ibid., p. 100.
28. Frazer, 1890, p. 987.
29. Maclure, W., Tanner, H.S., and Lewis, S., 1809, Observations on the geology of the United States, explanatory of a geological map: American Philosophical Society Transactions, v. 6, 411 p. and map.
30. King and Biekman, 1974.

31. Maclure, W., 1817, Observations on the Geology of the United States of America, with some remarks on the effect produced on the nature and fertility of soils, by the decomposition of the different classes of rocks; and an application to the fertility of every State in the Union, in reference to the accompanying geologic map: American Philosophical Society Transactions, Memoire, 2 plates, 127 p.
32. Stose, G.W., and Ljungstedt, O.A., 1932, Geologic map of the United States: U.S. Geological Survey, scale 1:2,500,000.
33. King, P.B., Beikman, H.M., and Edmonston, G.J., 1974, Geologic map of the United States (exclusive of Alaska and Hawaii): US Geological Survey, Scale 1: 2,500,000, 2 Plates: 40.75 × 52.50 inches and 40.63 × 52.49 inches; Legend.
34. Cuvier, G., and Brongniart, A., 1811, Essai sur la géographie minéralogique des environs de Paris, avec une carte géognostique, et des coupes de terrain (2 volumes): Paris, Baudouin, 271 Pl. I: Fig. 1 [of 11].
35. Klauk, E., n.d., Geology and physiography of the Crow Reservation, integrating research and education, impact of resource development on American Indian Lands, https://serc.carleton.edu/research_education/nativelands/crow/geology.html.
36. Gould, S.J., 1987, Time's Arrow, Time's Cycle: Cambridge, Mass., Harvard University Press, p. 10, 14.

Chapter 5

1. Popper, K.R., 1959, Back to the pre-Socratics: Proceedings of the Aristotelian Society, v. 59, no. 1, p. 8.
2. Romm, J., 1994, A new forerunner for continental drift: Nature, v. 367, p. 407.
3. Lamarck, J. B., 1802, Hydrogéologie ou recherches sur l'influence qu'ont les eaux sur la surface du globe terrestre; sur les causes de l'existence du bassin des mers, de son déplacement et de son transport successif sur les différens points de la surface de ce globe; enfin sur les changemens que les corps vivans exercent sur la nature et l'état de cette surface: Paris, An X, 268 p.
4. Carozzi, A.V., 1964, Lamarck's theory of the Earth: Hydrogeologie: Isis, v. 55, no. 3, p. 293–307.
5. Snider-Pellegrini, A., 1858, La création et ses mystères dévoilés; ouvrage où l'on expose clairement la nature de tous les êtres, les éléments dont ils sont composés et leurs rapports avec le globe et les astres, la nature et la situation du feu du soleil, l'origine de l'Amérique, et de ses habitants primitifs, la formation forcée de nouvelles planètes, l'origine des langues et les causes de la variété des physionomies, le compte courant de l'homme avec la terre, etc.: Paris, A. Franck, 487 p., plates no. 9 and 10..
6. Oreskes, N., 2003, From continental drift to plate tectonics, *in* Oreskes, N., ed., Plate Tectonics: An Insider's History to the Modern Theory of the Earth: Boulder, Westview Press, p. 5.

7. Suess, E., 1885, Das antlitz der Erde, Bd. 1: Vienna, F. Tempsky; Leipzig, G. Freytag, 778 p.
8. Suess, E., 1909, The Face of the Earth (translated by Sollas, H.B.C.): Oxford, Clarendon Press, 673 p.
9. Arber, E.A.N., 1905, Catalogue of the Fossil Plants of the Glossopteris Flora in the Department of Geology, British Museum (Natural History); Being a Monograph of the Permo-Carboniferous Flora of India and the Southern Hemisphere: Hertford, Stephen Austin, 255 p.
10. Scott, R.F., 1913, Scott's Last Expedition (Vol. 1): New York, Dodd, Mead and Co, p. 388–389.
11. Stillwell, J.D., and Long, J.A., 2011, Frozen in Time: Prehistorical Life in Antarctica: Clayton, Australia, CSIRO, 248 p.
12. Seward, A.C., 1914, Antarctic fossil plants: British Museum (Natural History) report, British Antarctic ("Terra Nova") report, 1910: Natural History Report, Geological Studies: London, Printed by order of the Trustees of the British Museum, 1914–1964, v. 1, p. 1–49.
13. Ibid., p. 44.
14. Greene, M.T., 2015, Alfred Wegener: Science, Exploration and the Theory of Continental Drift: Baltimore, Johns Hopkins University Press, p. 327.
15. Wegener, A., 1912, Die Herausbildung der Grossformen der Erdrinde (Kontinente und Ozeane), auf geophysikalischer Grundlage: Petermanns Geographische Mitteilungen, v. 63, p.185–195.
16. Oreskes, N., 1999, The Rejection of Continental Drift: New York, Oxford University Press, 420 p.
17. Oreskes, 2003, p. 7.
18. Geodesy is the science of measuring the Earth's precise geometric shape, orientation in space, and gravity field. National Oceanographic and Atmospheric Administration, https://oceanservice.noaa.gov/facts/geodesy.html.
19. Green, 2015, p. 545.
20. Wegener, 1966 [1915].
21. Brink, A.S., 1951, On the genus *Lystrosaurus* Cope: Transactions of the Royal Society of South Africa, v. 33, no. 1, p. 107–120.
22. Williston, S.W., 1925, The Osteology of the Reptiles (Society for the Study of Amphibians and Reptiles), W.K. Gregory, ed.: Cambridge, Massachusetts, Harvard University Press, 324 p.
23. Seeley, H.G., 1895, Researches on the structure, organization, and classification of the fossil Reptilia, Part IX, Section 5, On the skeleton in the New Cynodontia from the Karroo Rocks: Philosophical Transactions of the Royal Society of London, B 186, p. 59.
24. Ibid., p. 63.
25. Piñeiro, G., Ferigolo, J., Menechel, M., and Laurin, M., 2012, The oldest known amniotic embryos suggest viviparity in mesosaurs: Historical Biology, v. 24, no. 6, p. 620–630.

26. Colbert, E.H., 1973, Wandering Lands and Animals: New York, Dutton, 323 p.
27. Holmes, A., 1941, Principles of Physical Geology: London, Thomas Nelson, p. 505.
28. Du Toit, A., 1937, Our Wandering Continents: An Hypothesis on Continental Drifting: Edinburgh, Oliver and Boyd, 366 p.
29. Theberge, A.E., 2012, The myth of the telegraphic plateau: Hydro International, May 10, 2 p., https://www.hydro-international.com/content/article/the-myth-of-the-telegraphic-plateau.
30. Wertenbaker, W., 2000, Rock stars: William Maurice Ewing: Pioneer explorer of the ocean floor and architect of Lamont: GSA Today, October, p. 28–29.
31. Heezen, B.C., and Tharp, M., 1977, World ocean floor panorama (map, painted by Berann, H.): Columbia University, Office of Naval Research.
32. Bullard, E.C., Maxwell, A.E, and Revelle, R., 1956, Heat flow through the deep ocean floor: Advances in Geophysics, v. 3, p. 153–181.
33. Von Herzen, R.P., 1959, Heat-flow values from the South-Eastern Pacific: Nature, v. 183, p. 882–883.
34. Von Herzen, R.P., and Uyeda, S., 1963, Heat flow through the Eastern Pacific ocean floor: Journal of Geophysical Research, v. 68, no. 14, p. 4219–4450.
35. Hess, H.H., 1962, History of ocean basins: Petrologic Studies, November, p. 590–620.
36. Dietz, R.S., 1961, Continent and ocean basin evolution by spreading of the sea floor: Nature, v. 190, p. 854–857.
37. Picard, M.D., 1989, Harry Hammond Hess and the theory of sea-floor spreading: Journal of Geological Education, v. 37, p. 346–349.
38. Dietz, R.S., 1968, Reply [to "Arthur Holmes: Originator of Spreading Ocean Floor Hypothesis"]: Journal of Geophysical Research, v. 73, p. 6567.
39. Menard, H.W., 1955, Deformation of the northeastern Pacific Basin and the west coast of North America: Geological Society of America Bulletin, v. 66, p. 1149–1198.
40. Vine, F.J., and Matthews, D.H., 1963, Magnetic anomalies over oceanic ridges: Nature, v. 199, p. 947–949.
41. Mason, R.G., 1958, A magnetic survey off the west coast of the United States between latitudes 32° and 36° N. and longitudes 121° and 128° W.: Geophysical Journal of the Royal Astronomical Society, v. 1, p. 320–329.
42. Runcorn, S.K., 1959, Rock magnetism: The magnetization of ancient rocks bears on the questions of polar wandering and continental drift: Science, v. 129, no. 3355, p. 1002–1012.
43. Wilson J.T., 1965, A new class of faults and their bearing on continental drift: Nature, v. 207, p. 343–347.
44. Wilson, J.T., 1963, A possible origin of the Hawaiian Islands: Canadian Journal of Physics, v. 41, no. 6, p. 863–870.
45. Quennell, A.M., 1958, The structural and geomorphic evolution of the Dead Sea Rift: Journal of the Geological Society of London, v. 114, pp. 1–24.
46. Wilson, J.T., 1968, Static or mobile earth: The current scientific revolution: Proceedings of the American Philosophical Society, v. 112, no. 5, p. 312.

47. Bullard, E., Everett, J.E., and Smith, A.G., 1965, The fit of the continents around the Atlantic: Philosophical Transactions of the Royal Society of London, Series A, Mathematical and Physical Sciences, v. 258, no 1088, p. 41–51.
48. Bullard, E.C., 1975, The emergence of plate tectonics: a personal view: Annual Review of Earth and Planetary Sciences, v. 3, no. 1, p. 21.
49. Krill, A., 2011, Fixists vs. mobilists in the geological contest of the century, 1844–1969: Trondheim, Fixists.com.
50. Gordon, R.G., 2000, Diffuse oceanic plate boundaries: Strain rates, vertically averaged rheology, and comparison with narrow plate boundaries and stable plate interiors, *in* Richards, M., Gordon, R.G., and Van Der Hilst, R.D., eds., History and Dynamics of Global Plate Margins: Geophysical Monographs 121, American Geophysical Union, p. 143–159.
51. Forsyth, D., and Uyeda, S., 1975, On the relative importance of the driving forces of plate motion: Geophysical Journal International, v. 43, no. 1, p. 163–200.

Chapter 6

1. Richter, C.F., 1935, An instrumental earthquake magnitude scale: Bulletin of the Seismological Society of America, v. 25, no. 1, p. 1–32.
2. Hanks, T.C., and Kanamori, H., 1979, A moment magnitude scale: Journal of Geophysical Research, Solid Earth, v. 84, no. B5, p. 2348–2350.
3. For more on earthquake magnitude as measured by the moment magnitude scale, see https://sos.noaa.gov/datasets/earthquake-magnitude-perspective/.
4. https://www.usgs.gov/natural-hazards/earthquake-hazards/national-earthquake-information-center-neic.
5. Vigil, J.F., n.d., This dynamic planet: The US Geological Survey, the Smithsonian Institution, and the US Naval Research Laboratory, https://pubs.usgs.gov/gip/earthq1/plate.html.
6. Frank, F.C., 1968, Curvature of island arcs: Nature, v. 220, no. 5165, p. 363.
7. De Vries, M.V.W., Bingham, R.G., and Hein, A.S., 2017, A new volcanic province: An inventory of subglacial volcanoes in West Antarctica: Geological Society of London Special Publications, v. 461, no. 1, p. 231–248.
8. Jamieson, A.J., Malkocs, T., Piertney, S.B., Fujii, T., and Zhang, Z., 2017, Bioaccumulation of persistent organic pollutants in the deepest ocean fauna: Nature Ecology and Evolution, v. 1, article 0051.
9. Cavallo, E.A., Powell, A., and Becerra, O., 2010, Estimating the direct economic damage of the earthquake in Haiti, IDB Working Paper IBD-WP-163: Washington, D.C., Inter-American Development Bank.
10. Torsvik, T.H., Doubrovine, P.V., Steinberger, B., Gaina, C., Spakman, W., and Domeier, M., 2017, Pacific plate motion change caused the Hawaiian-Emperor Bend, Nature Communications, v. 8, no. 15660, 12 p.

11. Huang, H.H., Lin, F.C., Schmandt, B., Farrell, J., Smith, R.B., and Tsai, V.C., 2015, The Yellowstone magmatic system from the mantle plume to the upper crust: Science, v. 348, no. 6236, p. 773–776.
12. Smith, R.B., Jordan, M., Steinberger, B., Puskas, C.M., Farrell, J., Waite, G.P., Husen, S., Chang, W.L., and O'Connell, R., 2009, Geodynamics of the Yellowstone hotspot and mantle plume: Seismic and GPS imaging, kinematics, and mantle flow: Journal of Volcanology and Geothermal Research, v. 188, p. 26–56.
13. Mastin, L.G., Van Eaton, A.R., and Lowenstern, J.B., 2014, Modeling ash fall distribution from a Yellowstone supereruption: Geochemistry, Geophysics, Geosystems, v. 15, no. 8, p. 3459–3475.
14. Gordon, 2000, p 143.
15. Pliny the Younger, 2016, Letters of Pliny (Bosanquet, F.C.T., ed.; Melmoth, W., trans.): Project Gutenberg Ebook, Letter LXV.
16. Ibid., Letter LXVI.
17. UNESCO, 2015, Tsunami warning and mitigation systems to protect coastal communities, Indian Ocean Tsunami Warning and Mitigation System (IOTWS) 2005–2015, Fact Sheet, May: Paris, Intergovernmental Oceanographic Commission.
18. Breuer, D., 2011, Stagnant lid convection, *in* Gargaud, M., et al., eds., Encyclopedia of Astrobiology: Berlin, Springer, p. 125.
19. Nagel, T.J., Hoffmann, J.E., and Münker, C., 2012, Generation of Eoarchean tonalite-trondhjemite-granodiorite series from thickened mafic arc crust: Geology, v. 40, no. 4, p. 375–378.
20. Johnson, T.E., Brown, M., Gardiner, N.J., Kirkland, C.L., and Smithies, R.H., 2017, Earth's first stable continents did not form by subduction: Nature, v. 543, March 9, p. 239–242.
21. Hatcher, R.P., Jr., 2010, The Appalachian orogeny: A brief summary, *in* Tollo, R.P., Bartholomew, M.J., Hibbard, J.P., and Karabinos, P.M., eds., From Rodinia to Pangea: The Lithotectonic Record of the Appalachian Region: Geological Society of America Memoire 206, p. 1–20.
22. Scotese, C.R., 2009, Late Proterozoic plate tectonics and palaeogeography: A tale of two supercontinents, Rodinia and Pannotia: Geological Society of London Special Publications, v. 326, no. 1, p. 67–83.

Chapter 7

1. Raup, D.M., 1991, Extinction: Bad Genes or Bad Luck?: New York, Norton, p. 4.
2. Tashiro, T., Ishida, A., Hori, M., Igisu, M., Koike, M., Méjean, P., Takahata, N., Sano, Y., and Komiya, T., 2017, Early trace of life from 3.95 Ga sedimentary rocks in Labrador, Canada: Nature, v. 549, no. 7673, p. 516–518.
3. Javaux, E.J., and Marshall, C.P., 2005, Tracking the record of early life: Carnets de Géologie, M02, Abstract05.
4. Bowler, P.J., 2002, Charles Darwin: The Man and His Influence: London, Cambridge University Press, 264 p.

5. Darwin, C.R., 1835, Letter 282, Darwin Correspondence Project, https://www.darwinproject.ac.uk/letter/DCP-LETT-282.xml, accessed February 25, 2020.
6. Herbert, S., 2005, Charles Darwin, Geologist: Ithaca, Cornell University Press, p. xv.
7. Eiseley, L.C., 1959, Charles Lyell: Scientific American, v. 201, no. 2, p. 98.
8. Wilkins, J.S., 2009, Species: A History of the Idea: Berkeley, University of California Press, 320 p.
9. Gould, J., 1837, Remarks on a group of ground finches from Mr. Darwin's collection, with characters of the new species: Proceedings of the Zoological Society of London, v. 5, p. 4–7.
10. Darwin, C.R., 1859, On the Origin of Species by Means of Natural Selection, or, the Preservation of Favoured Races in the Struggle for Life: London, John Murray, 564 p.
11. Darwin, C.R., 1860, On the Origin of Species by Means of Natural Selection, or the Preservation of Favoured Races in the Struggle for Life: New York, Appleton, 474 p.
12. Penny, D., 2011, Darwin's theory of descent with modification, versus the biblical tree of life: PLOS Biology, v. 9, no. 7, p. 1.
13. Darwin, 1860, p. 119.
14. Lovejoy, A.O., 2001, The Great Chain of Being: A Study of the History of an Idea: Abingdon, Routledge, p. 21.
15. Lombardo, P.A., 1982, The great chain of being and the limits to the Machiavellian cosmos: Journal of Thought, v. 17, no. 1, p. 39.
16. Switek, B., 2011, Written in Stone: London, Icon Books, p. 20.
17. Matthews, W.H., III, 1969, The Geologic Story of the Palo Duro Canyon: Austin, Bureau of Economic Geology, https://www.gutenberg.org/files/52179/52179-h/52179-h.htm#fig6, figure 6.
18. Mayr, E., 1977, Darwin and natural selection: How Darwin may have discovered his highly unconventional theory: American Scientist, v. 65, no. 3, p. 321.
19. Prothero, D.R., 2013, Bringing Fossils to Life: An Introduction to Paleobiology: New York, Columbia University Press, p. 7.
20. Leidy, J., 1865, Cretaceous Reptiles of the United States (Smithsonian Contributions to Knowledge 192): New York, Appleton, p. 76–77.
21. Everhart, M., 2002, The tale of a tail: Or how easy it was to put the head on the wrong end of *Elasmosaurus platyurus* Cope 1868, http://oceansofkansas.com/tale-tail.html, accessed March 29, 2020.
22. Leidy J., 1870, Remarks on *Elasmosaurus platyurus*: Proceedings of the Academy of Natural Sciences of Philadelphia, v. 22, p. 9–10.
23. Davidson, J.P., 2002, Bonehead mistakes: The background in scientific literature and illustrations for Edward Drinker Cope's first restoration of *Elasmosaurus platyurus*: Proceedings of the Academy of Natural Sciences of Philadelphia, v. 152, no. 1, p. 215–240.
24. Huntington, T., 1998, The great feud: American History, v. 33, no. 3, p. 17–18.
25. Marsh, O.C., 1875, Odontornithes, or birds with teeth: American Naturalist, v. 9, no. 12, p. 625–631.
26. Marsh, O.C., 1883, Birds with Teeth, 3rd Annual Report of the Secretary of the Interior: Washington, D.C., U.S. Government Printing Office, v. 3, p. 43–88.

27. Miko, I., 2008, Gregor Mendel and the principles of inheritance: Nature Education, v. 1, no. 1, p. 134–137.
28. Johannsen, W., 1909, Elemente der exakten Erblichkeitslehre [Elements of the Exact Theory of Inheritance]: Jena, Gustav Fischer.
29. Dahm, R., 2008, Discovering DNA: Friedrich Miescher and the early years of nucleic acid research: Human Genetics, v. 122, p. 565–581, doi: 10.1007/s00439-007-0433-0.
30. Avery, O.T., MacLeod, C.M., and McCarty, M., 1944, Studies on the chemical nature of the substance inducing transformation of pneumococcal types: Induction of transformation by a desoxyribonucleic acid fraction isolated from pneumococcus type III: Journal of Experimental Medicine, v. 79, no. 2, p. 137–158.
31. Watson, J.D., and Crick, F.H.C., 1953b, The structure of DNA: Cold Spring Harbor Symposia on Quantitative Biology, v. 18, p. 123–131.
32. Watson, J.D., and Crick, F.H.C., 1953a, Molecular structure of nucleic acids: Nature, v. 171, no. 4356, p. 737–738.
33. National Academy of Sciences, 1999, Science and Creationism: A View from the National Academy of Sciences (second edition): Washington, D.C., National Academy Press, p. 3–4.
34. International Human Genome Sequencing Consortium, 2001, Initial sequencing and analysis of the human genome: Nature, v. 409, February 15, p. 860–921, doi: 10.1038/35057062.
35. Morlon, H., Parsons, T.L., and Plotkin, J.B., 2011, Reconciling molecular phylogenies with the fossil record: Proceedings of the National Academy of Sciences, v. 108, no. 39, p. 16327–16332, https://doi.org/10.1073/pnas.1102543108.
36. Hunter, P., 2013, Molecular fossils probe life's origins: European Molecular Biology Organization Reports, v. 14, no. 11, p. 964–967, https://dx.doi.org/10.1038%2Fembor.2013.162.
37. Singh, V., and Singh, K., 2018, Modern synthesis, *in* Vonk, J., and Shackelford, T.K., eds., Encyclopedia of Animal Cognition and Behavior: New York, Springer, p. 1–5.
38. Huxley, J., 1942, Evolution: The Modern Synthesis: London, George Allen & Unwin, 645 p.
39. Simpson, G.G., 1944, Tempo and Mode in Evolution: New York, Columbia University Press, 237 p.
40. Olson, E.C., 1991, George Gaylord Simpson: June 16, 1902–October 6, 1984: National Academy of Sciences Biographical Memoirs, v. 60, p. 332, https://doi.org/10.17226/6061.
41. Pigliucci, M., and Muller, G., 2010, Evolution: The Extended Synthesis: Cambridge, Massachusetts, MIT Press, 504 p.
42. Eldredge, N., and Gould, S.J., 1972, Punctuated equilibria: An alternative to phyletic gradualism, in Schopf, T.J.M., ed., Models in Paleobiology: San Francisco, Freeman Cooper, p. 82–115.
43. Gould, S.J., and Eldredge, N., 1977, Punctuated equilibria: The tempo and mode of evolution reconsidered: Paleobiology, v. 3, no. 2, p. 115–151.
44. Gould, S.J., and Eldredge, N., 1993, Punctuated equilibrium comes of age: Nature, v. 366, no. 6452, p. 223–227.

45. Saitta, D., 2003, Stephen Jay Gould: In memoriam: Rethinking Marxism, v. 15, no. 4, p. 445–449.
46. Yoon, C.K., 2002, Stephen Jay Gould, evolution theorist, dies at 60: New York Times, May 21, https://www.nytimes.com/2002/05/21/us/stephen-jay-gould-60-is-dead-enlivened-evolutionary-theory.html.
47. MacFadden, B.J., 2005, Fossil horses: Evidence for evolution: Science, v. 308, p. 1728–1730.
48. Cuvier, G., 1804, Sur les espèces d'animaux dont proviennent les os fossiles: Annales du Muséum Nationale d'Histoire Naturelle. (Paris), tome troisième, p. 276.
49. Owen, R., 1841a, Description of the fossil remains of a mammal (*Hyracotherium leporinum*) and of a bird (*Lithornis vulturinus*) from the London Clay: Transactions of the Geological Society of London, v. 2, no. 1, p. 203–208.
50. Leidy, J., 1847, On the fossil horse of America: Proceedings of the Academy of Natural Sciences of Philadelphia, v. 3, p. 262–266.
51. Mitchill, S.L., 1826, Catalogue of Organic Remains Presented to the New York Lyceum of Natural History: New York, Seymour, p. 7.
52. Guthrie, R.D., 2003, Rapid body size decline in Alaskan Pleistocene horses before extinction: Nature, v. 426, no. 6963, p. 169–171.
53. Leidy, J., and Gibbes, R.W., 1847, On the fossil horse of America: Description of new species of Squalides from the Tertiary Beds of South Carolina: Proceedings of the Academy of Natural Sciences of Philadelphia, v. 3, no. 11, p. 263.
54. Glassman, S., Bolt, E.A., Jr., and Spamer, E.E., 1993, Joseph Leidy and the "Great Inventory of Nature": Proceedings of the Academy of Natural Sciences of Philadelphia, v. 144, p. 1–19.
55. Prothero, D.R., 2016, The Princeton Field Guide to Prehistoric Mammals (Vol. 112): Princeton, Princeton University Press, p. 187.
56. Marsh, O.C., 1876, Notice of new Tertiary mammals, V: American Journal of Science, v. 71, p. 401–404.
57. MacFadden, 2005.
58. MacFadden, B.J., Oviedo, L.H., Seymour, G.M., and Ellis, S., 2012, Fossil horses, orthogenesis, and communicating evolution in museums: Evolution, Education and Outreach, v. 5, p. 30, doi: 10.1007/s12052-012-0394-1.
59. Broo, J., and Mahoney, J., 2015, Chewing on Change: Exploring the Evolution of Horses in Response to Climate Change: Gainesville, University of Florida, p. 4.
60. Millar, C.D., and Lambert, D.M., 2013, Ancient DNA: Towards a million-year-old genome: Nature, v. 499, no. 7456, p. 34–35, doi: 10.1038/nature12263.
61. Cappellini, E., et al., 2019, Early Pleistocene enamel proteome from Dmanisi resolves *Stephanorhinus* phylogeny: Nature, v. 574, no. 7776, p. 103–107.
62. Kettlewell, H.B.D., 1955, Selection experiments on industrial melanism in the Lepidoptera: Heredity, v. 9, p. 323–342.
63. Majerus, M.E.N., 1998, Melanism: Evolution in Action: Oxford, Oxford University Press, 364 p.
64. Shubin, N.H., Daeschler, E.B., and Jenkins, F.A., 2006, The pectoral fin of *Tiktaalik roseae* and the origin of the tetrapod limb: Nature, v. 440, no. 7085, p. 764–771.

65. Cloutier, R., Clement, A.M., Lee, M.S., Noël, R., Béchard, I., Roy, V., and Long, J.A., 2020, Elpistostege and the origin of the vertebrate hand: Nature, v. 579, p. 549–554.
66. Cuvier, G., 1818, Essay on the theory of the Earth: New York, Kirk & Mercein, 431 p.
67. Twitchett, R.J., 2006, The palaeoclimatology, palaeoecology and palaeoenvironmental analysis of mass extinction events: Palaeogeography, Palaeoclimatology, Palaeoecology, v. 232, no. 2–4, p. 190.
68. Burgess, S.D., Muirhead, J.D., and Bowring, S.A., 2017, Initial pulse of Siberian Traps sills as the trigger of the end-Permian mass extinction: Nature Communications, v. 8, no. 1, p. 1.
69. Kielan-Jaworowska, Z., Hurum, J.H., and Lopatin, A.V., 2005, Skull structure in *Catopsbaatar* and the zygomatic ridges in multituberculate mammals: Acta Palaeontologica Polonica, v. 50, no. 3, p. 492; Kielan-Jaworowska, Z., and Hurum, J.H., 2006, Limb posture in early mammals: Sprawling or parasagittal: Acta Palaeontologica Polonica, v. 51, no. 3, p. 397.
70. Van Valen, L., and Sloan, R.E., 1966, The extinction of the multituberculates: Systematic Zoology, v. 15, no. 4, p. 261–278.
71. Hodgskiss, M.S., Crockford, P.W, Peng, Y., Wing, B.A., and Horner, T.J., 2019, A productivity collapse to end Earth's great oxidation: Proceedings of the National Academy of Sciences, v. 116, no. 35, p. 17207.
72. Darroch, S.A., Boag, T.H., Racicot, R.A., Tweedt, S., Mason, S.J., Erwin, D.H., and Laflamme, M., 2016, A mixed Ediacaran-metazoan assemblage from the Zaris Sub-basin, Namibia: Palaeogeography, Palaeoclimatology, Palaeoecology, v. 459, p. 198–208.
73. Stanley, S.M., 2016, Estimates of the magnitudes of major marine mass extinctions in earth history: Proceedings of the National Academy of Sciences, v. 113, no. 42, p. E6325–E6334.
74. Kennett, J.P., and Stott, L.D., 1991, Abrupt deep-sea warming, palaeoceanographic changes and benthic extinctions at the end of the Palaeocene: Nature, v. 353, no. 6341, p. 225–229.
75. Röhl, U., Westerhold, T., Bralower, T.J., and Zachos, J.C., 2007, On the duration of the Paleocene-Eocene thermal maximum (PETM): Geochemistry, Geophysics, Geosystems, v. 8, no. 12, p. 1–13.
76. Gingerich, P.D., 2006, Environment and evolution through the Paleocene–Eocene thermal maximum: Trends in Ecology & Evolution, v. 21, no. 5, p. 246–253.
77. Barnosky, A.D., Carrasco, M.A., and Davis, E.B., 2005, The impact of the species-area relationship on estimates of paleodiversity: PLOS Biology, v. 3, no. 8, p. E266.
78. Bokulich, A., 2018, Using models to correct data: Paleodiversity and the fossil record: Synthese, p. 1–22.
79. Sepkoski, J.J., 1984, A kinetic model of Phanerozoic taxonomic diversity, III, Post-Paleozoic families and mass extinctions: Paleobiology, v. 10, no. 2, p. 246–267.
80. Sepkoski, J.J., 1992, A compendium of fossil marine animal families: Contributions in Biology and Geology, v. 83, p. 1–156.
81. Benton, M.J., 1985, Mass extinction among non-marine tetrapods: Nature, v. 316, no. 6031, p. 811–814; Padian, K., Clemens, W.A., and Valentine, J.W., 1985, Terrestrial

vertebrate diversity: Episodes and insights, *in* Valentine, J.W., ed., Phanerozoic Diversity Patterns: Profiles in Macroevolution: Princeton, Princeton University Press, p. 41–96.

82. Knoll, A.H., Niklas, K.J., and Tiffney, B.H., 1979, Phanerozoic land-plant diversity in North America: Science, v. 206, no. 4425, p. 1400–1402.
83. Rohde, R.A., and Muller, R.A., 2005, Cycles in fossil diversity: Nature, v. 434, no. 7030, p. 209. The figure has been adapted to show geologic time from oldest Phanerozoic Eon to youngest (from left to right).
84. National Ocean Service, 2014, How much of the ocean have we explored? National Oceanographic and Atmospheric Administration, http://oceanservice.noaa.gov/facts/exploration.html.

Chapter 8

1. Reference to dates in this section and beyond, related to the geologic time scale, follow the International Commission on Stratigraphy guidelines. Dates are given in 100s of millions or 1,000s of millions (the latter equivalent to billions) of year. See https://stratigraphy.org/ICSchart/ChronostratChart2020-03.pdf.
2. Plumb, K.A., 1991, New Precambrian time scale: Episodes, v. 14, no. 2, p. 139–140.
3. Valley, J.W., Peck, W.H., King, E.M., and Wilde, S.A., 2002, A cool early Earth: Geology, v. 30, no. 4, p. 351–354; Charnay, B., Le Hir, G., Fluteau, F., Forget, F. and Catling, D.C., 2017, A warm or a cold early Earth? New insights from a 3-D climate-carbon model: Earth and Planetary Science Letters, v. 474, p. 97–109.
4. Bottke, W.F., and Norman, M.D., 2017, The Late Heavy Bombardment: Annual Review of Earth and Planetary Sciences, v. 45, p. 619–647.
5. Compston, W., and Pidgeon, R.T., 1986, Jack Hills, evidence of more very old detrital zircons in Western Australia: Nature, v. 321, no. 6072, p. 766.
6. Bowring, S.A., Williams, I.S., and Compston, W., 1989, 3.96 Ga gneisses from the slave province, Northwest Territories, Canada: Geology, v. 17, no. 11, p. 971–975.
7. Kerr, R.A., 1984, Making the moon from a big splash: Science, v. 226, p. 1060–1062.
8. Lock, S.J., and Stewart, S.T., 2017, The structure of terrestrial bodies: Impact heating, corotation limits, and synestias: Journal of Geophysical Research: Planets, v. 122, no. 5, p. 950–982.
9. Lock, S.J., Stewart, S.T., Petaev, M.I., Leinhardt, Z., Mace, M.T., Jacobsen, S.B., and Cuk, M., 2018, The origin of the moon within a terrestrial synestia: Journal of Geophysical Research: Planets, v. 123, no. 4, p. 910–951.
10. Tarduno, J.A., Cottrell, R.D., Davis, W.J., Nimmo, F., and Bono, R.K., 2015, A Hadean to Paleoarchean geodynamo recorded by single zircon crystals: Science, v. 349, no. 6247, p. 521–524.
11. Singer, B.S., Jicha, B.R., Mochizuki, N., and Coe, R.S., 2019, Synchronizing volcanic, sedimentary, and ice core records of Earth's last magnetic polarity reversal: Science Advances, v. 5, no. 8, p. eaaw4621.

12. NASA, 2012, Magnetic pole reversal happens all the (geologic) time, November 30, https://www.nasa.gov/topics/earth/features/2012-poleReversal.html.
13. Piper, J.D., 2013. A planetary perspective on Earth evolution; lid tectonics before plate tectonics: Tectonophysics, v. 589(C), p. 44–56.
14. O'Neill, C., and Debaille, V., 2014, The evolution of Hadean–Eoarchaean geodynamics: Earth and Planetary Science Letters, v. 406, p. 49–58.
15. Lammer, H., et al., 2018, Origin and evolution of the atmospheres of early Venus, Earth and Mars: Astronomy and Astrophysics Review, v. 26, no. 1, p. 1–72.
16. Bell, E.A., Boehnke, P., Harrison, T.M., and Mao, W.L., 2015, Potentially biogenic carbon preserved in a 4.1 billion-year-old zircon: Proceedings of the National Academy of Sciences, v. 112, no. 47, p. 14518–14521.
17. House, C.H., 2015, Penciling in details of the Hadean: Proceedings of the National Academy of Sciences, v. 112, no. 47, p. 14410–14411.
18. Robb, L.J., Knoll, A.H., Plumb, K.A., Shields, G.A., Strauss, H., and Veizer, J., 2004, The Precambrian: The Archean and Proterozoic Eons, *in* Gradstein, F.M. and Ogg, J.G., eds. A Geologic Time Scale: Cambridge, Cambridge University Press, p. 131.
19. Gomes, R., Levison, H.F., Tsiganis, K., and Morbidelli, A., 2005, Origin of the cataclysmic Late Heavy Bombardment period of the terrestrial planets: Nature, v. 435, no. 7041, p. 466.
20. Mikhail, S., and Sverjensky, D.A., 2014, Nitrogen speciation in upper mantle fluids and the origin of Earth's nitrogen-rich atmosphere: Nature Geoscience, v. 7, no. 11, p. 816–819.
21. Robert, F., 2001, The origin of water on Earth: Science, v. 293, no. 5532, p. 1056–1058.
22. Sarafian, A.R., Nielsen, S.G., Marschall, H.R., McCubbin, F.M., and Monteleone, B.D., 2014, Early accretion of water in the inner solar system from a carbonaceous chondrite–like source: Science, v. 346, no. 6209, p. 623–626.
23. NASA, Why do we have oceans? https://oceanservice.noaa.gov/facts/why_oceans.html.
24. Rosing, M.T., 1999, 13C-depleted carbon microparticles in >3700-Ma sea-floor sedimentary rocks from West Greenland: Science, v. 283, no. 5402, p. 674–676.
25. Nutman, A.P., Bennett, V.C., Friend, C.R., Van Kranendonk, M.J., and Chivas, A.R., 2016, Rapid emergence of life shown by discovery of 3,700-million-year-old microbial structures: Nature, v. 537, no. 7621, p. 535.
26. Messing, C.G., Neumann, A.C., and Lang, J.C., 1990, Biozonation of deep-water lithoherms and associated hardgrounds in the northeastern Straits of Florida: Palaios, v. 5, no. 1, p. 15–33.
27. Gauger, T., Konhauser, K., and Kappler, A., 2015, Protection of phototrophic iron(II)-oxidizing bacteria from UV irradiation by biogenic iron(III) minerals: Implications for early Archean banded iron formation: Geology, v. 43, no. 12, p. 1067–1070.
28. Campbell, I.H., and Allen, C.M., 2008, Formation of supercontinents linked to increases in atmospheric oxygen: Nature Geoscience, v. 1, no. 8, p. 554.
29. Ibid.
30. Robb, Knoll, Plumb, Shields, Strauss, and Veizer, 2004, p. 132.

31. Næraa, T., Scherstén, A., Rosing, M.T., Kemp, A.I.S., Hoffmann, J.E., Kokfelt, T.F., and Whitehouse, M.J., 2012, Hafnium isotope evidence for a transition in the dynamics of continental growth 3.2 Gyr ago: Nature, v. 485, no. 7400, p. 627.
32. Evans, D.A.D., and Pisarevsky, S.A., 2008, Plate tectonics on early Earth? Weighing the paleomagnetic evidence, *in* Condie, K.C., and Pease, V., eds., When Did Plate Tectonics Begin on Planet Earth?: Geological Society of America Special Paper 440, p. 249–263.
33. McLelland, J., Daly, J.S., and McLelland, J.M., 1996, The Grenville orogenic cycle (ca. 1350–1000 Ma): An Adirondack perspective: Tectonophysics, v. 265, nos. 1–2, p. 1–28.
34. Mosher, S., 1998, Tectonic evolution of the southern Laurentian Grenville orogenic belt: Geological Society of America Bulletin, v. 110, no. 11, p. 1357–1375.
35. Scotese, 2009.
36. Gumsley, A.P., Chamberlain, K.R., Bleeker, W., Söderlund, U., de Kock, M.O., Larsson, E.R., and Bekker, A., 2017, Timing and tempo of the Great Oxidation Event: Proceedings of the National Academy of Sciences, v. 114, no. 8, p. 1811–1816.
37. Ogg, Ogg, and Gradstein, 2016, p. 23.
38. Ibid., p. 30.
39. Hoffman, P.F., et al., 2017, Snowball Earth climate dynamics and Cryogenian geology-geobiology: Science Advances, v. 3, no. 11, p. e1600983.
40. Sohl, L.E., Chandler, M.A., Jonas, J., and Rind, D.H., 2014, Energy and heat transport constraints on tropical climates of the Sturtian Snowball Earth: AGU Fall Meeting Abstracts, PP43C-1487.
41. Narbonne, G.M., and Gehling, J.G., 2003, Life after snowball; the oldest complex Ediacaran fossils: Geology, v. 31, no. 1, p. 27–30.
42. Matthews, S.C., and Missarzhevsky, V., 1975, Small shelly fossils of late Precambrian and early Cambrian age; a review of recent work: Journal of the Geological Society of London, v. 131, p. 289–304.

Chapter 9

1. McKerrow, W.S., Mac Niocaill, C., and Dewey, J.F., 2000, The Caledonian orogeny redefined: Journal of the Geological Society of London, v. 157, p. 1151.
2. Beasecker, J., Chamberlin, Z., Lane, N., Reynolds, K., Stack, J., Wahrer, K., Wolff, A., Devilbiss, J., Wahr, C., Durbin, D., and Garneau, H., 2020, It's time to defuse the Cambrian "explosion": GSA Today, v. 30, no. 12, p. 27.
3. Fox, D., 2016, What sparked the Cambrian explosion?: Nature, v. 530, no. 7590, p. 268–270.
4. Yochelson, E.L., 1996, Discovery, collection, and description of the Middle Cambrian Burgess Shale biota by Charles Doolittle Walcott: Proceedings of the American Philosophical Society, v. 140, no. 4, p. 469.
5. Gould, 1989, p. 13.

6. Yochelson, 1996, p. 469.
7. Simonetta, A.M., 1975, The Cambrian non trilobite arthropods from the Burgess Shale of British Columbia: A study of their comparative morphology taxonomy and evolutionary significance: Palaeontographia Italica, v. 69, tabs. I–LXI, p. 1–37.
8. Gould, 1989, p. 14.
9. Whittington, H.B., and Briggs, D.E.G., 1985, The largest Cambrian animal, *Anomalocaris*, Burgess Shale, British-Columbia: Philosophical Transactions of the Royal Society of London B, Biological Sciences, v. 309, no. 1141, p. 571.
10. Aitken, J.D., and McIlreath, I.A., 1984, The Cathedral Reef Escarpment, a Cambrian great wall with humble origins (British Columbia, Canada): Geos, v. 13, no. 1, p. 17–19.
11. Morris, S.C., and Whittington, H.B., 1985, Fossils of the Burgess Shale: A national treasure in Yoho National Park, British Columbia (Vol. 43): Natural Resources Canada, p. 21.
12. Gould, 1989.
13. Hennig, W., 1966, Phylogenetic Systematics: Urbana, University of Illinois Press, 263 p.
14. McKerrow, Mac Niocaill, and Dewey, 2000.
15. Oldroyd, D.R., and McKenna, G., 1995. A note on Andrew Ramsay's unpublished report on the St. David's area, recently discovered: Annals of Science, v. 52, no. 2, p. 196.
16. Zalasiewicz, J.A., Taylor, L., Rushton, A.W.A., Loydell, D.K., Rickards, R.B., and Williams, M., 2009, Graptolites in British stratigraphy: Geological Magazine, v. 146, no. 6, p. 785.
17. Szaniawski, H., 2009, The earliest known venomous animals recognized among conodonts: Acta Palaeontologica Polonica, v. 54, no. 4, p. 669–676.
18. Pander, C.H., 1856, Monographie der fossilen Fische des Silurischen Systems der russisch–baltischen Gouvernements: St. Petersburg, Akademie der Wissenschaften, 91 p.
19. Ulrich, E.O., and Bassler, R.S., 1926, A classification of the toothlike fossils, conodonts, with descriptions of American Devonian and Mississippian species: U.S. National Museum Proceedings, v. 68, no. 2613, article 12, p. 63, 11 plates.
20. Ogg, Ogg, and Gradstein, 2016, p. 59; Beasecker et al., 2020.
21. Murchison, 1839, p. 11.
22. Lapworth, C., 1879.
23. McKerrow, Mac Niocaill, and Dewey, 2000.
24. Barclay, W.J., 2005, Introduction to the Old Red Sandstone of Great Britain, *in* Barclay, W.J., Browne, M.A.E., McMillan, A.A., Pickett, E.A., Stone, P., and Wilby, P.R., eds., The Old Red Sandstone of Great Britain (Geological Conservation Review Series 31): Peterborough, Joint Nature Conservation Committee, p. 14–15.
25. Murchison, 1839, p. 587 and plate I.
26. Raup, 1991, p. 29; Gould, 1989.
27. House, M.R., 2002, Strength, timing, setting and cause of mid-Palaeozoic extinctions: Palaeogeography, Palaeoclimatology, Palaeoecology, v. 181, p. 21.
28. Glasspool, I.J., and Scott, A.C., 2010, Phanerozoic concentrations of atmospheric oxygen reconstructed from sedimentary charcoal: Nature Geoscience, v. 3, p. 628.

29. Ogg, Ogg, and Gradstein, 2016, p. 104.
30. Anderson, J.S., Smithson, T., Mansky, C.F., Meyer, T., and Clack, J., 2015, A diverse tetrapod fauna at the base of "Romer's Gap": PLOS ONE, v. 10, no. 4. p. 23–24.
31. Merck, J., 2019, The reptilian stem: Sauropsida, Eureptilia, Diapsida, https://www.geol.umd.edu/~jmerck/geol431/lectures/17sauropsida.html, accessed May 30, 2020.
32. Arber, 1905, p. xix.
33. Seward, 1914, p. 26–28.
34. Ogg, Ogg, and Gradstein, 2016, p. 121.
35. Lozovsky, V.R., 2005, Olson's Gap or Olson's Bridge, that is the question, *in* Lucas, S.G., and Zeigler, K.E., eds., The Nonmarine Permian, New Mexico Museum of Natural History and Science Bulletin, no. 30, p. 179–184; Lucas, S.G., 2005, Olsen's Gap or Olsen's Bridge, an answer, *in* Lucas, S.G., and Zeigler, K.E., eds., *The Nonmarine Permian: New Mexico Museum of Natural History and Science Bulletin*, no. 30, p. 185–186; Benton, M.J., 2012, No gap in the Middle Permian record of terrestrial vertebrates: Geology, v. 40, p. 339–342.
36. Lucas, S.G., 2013, No gap in the Middle Permian record of terrestrial vertebrates (forum comment): Geological Society of America, September, p. e293.
37. Ogg, Ogg, and Gradstein, 2016, p. 121.
38. Sánchez-Villagra, M.R., 2010, Developmental palaeontology in synapsids: The fossil record of ontogeny in mammals and their closest relatives: Proceedings of the Royal Society of Britain, v. 277, no. 1685, p. 1139.
39. Shen, S.Z., and Bowring, S.A., 2014, The end-Permian mass extinction: A still-unexplained catastrophe: National Science Review, v. 1, no. 4, p. 492.
40. Reichow, M.K., et al., 2009, The timing and extent of the eruption of the Siberian Traps large igneous province: Implications for the end-Permian environmental crisis: Earth and Planetary Science Letters, v. 277, nos. 1–2, p. 9; Renne, P., and Basu, A.R., 1991, Rapid eruption of the Siberian Traps flood basalts at the Permo-Triassic boundary: Science, v. 253, no. 5016, p. 176–179.
41. Burgess, Muirhead, and Bowring, 2017, Initial pulse of Siberian Traps sills as the trigger of the end-Permian mass extinction: Nature Communications, v. 8, no. 1, p. 1–6.
42. Shen, S.Z., and Bowring, S.A., 2014, The end-Permian mass extinction: A still unexplained catastrophe: National Science Review, v. 1, no. 4, p. 494.
43. Rothman, D.H., Fourier, G.P., French, K.L., Alm, E.J., Boyle, E.A., Cao, C., and Summons, R.E., 2014, Methanogenic burst in the end-Permian Carbon Cycle: Proceedings of the National Academy of Sciences, v. 111, no. 15, p. 5462.
44. Shen, S.Z., et al., 2011, Calibrating the end-Permian mass extinction: Science, v. 334, no. 6061, p. 1367–1372.

Chapter 10

1. Lucas, S.G., 2010, The Triassic time scale: An introduction, *in* Lucas, S.G., ed., The Triassic Time Scale: Geological Society of London Special Publications, v. 334, p. 2.
2. Ibid., p. 9.

3. Currie, B.S., Colombi, C.E., Tabor, N.J., Shipman, T.C., and Montañez, I.P., 2009, Stratigraphy and architecture of the Upper Triassic Ischigualasto Formation, Ischigualasto Provincial Park, San Juan, Argentina: Journal of South American Earth Sciences, v. 27, p. 74–87.
4. Seeley, H.G., 1887, On a sacrum apparently indicating a new type of bird, *Ornithodesmus cluniculus* Seeley: Quarterly Journal of the Geological Society of London, v. 43, p. 206–211; Seeley, H.G., 1888, On *Thecospondylus daviesi* (Seeley), with some remarks on the classification of the *Dinosauria*: Quarterly Journal of the Geological Society of London, v. 44, p. 79–87.
5. Marsh, O.C., 1881, Principal characters of American Jurassic dinosaurs, Part V: American Journal of Science, series 3, v. 21, May, p. 417–423.
6. Baron, M.G., Norman, D.B., and Barrett, P.M., 2017, A new hypothesis of dinosaur relationships and early dinosaur evolution: Nature, v. 543, no. 7646, p. 501–506.
7. Brusatte, S.L., O'Connor, J.K., and Jarvis, E.D., 2015, The origin and diversification of birds: Current Biology, v. 25, no. 19, p. R888.
8. Sereno, P.C., 2013, Basal sauropodomorphs and the vertebrate fossil record of the Ischigualasto Formation (Late Triassic: Carnian-Norian) of Argentina: Journal of Vertebrate Paleontology, v. 32, no. sup1: Memoir 12: p. 1–9.
9. Sereno, P.C., Forster, C.A., Rogers, R.R., and Monetta, A.M., 1993, Primitive dinosaur skeleton from Argentina and the early evolution of *Dinosauria*: Nature, v. 361, no. 6407, p. 64–66.
10. Sues, H.D., Nesbitt, S.J., Berman, D.S., and Henrici, A.C., 2011, A late-surviving basal theropod dinosaur from the latest Triassic of North America: Proceedings of the Royal Society, v. B 278, no. 1723, p. 3459.
11. Ibid., p. 3463.
12. Jiang, D.Y., et al., 2016, A large aberrant stem ichthysauriform indicating early rise and demise of ichthysauromorphs in the wake of the end-Permian extinction: Scientific Reports, v. 6, no. 26232, p. 1.
13. Ogg, Ogg, and Gradstein, 2016, p. 151.
14. Ward, P.D., Garrison, G.H., Haggart, J.W., Kring, D.A., and Beattie, M.J., 2004, Isotopic evidence bearing on the late Triassic extinction events, Queen Charlotte Islands, British Columbia, and implications for the duration and cause of the Triassic–Jurassic mass extinction: Earth and Planetary Science Letters, v. 224, p. 599.
15. Woodford, A.O., 1971, Catastrophism and evolution: Journal of Geological Education, v. 19, no. 5, p. 229.
16. D'Orbigny, A., 1849–1852, Cours élémentaire de paléontologic et de Géologie Stratigraphies (2 volumes): Paris, Masson, 1146 p.
17. Oppel, A., 1856–1858, Die Juraformation Englands, Frankreichs und das Südwestlichen Deutschlands: Stuttgart, Ebner & Serebert, 857 p.
18. Ogg, Ogg, and Gradstein, 2016, p. 156.
19. Martill, D.M., Vidovic, S.U., Howells, C., and Nudds, J.R., 2016, The oldest Jurassic dinosaur: A basal neotheropod from the Hettangian of Great Britain: PLOS One, v. 11, no. 1, p. e0154352.

20. Conybeare, W.D., 1822, Additional notices on the fossil genera *Ichthyosaurus* and *Plesiosaurus*: Transactions of the Geological Society of London, series 2, v. 1, p. 103–123.
21. Barthel, K.W., Swinburne, N.H.M., and Morris, S.C., 1990, Solnhofen: A Study of Mesozoic Palaeontology: Cambridge, Cambridge University Press, p. 197.
22. Van Valen and Sloan, 1966, p. 261.
23. Ogg, Ogg, and Gradstein, 2016, p. 167.
24. Hickey, L.J., and Doyle, J.A., 1977, Early Cretaceous fossil evidence for angiosperm evolution: Botanical Review, v. 43, no. 1, p. 3–4.
25. Xing Xu, Zhonghe Zhou and Xiaolin Wang, 2000, The smallest known non-avian theropod dinosaur: Nature, v. 408, no. 7, p. 705–708.
26. Lawson, D.A., 1975, Pterosaur from the latest Cretaceous of West Texas: Discovery of the largest flying creature: Science, v. 187, no. 4180, p. 947.
27. Frey, E., and Martill, D.M., 1996, A reappraisal of *Arambourgiania* (Pterosauria, Pterodactyloidea): One of the world's largest flying animals: Neues Jahrbuch für Geologie und Paläontologie-Abhandlungen, p. 221–247.
28. Harrell, T. L., Jr., Gibson, M.A., and Langston, W., Jr., 2016, A cervical vertebra of *Arambourgiania philadelphiae* (Pterosauria, Azhdarchidae) from the late Campanian micaceous facies of the Coon Creek Formation in McNairy County, Tennessee, USA: Bulletin of the Alabama Museum of Natural History, v. 33, no. 2, p. 94–103.
29. Krause, D.W., et al., 2020, Skeleton of a Cretaceous mammal from Madagascar reflects long-term insularity: Nature, v. 581, p. 1–7.
30. Sloan, R.E., and Van Valen, L., 1965, Cretaceous mammals from Montana: Science, v. 148, no. 3667, p. 220–227.
31. Alvarez, L.W., Alvarez, W., Asaro, F., and Michel, H.V., 1980, Extraterrestrial cause of the Cretaceous–Tertiary extinction: Science, v. 208, no. 4448, p. 1095–1108.
32. Hildebrand, A.R., Penfield, G.T., Kring, D.A., Pilkingotn, M., Camargo, Z.A., Jacobsen, S.B., and Boynton, W.V., 1991, Chicxulub Crater: A possible Cretaceous/Tertiary impact crater in the Yucatán Peninsula, Mexico: Geology, v. 19, p. 867–871.
33. Sanford, J.C., Snedden, J.W., and Gulich, S.P.S., 2016, The Cretaceous–Paleogene boundary deposit in the Gulf of Mexico: Large scale oceanic basin response to the Chicxulub impact: Journal of Geophysical Research, Solid Earth, v. 121, no. 3, p. 1240–1261.
34. Ibid., p. 1257.
35. Morgan, J.V., et al., 2016, The formation of peak rings in large impact craters: Science, v. 354, no. 6314, p. 878–882 .
36. Artemieva, N., Morgan, J., and the Expedition 364 Science Party, 2017, Quantifying the release of climate-active gases by large meteorite impacts with a case study of Chicxulub: Geophysical Research Letters, v. 44, no. 20, p. 10, 180.
37. Kunio, K., and Oshima, N., 2017, Site of asteroid impact changed the history of life on Earth: The low probability of mass extinction: Scientific Reports, v. 7, no. 1, p. 14855.
38. Jolley, D., Gilmour, I., Gurov, E., Kelley, S., and Watson, J., 2010, Two large meteorite impacts at the Cretaceous–Paleogene Boundary: Geology, v. 38, no. 9, p. 835–838.

39. Keller, G., Adatte, T., Pardo, J.A., and Lopez-Oliva, J.G., 2009, New evidence concerning the age and biotic effects of the Chicxulub impact in NE Mexico: Journal of the Geological Society of London, v. 166, no. 3, p. 393–411.
40. Schoene, B., Eddy, M.P., Samperton, K.M., Keller, C.B., Keller, G., Adatte, T., and Khadri, S.F., 2014, U-Pb geochronology of the Deccan Traps and relation to the end-Cretaceous mass extinction: Science, v. 363, no. 6429, p. 862–866.
41. Schulte, P., et al., 2010, The Chicxulub asteroid impact and mass extinction at the Cretaceous–Paleogene boundary: Science, v. 327, no. 5970, p. 1214–1218.

Chapter 11

1. Cohen, K.M., Finney, S.C., Gibbard, P.L., and Fan, J.X., 2013 (updated 2020), The ICS international chronostratigraphic chart: Episodes, v. 36, p. 199–204.
2. U.S. Geological Survey Geologic Names Committee, 2018, Divisions of geologic time—major chronostratigraphic and geochronologic units: U.S. Geological Survey Fact Sheet 2018–3054, 2 p.
3. Walker, J.D., Geissman, J.W., Bowring, S.A., and Babcock, L.E., 2013, The Geological Society of America geologic time scale: GSA Bulletin, v. 125, no. 3–4, pp. 259–272.
4. U.S. Geological Survey Geologic Names Committee, 2018.
5. Fuentes, A.J., Clyde, W.C., Weissenburger, K., Bercovici, A., Lyson, T.R., Miller, I.M., Ramezani, J., Isakson, V., Schmitz, M.D., and Johnson, K.R., 2019, Constructing a time scale of biotic recovery across the Cretaceous–Paleogene boundary, Corral Bluffs, Denver Basin, Colorado, USA: Rocky Mountain Geology, v. 54, no. 2, p. 133–153.
6. Gazin, C.L., 1941, Paleocene mammals from the Denver Basin, Colorado: Journal of the Washington Academy of Sciences, v. 31, no. 7, p. 289–295.
7. Grand Junction Daily Sentinel, 2019, Moment of extinction: How a Grand Junction geologist got a closer look at the K-T boundary, November 18, https://www.gjsentinel.com/news/western_colorado/moment-of-extinction-how-a-grand-junction-geologist-got-a/article_c5620344-09c3-11ea-bf09-20677ce07cb4.html.
8. Fuentes, Clyde, Weissenburger, Bercovici, Lyson, Miller, Ramezani, Isakson, Schmitz, and Johnson, 2019.
9. Denver Museum of Nature and Science, 2019, The mammals, https://coloradosprings.dmns.org/the-mammals/.
10. Harlan, R., 1834, Notice of fossil bones found in the Tertiary in the State of Louisiana: Transactions of the American Philosophical Society, v. 4, p. 397–403.
11. Gingerich, P.D., Arif, M., Bhatti, M.A., Anwar, M., and Sanders, W.J., 1997, *Basilosaurus drazindai* and *Basiloterus hussaini*, new *Archaeoceti* (Mammalia, Cetacea) from the middle Eocene Drazinda Formation, with a revised interpretation of ages of whale-bearing strata in the Kirthar Group of the Sulaiman Range, Punjab (Pakistan): Contributions from the Museum of Paleontology, University of Michigan, v. 30, no. 2, October 1, p. 55–81.

12. Gingerich, P.D., 2012, Evolution of whales from land to sea: Proceedings of the American Philosophical Society, v. 156, September 3, p. 312.
13. Sarich, V.M., 1985, Molecular clocks and eutherian phylogeny (paper): Fourth International Theriological Congress, Edmonton, August 13–20.
14. De Muizon, C., 2001, Walking with whales: Nature, v. 413, no. 6853, p. 259-260.
15. Lihoreau, F., Boisserie, J.R., Manthi, F.K., and Ducrocq, S., 2015, Hippos stem from the longest sequence of terrestrial cetartiodactyl evolution in Africa: Nature Communications, v. 6, no. 1, p. 1–8.
16. Gingerich, 2012, p. 315.
17. Meyer, H.W., 2003, The Fossils of Florissant: Washington, D.C., Smithsonian Books, 258 p.
18. McInerney, F.A., and Wing, S.L., 2011, The Paleocene–Eocene thermal maximum: A perturbation of carbon cycle, climate, and biosphere with implications for the future: Annual Review of Earth and Planetary Sciences, v. 9, May, p. 489–516.
19. Foster, G.L., Lear, C.H., and Rae, J.B.W., 2012, The evolution of pCO2, ice volume and climate during the middle Miocene: Earth and Planetary Science Letters, v. 311–344, August, p. 243–254.
20. Böhme, M., 2003, The Miocene Climatic Optimum: Evidence from ectothermic vertebrates of Central Europe: Palaeogeography, Palaeoclimatology, Palaeoecology, v. 195, p. 389–401.
21. Hsu, K.J., Montadert, L., Bernoulli, D., Cita, M.B., Erikson, A., Garrison, R.E., Kidd, R.B., Melières, F., Müller, C., and Wright, R., 1977, History of the Mediterranean salinity crisis: Nature, v. 267, p. 399.
22. Ryan, W.B.F., 2009, Decoding the Mediterranean salinity crisis: Sedimentology, v. 56, p. 95–136.
23. Hsu, Montadert, Bernoulli, Cita, Erikson, Garrison, Kidd, Melières, Műller, and Wright, 1977, p. 402.
24. Krijgsman, W., Hilgen, F.J., Raffi, I., Sierro, F.J., and Wilson, D.S., 1999, Chronology, causes and progression of the Messinian salinity crisis: Nature, v. 400, August 12, p. 652–655.
25. Garcia-Castellanos, D., Estrada, F., Jiménez-Munt, I., Gorinin, C., Fernàndez, M., Vergés, J., and De Vincente, R., 2009, Catastrophic flood of the Mediterranean after the Messinian salinity crisis: Nature, v. 462, p. 778–782, and supplement, 4 p.
26. Clavel, J., and Morlon, H., 2017, Accelerated body size evolution during cold climatic periods in the Cenozoic: Proceedings of the National Academy of Sciences, v. 114, no. 16, p. 4183–4188.
27. Bergmann, C., 1847, Ueber die verhaltnisse der warmeokono-mie der thiere zu ihrer grosse (About the relationships between heat conservation and body size of animals): Gottinger Studien, v. 1, p. 595–708.
28. Ashton, K.G., Tracy, M.C., and Queiroz, A.D., 2000, Is Bergmann's rule valid for mammals?: American Naturalist, v. 156, no. 4, p. 390–415.
29. Federov, A.V., Brierley, C.M., Lawrence, K., Liu, Z., Dekens, P.S., and Ravelo, A.C., 2013, Patterns and mechanisms of early Pliocene warmth: Nature, v. 496, no. 7443, p. 43–49.

30. Gibbons, A., 2002, In search of the first hominids: Science, v. 295, no. 5558, p. 1214–1219.
31. Johanson, D.C., and White, T.D., 1979, A systematic assessment of early African hominids: Science, v. 203, no. 4378, p. 321–329.
32. Leakey, M.D., and Hay, R.L., 1979, Pliocene footprints in the Laetolil Beds at Laetoli, northern Tanzania: Nature, v. 278, p. 323.
33. Denton, W., 1875, On the asphalt bed near Los Angeles, California: Proceedings of the Boston Society of Natural History, v. 18, p. 185–186.
34. Orcutt, M.L., 1954, The discovery in 1901 of the La Brea Fossil Beds: Historical Society of Southern California Quarterly, v. 36, no. 4, p. 338–341.
35. Holden, A.R., Koch, J.B., Griswold, T., Erwin, D.M., and Hall, J., 2014, Leafcutter bee nests and pupae from the Rancho La Brea Tar Pits of southern California: Implications for understanding the paleoenvironment of the late Pleistocene: PLOS One, v. 9, no. 4, p. e94724.
36. Broecker, W.S., Andree, M., Wolfli, W., Oeschger, H., Bonani, G., Kennett, J., and Peteet, D., 1988, The chronology of the last deglaciation: Implications to the cause of the Younger Dryas event: Paleoceanography, v. 3, no. 1, p. 1–19.
37. Lea, D.W., Pak, D.K., Peterson, L.C., and Hughen, K.A., 2003, Synchroneity of tropical and high-latitude Atlantic temperatures over the last glacial termination: Science, v. 301, no. 5638, p. 1361–1364.
38. Firestone, R.B., et al., 2007, Evidence for an extraterrestrial impact 12,900 years ago that contributed to the megafaunal extinctions and the Younger Dryas cooling: Proceedings of the National Academy of Sciences, v. 104, no. 41, p. 16016–16021.
39. Pino, M., et al., 2019, Sedimentary record from Patagonia, southern Chile supports cosmic-impact triggering of biomass burning, climate change, and megafaunal extinctions at 12.8 ka: Scientific Reports, v. 9, no. 1, p. 1–27.
40. Sandoval-Castellanos, E., Wutke, S., Gonzalez-Salazar, C., and Ludwig, A., 2017, Coat colour adaptation of post-glacial horses to increasing forest vegetation: Nature Ecology & Evolution, v. 1, p. 1816–1819.

Chapter 12

1. Hume, D., 1848 [1777], An Enquiry Concerning Human Understanding, *in* Philosophical Essays Concerning Human Understanding: London, A. Millan, p. 36–37; Hume, D., 2000 [1777], An Enquiry Concerning Human Understanding: A Critical Edition (Vol. 3): Oxford, Oxford University Press.
2. Santayana, G., 2013, Life of Reason: Amherst, New York, Prometheus Books, p. 82.
3. Hoegh-Guldberg, O., et al., 2018, Impacts of 1.5°C global warming on natural and human systems, *in* Masson-Delmotte, V., et al., eds., Global Warming of 1.5°C: An IPCC Special Report on the impacts of global warming of 1.5°C above pre-industrial levels and related global greenhouse gas emission pathways, in the context of

strengthening the global response to the threat of climate change, sustainable development, and efforts to eradicate poverty, p. 262.

4. Gladwell, M., 1996, The tipping point: New Yorker Magazine, June 3, 1996, https://www.newyorker.com/magazine/1996/06/03/the-tipping-point.
5. Lorenz, E., 2000, The butterfly effect: World Scientific Series on Nonlinear Science, Series A, v. 39, p. 91.
6. Hoegh-Guldberg, O.D., et al., 2018, p. 257.
7. United Nations, Department of Economic and Social Affairs, 2015, World Population Projected to Reach 9.7 Billion by 2050, July 29, http://www.un.org/en/development/desa/news/population/2015-report.html, accessed February 25, 2017.
8. U.S. Census Bureau, 2017, U.S. and world population clock, February 24, https://www.census.gov/popclock/.
9. Zahid, H.T., Robinson, E., and Kelly, R.L., 2016, Agriculture, population growth, and statistical analysis of the radiocarbon record: Proceedings of the National Academy of Sciences, v. 114, no. 4, p. 931–935; National Academy of Sciences. 2016, Correction for Zahid et al., Agriculture, population growth, and statistical analysis of the radiocarbon record: Proceedings of the National Academy of Sciences, v. 113, no. 8, p. E 2546.
10. Pingali, P.L., 2012, Green revolution: Impacts, limits and the path ahead: Proceedings of the National Academy of Sciences, v. 109, no. 31, p. 12302–12308.
11. Warren, S.G., 2015, Can human population be stabilized?: Earth's Future, v. 3, p. 82–94.
12. Venter, O., et al., 2016, Sixteen years of change in the global terrestrial footprint and implications for biodiversity conservation: Nature Communications, v. 7, no. 1, p. 1–11.
13. Thomas, L., 2013, Coal Geology: Hoboken, New Jersey, Wiley-Blackwell, p. 54.
14. Heath, R.C., 1998, Basic ground water hydrology: Water Supply Paper 2220, U.S. Department of the Interior, USGS, 86 p.
15. Gornitz, V., Rosenzweig, C., and Hillel, D., 1997, Effects of anthropogenic intervention in the land hydrologic cycle on global sea level rise: Global and Planetary Change, v. 14, nos. 3–4, p. 147–161.
16. United Nations, 2019, World Water Development Report 2019: "Leaving no one behind," https://www.unwater.org/world-water-development-report-2019-leaving-no-one-behind/, accessed May 20, 2020.
17. U.S. Geological Survey, Ground water quality, https://www.usgs.gov/special-topic/water-science-school/science/groundwater-quality?qt-science_center_objects=0#qt-science_center_objects, accessed May 20, 2020.
18. U.S. Geological Survey, Contamination of ground water, https://www.usgs.gov/special-topic/water-science-school/science/contamination-groundwater?qt-science_center_objects=0#qt-science_center_objects, accessed May 20, 2020.
19. Hertel, T.W., and Liu, J., 2016, Implications of water scarcity for economic growth: OECD Environment Working Papers 109, Paris.
20. Heath, 1998.
21. Barlow, P.M., 2003, Ground water in freshwater-saltwater environments of the Atlantic coast (Vol. 1262): U.S. Geological Survey, p. 5.

22. Balog, J., 2009, TED Talk, https://www.ted.com/talks/james_balog_time_lapse_proof_of_extreme_ice_loss#t-319446.
23. Blunden, J., and Arndt, D., 2019, State of the climate in 2018: Bulletin of the American Meteorological Society, v. 100, no. 9, p. S55.
24. Parnreiter, C., 2009, Megacities in the geography of global economic governance: Die Erde, v. 140, no. 4, p. 371.
25. Leakey, R.E., and Lewin, R., 1996, The Sixth Extinction: Patterns of Life and the Future of Humankind: New York, Anchor.
26. Menne, M.J., Williams, C.N., Jr., and Vose, R.S., 2009, The U.S. historical climatology network monthly temperature data, version 2: New bias adjustments reduce uncertainty in temperature trends for the United States: Bulletin of the American Meteorology Society, v. 90, no. 7, p. 993–1007.
27. Camuffo, D., and Bertolin, C., 2012, The earliest temperature observations in the world: The Medici Network: Climatic Change, v. 111, no. 2, p. 335–363.
28. Mann, M.E., Bradley, R.S., and Hughes, M.K., 1998, Global-scale temperature patterns and climate forcing over the past six centuries: Nature, v. 392, no. 6678, p. 779–787.
29. Folland, C.K., Karl, T.R., Christy, J.R., Clarke, R.A., Gruza, G.V., Jouzel, J., Mann, M.E., Oerlemans, J., Salinger, M.J., and Wang, S.W., 2001, Observed climate variability and change, *in* Houghton, J.T., Ding, Y. Griggs, D.J., Noguer, M., van der Linden, P.J., Dai, X., Maskell, K., and Johnson, C.A., eds., Climate Change 2001: The Scientific Basis: Cambridge, Cambridge University Press, p. 134.
30. Ibid.; Mann, M.E., Bradley, R.S., and Hughes, M.K., 1999, Northern hemisphere temperatures during the last millennium: Inferences, uncertainties, and limitations: Geophysical Research Letters, v. 26, p. 759–762.
31. Briffa, K.R., 2000, Annual climate variability in the Holocene: Interpreting the message of ancient trees: Quaternary Science Reviews, v. 19, p. 87–105; Briffa, K.R., Jones, P.D., Schweingruber, F.H., Shiyatov, S.G., and Cook, E.R., 2002, Unusual twentieth-century summer warmth in a 1,000-year temperature record from Siberia: Nature, v. 376, no. 6536, p. 156–159; Jones, P.D., Briffa, K.R., Barnett, T.P., and Tett, S.F.B., 1998, High-resolution paleoclimate records for the last millennium: Interpretation, integration and comparison with general circulation model control-run temperatures: Holocene, v. 8, p. 455–471.
32. Worster, D., 1979, The Dust Bowl: The Southern Plains in the 1930s: New York, Oxford University Press, 304 p.
33. Donarummo, J., Jr., Ram, M., and Stoermer, E.F., 2003, Possible deposit of soil dust from the 1930's U.S. Dust Bowl identified in Greenland ice: Geophysical Research Letters, v. 30, no. 6, 4 p.
34. Trenbreth, K., 2007, Can climate change explain odd weather: National Public Radio (interview by Ira Flatow), January 19, http://www.npr.org/templates/story/story.php?storyId=6921972.
35. Bush, E., and Lemmen, D.S., eds., 2019, Canada's changing climate report: Government of Canada, Ottawa, 444 p.

36. NOAA National Centers for Environmental Information, 2020, Climate at a Glance: Global Time Series, https://www.ncdc.noaa.gov/cag/, accessed May 30, 2020.
37. NASA, 2020, 2019 was the second warmest year on record, https://earthobservatory.nasa.gov/images/146154/2019-was-the-second-warmest-year-on-record, accessed April 12, 2020.
38. Intergovernmental Panel on Climate Change, 2015, Climate change 2014 synthesis report, *in* Pachauri, R.K., et al., eds., Contribution of Working Groups I, II and III to the Fifth Assessment Report of the Intergovernmental Panel on Climate Change: Geneva, IPCC, 151 p.
39. NASA, 2020, Milankovitch (orbital) cycles and their role in Earth's climate, https://climate.nasa.gov/news/2948/milankovitch-orbital-cycles-and-their-role-in-earths-climate/, accessed May 25, 2020.
40. Milankovitch, M., 1930, Mathematische Klimalehre und astronomische Theorie der Klimaschwankungen: Handbuch der Klimatologie 1: Berlin, Borntraeger, 176 p.
41. Buis, A., 2020, Why Milankovitch (orbital) cycles can't explain Earth's current warming, NASA Jet Propulsion Lab Blog, February 27, https://climate.nasa.gov/blog/2949/why-milankovitch-orbital-cycles-cant-explain-earths-current-warming/, accessed May 26, 2020.
42. NOAA, 2019, Carbon dioxide levels in atmosphere hit record high in May: Monthly average surpassed 414 ppm at NOAA's Mauna Loa Observatory in Hawaii: ScienceDaily, June 4, www.sciencedaily.com/releases/2019/06/190604140109.htm, accessed May 18, 2020.
43. NOAA, 2020, Monthly Average Mauna Loa CO2, Retrieved May 18, 2020 from https://www.esrl.noaa.gov/gmd/ccgg/trends/
44. Le Quéré, C., et al., 2020, Temporary reduction in daily global CO_2 emissions during the COVID-19 forced confinement: Nature Climate Change, v. 10, p. 1.
45. NOAA, 2020, NOAA exploring impact of COVID-19 response on the environment, May 6, 2020, https://research.noaa.gov/article/ArtMID/587/ArticleID/2617/NOAA-exploring-impact-of-coronavirus-response-on-the-environment, accessed May 25, 2020.
46. Scripps Institution for Oceanography, 2020, https://scripps.ucsd.edu/programs/keelingcurve/.
47. Monroe, R., 2020, What does it take for the coronavirus (or other major economic events) to affect global carbon dioxide readings? https://scripps.ucsd.edu/programs/keelingcurve/2020/03/11/what-does-it-take-for-the-coronavirus-or-other-major-economic-events-to-affect-global-carbon-dioxide-readings/, accessed June 5, 2020.
48. Zemp, M., et al., 2015, Historically unprecedented global glacial decline in the early 21st century: Journal of Glaciology, v. 61, no. 228, p. 745–762.
49. Field, C.B., et al., eds., 2014, Climate change 2014: Impacts, adaptation, and vulnerability, Part A, Global and sectoral aspects, *in* Contribution of Working Group II to the Fifth Assessment Report of the Intergovernmental Panel on Climate Change: Cambridge, Cambridge University Press, 1150 p. www.cambridge.org/9781107641655. The IPCC is preparing the Sixth Assessment Report

AR6 Climate Change, to be printed later in 2021, https://www.ipcc.ch/report/sixth-assessment-report-working-group-ii/.

50. Trenberth, K.E., 2011, Changes in precipitation with climate change: Climate Research, v. 47, nos. 1–2, p. 123–138.
51. Herring, S.C., Hoell, A., Hoerling, M.P., Kossin, J.P., Schreck, C.J., III, and Stott, P.A., eds., 2016, Explaining extreme events of 2015 from a climate perspective: Bulletin of the American Meteorological Society, v. 97, no. 12, p. S1–S145.
52. Yochum, S.E., 2015, Colorado Front Range flood of 2013: Peak flows and flood frequencies: Proceedings of the 3rd Joint Federal Interagency Conference on Sedimentation and Hydrologic Modeling, April 19–23, Reno, p. 537–548.
53. Colorado Water Conservation Board, 2014, CDOT/CWCB hydrology investigation phase one, 2013 peak flow determinations: State of Colorado Technical Memorandum, Denver, 8 p.
54. CIRES, 2013, Severe flooding on the Colorado Front Range, September 2013, preliminary assessment: University of Colorado, 4 p.
55. Westerling, A.L., Hidalgo, H.G., Cayan, D.R., and Swetnam, T.W., 2006, Warming and earlier spring increase western U.S. forest wildfire activity: Science, August 18, p. 940–943.
56. Evans, D., et al., 2018, Eocene greenhouse climate revealed by coupled clumped isotope-Mg/Ca thermometry: Proceedings of the National Academy of Sciences, v. 115, no. 6, p. 1174–1179, .
57. Steele, A., 2012, Looking backwards, looking forwards: A consideration of the foibles of action research within teacher work: The Canadian Journal of Action Research, v. 13, no. 2, p. 17.

Bibliography

Aalto, K.R., 2004, Rock stars: Clarence King (1842–1901): Pioneering geologist of the West: GSA Today, February, p. 18–19.

Agassiz, L., 1860, [Review of] On the origin of species: American Journal of Science and Arts, series 2, July 30, p. 142–154, doi: 10.7135/UPO9780857286512.023.

Agassiz, L., 1866 [1867], Geological Sketches (Vol. 2): Boston, Ticknor and Fields, 311 p., doi: 10.5962/bhl.title.166203.

Agassiz, L., and Bettannier, J., 1840, Études sur les glaciers (Studies on glaciers). Neuchâtel: Jent et Gassmann, 346 p., doi: 10.5962/bhl.title.151173.

Agricola, Georgius, 1546, De natura fossilium (Textbook of Mineralogy), translated by M.C. Bandy and J.A. Bandy, 1955, Mineola, New York: The Geological Society of America, 256 p., doi: 10.1130/SPE63.

Aitken, J.D., and McIlreath, I.A., 1984, The Cathedral Reef Escarpment, a Cambrian great wall with humble origins (British Columbia, Canada): Geos, v. 13, no. 1, p. 17–19.

Alberti, F.A. von, 1834, Beitrag zu einer Monographie des bunten Sandsteins, Muschelkalks und Keupers: Und die Verbindung dieser gebilde zu einer Formation. Tubingen: Stuttgart University, 366 p.

Alvarez, L.W., Alvarez, W., Asaro, F., and Michel, H.V., 1980, Extraterrestrial cause of the Cretaceous–Tertiary extinction: Science, v. 208, no. 4448, p. 1095–1108, doi: 10.1126/science.208.4448.1095.

Anderson, J.S., Smithson, T., Mansky, C.F., Meyer, T., and Clack, J., 2015, A diverse tetrapod fauna at the base of "Romer's Gap": PLOS ONE, v. 10, no. 4. p. 23–24, doi: 10.1371/journal.pone.0125446.

Anning, M., 1839, Note on the supposed frontal spine in the genus Hybodus: Magazine of Natural History, v. 12, p. 605.

Anon., 1817, Memoir of Abraham Gottlob Werner, late professor of mineralogy at Freiberg: Philosophical Magazine, series 1, v. 50, no. 233, p. 182–189.

Arber, E.A.N., 1905, Catalogue of the fossil plants of the Glossopteris Flora in the Department of Geology, British Museum (Natural History); Being a Monograph of the Permo-Carboniferous Flora of India and the Southern Hemisphere: Hertford, Stephen Austin, 255 p., doi: 10.5962/bhl.title.7567.

Archer, M.O., Hietala, H., Hartinger, M.D., Plaschke, F., and Angelopoulos, V., 2019, Direct observations of a surface eigenmode of the dayside magnetopause: Nature Communications, v. 10, no. 1, p. 1–11, doi: 10.1038/s41467-018-08134-5.

Arnold, L., 1999, Becoming a geologist: Florence Bascom in Wisconsin, 1874–1887: Earth Science History, v. 12, no. 2, p. 159–179, doi: 10.17704/eshi.18.2.m5500526x7331430.

Artemieva, N., Morgan, J., and the Expedition 364 Science Party, 2017, Quantifying the release of climate-active gases by large meteorite impacts with a case study of Chicxulub: Geophysical Research Letters, v. 44, no. 20, p. 10,180–10,188, doi:10.1002/2017GL074879.

Ashton, K.G., Tracy, M.C., and Queiroz, A.D., 2000, Is Bergmann's rule valid for mammals?: American Naturalist, v. 156, no. 4, p. 390–415, doi: 10.1086/303400.

Aton, J.M., 1994, John Wesley Powell (Western Writers Series 114): Boise, Boise State University Printing and Graphic Services, 55 p.

Aubry, M.P., Berggren, W.A., Van Couvering, J., McGowran, B., Pillans, B., and Hilgen, F., 2005, Quaternary: Status, rank, definition, and survival: Episodes, v. 28, no. 2, p. 118–120.

Avery, O.T., MacLeod, C.M., and McCarty, M., 1944, Studies on the chemical nature of the substance inducing transformation of pneumococcal types: Induction of transformation by a desoxyribonucleic acid fraction isolated from pneumococcus type III: Journal of Experimental Medicine, v. 79, no. 2, p. 137–158, doi: 10.1084/jem.79.2.137.

Barclay, W.J., 2005, Introduction to the Old Red Sandstone of Great Britain, *in* Barclay, W.J., Browne, M.A.E., McMillan, A.A., Pickett, E.A., Stone, P., and Wilby, P.R., eds., The Old Red Sandstone of Great Britain (Geological Conservation Review Series 31): Peterborough, Joint Nature Conservation Committee, p. 1–18.

Barlow, P.M., 2003, Ground water in freshwater-saltwater environments of the Atlantic coast (Vol. 1262): U.S. Geological Survey.

Barnosky, A.D., Carrasco, M.A., and Davis, E.B., 2005, The impact of the species-area relationship on estimates of paleodiversity: PLOS Biology, v. 3, no. 8, p. E266, doi: 10.1371/journal.pbio.0030266.

Baron, M.G., Norman, D.B., and Barrett, P.M., 2017, A new hypothesis of dinosaur relationships and early dinosaur evolution: Nature, v. 543, no. 7646, p. 501–506, doi: 10.1038/nature21700.

Barthel, K.W., Swinburne, N.H.M., and Morris, S.C., 1990, Solnhofen: A Study of Mesozoic Palaeontology: Cambridge, Cambridge University Press, 246 p.

Bartholomew, M., 1973, Lyell and evolution: An account of Lyell's response to the prospect of an evolutionary ancestry of man: British Journal for the History of Science, v. 26, no. 3, p. 261–303, doi: 10.1017/S0007087400016265.

Bascom, F., Clark, W.B., Darton, N.H., Knapp, G.N., Kuemmel, H.B., Miller, B.L., and Salisbury, R.D., 1909, Philadelphia folio, Norristown, Germantown, Chester and Philadelphia, Pennsylvania–New Jersey–Delaware: U.S. Geological Survey Folios of the Geologic Atlas 162, doi: 10.3133/gf162.

Becquerel, H.A., 1896, Sur les radiations émises par phosphorescence: Comptes Rendus des Séances de l'Académie des Sciences, v. 122, p. 420–421.

Bell, E.A., Boehnke, P., Harrison, T.M., and Mao, W.L., 2015, Potentially biogenic carbon preserved in a 4.1 billion-year-old zircon: Proceedings of the National Academy of Sciences, v. 112, no. 47, p. 14518–14521, doi: 10.1073/pnas.1517557112.

Benett, E., 1831, A catalogue of the organic remains of the County of Wiltshire: Warminster, Vardy, 9 p.

Benton, M.J., 1985, Mass extinction among non-marine tetrapods: Nature, v. 316, no. 6031, p. 811–814, doi: 10.1038/316811a0.

Benton, M.J., 2012, No gap in the middle Permian record of terrestrial vertebrates: Geology, v. 40, p. 339–342, doi: 10.1130/G32669.1.

Benton, M.J., Sennikov, A.G., and Newell, A.J., 2010, Murchison's first sighting of the Permian at Vzackniki in 1841: Proceedings of the Geologists' Association, v. 121, p. 313–318, doi:10.1016/j.pgeola.2010.03.005.

Bergmann, C., 1847, Ueber die verhaltnisse der warmeokono-mie der thiere zu ihrer grosse (About the relationships between heat conservation and body size of animals): Gottinger Studien, v. 1, p. 595–708.

Berry, W.B.N. 1987, Growth of a Prehistoric Time Scale (revised edition): Palo Alto, Blackwell Scientific, 202 p.

Blunden, J., and Arndt, D., 2019, State of the climate in 2018: Bulletin of the American Meteorological Society, v. 100, no. 9, p. Si–S306, doi: 10.1175/2019BAMSStateof theClimate.1.

Böhme, M., 2003, The Miocene Climatic Optimum: Evidence from ectothermic vertebrates of Central Europe: Palaeogeography, Palaeoclimatology, Palaeoecology, v. 195, p. 389–401, doi: 10.1016/S0031-0182(03)00367-5.

Bokulich, A., 2018, Using models to correct data: Paleodiversity and the fossil record: Synthese, p. 1–22, doi: 10.1007/s11229-018-1820-x.

Bottke, W.F., and Norman, M.D., 2017, The Late Heavy Bombardment: Annual Review of Earth and Planetary Sciences, v. 45, p. 619–647, doi: 10.1146/annurev-earth-063016-020131.

Bowler, P.J., 2002, Charles Darwin: The Man and His Influence: London, Cambridge University Press, 264 p.

Bowring, S.A., Williams, I.S., and Compston, W., 1989, 3.96 Ga gneisses from the slave province, Northwest Territories, Canada: Geology, v. 17, no. 11, p. 971–975, doi: 10.1130/0091-7613(1989)017<0971:GGFTSP>2.3.CO;2.

Boyer, P.J., 1984, William Buckland, 1784–1855: Scientific institutions, vertebrate palaeontology and Quaternary geology [Ph.D. thesis]: Leicester, University of Leicester, v. 1, 443 p.

Breuer, D., 2011, Stagnant lid convection, *in* Gargaud, M., et al., eds., Encyclopedia of Astrobiology: Berlin: Springer, p. 125, doi: 10.1007/978-3-642-11274-4_1499.

Brewer, W.H., 1902, John Wesley Powell: American Journal of Science, v. 14, p. 377–382, doi: 10.2475/ajs.s4-14.83.377.

Briffa, K.R., 2000, Annual climate variability in the Holocene: Interpreting the message of ancient trees: Quaternary Science Reviews, v. 19, p. 87–105, doi: 10.1016/S0277-3791(99)00056-6.

Briffa, K.R., Jones, P.D., Schweingruber, F.H., Shiyatov, S.G., and Cook, E.R., 2002, Unusual twentieth-century summer warmth in a 1,000-year temperature record from Siberia: Nature, v. 376, no. 6536, p. 156–159, doi: 10.1038/376156a0.

Brink, A.S., 1951, On the genus Lystrosaurus Cope: Transactions of the Royal Society of South Africa, v. 33, no. 1, p. 107–120, doi: 10.1080/00359195109519880.

Broecker, W.S., Andree, M., Wolfli, W., Oeschger, H., Bonani, G., Kennett, J., and Peteet, D., 1988, The chronology of the last deglaciation: Implications to the cause of the Younger Dryas event: Paleoceanography, v. 3, no. 1, p. 1–19, doi: 10.1029/PA003i001p00001.

Broo, J., and Mahoney, J., 2015, Chewing on Change: Exploring the Evolution of Horses in Response to Climate Change: Gainesville, University of Florida, 12 p.

Brusatte, S.L., O'Connor, J.K., and Jarvis, E.D., 2015, The origin and diversification of birds: Current Biology, v. 25, no. 19, p. R888–R898, doi: 10.1016/j.cub.2015.08.003.

Buckland, W., 1824, Notice on the Megalosaurus or great fossil lizard of Stonesfield: Transactions of the Geological Society of London, series 2, no. 1, p. 390–396, doi: 10.1144/transgslb.1.2.390.

Buckland, W., 1823, Reliquiae diluvianae; Or, Observations on the Organic Remains Contained in Caves, Fissures and Diluvial Gravel, and on Other Geological Phenomena, Attesting the Action of an Universal Deluge: London, John Murray, 279 p. Published online December 2011, Cambridge University Press, 303 p., doi: 10.1017/CBO9780511694820.

Buckland, W., 1836, The Bridgewater Treatises on the power, wisdom and goodness of God as manifested in the creation (Sammeltitel). Treatise 6, Geology and mineralogy considered with reference to natural theology: London, Pickering, 2 Vol., v. 2, 128 p., doi: 10.5962/bhl.title.125523.

Bullard, E.C., Everett, J.E., and Smith, A.G., 1965, The fit of the continents around the Atlantic: Philosophical Transactions of the Royal Society of London, Series A, Mathematical and Physical Sciences, v. 258, no. 1088, p. 41–51, doi: 10.1098/rsta.1965.0020.

Bullard, E.C., Maxwell, A.E, and Revelle, R., 1956, Heat flow through the deep ocean floor: Advances in Geophysics, v. 3, p. 153–181, doi: 10.1016/S0065-2687(08)60389-1.

Burgess, S.D., Muirhead, J.D., and Bowring, S.A., 2017, Initial pulse of Siberian Traps sills as the trigger of the end-Permian mass extinction: Nature Communications, v. 8, no. 1, p. 1–6, doi: 10.1038/s41467-017-00083-9.

Bush, E., and Lemmen, D.S., eds., 2019, Canada's changing climate report: Government of Canada, Ottawa, 444 p.

Campbell, I.H., and Allen, C.M., 2008, Formation of supercontinents linked to increases in atmospheric oxygen: Nature Geoscience, v. 1, no. 8, p. 554–558, doi: 10.1038/ngeo259.

Camuffo, D., and Bertolin, C., 2012, The earliest temperature observations in the world: The Medici Network: Climatic Change, v. 111, no. 2, p. 335–363, doi: 10.1007/s10584-011-0142-5.

Cappellini, E., et al., 2019, Early Pleistocene enamel proteome from Dmanisi resolves Stephanorhinus phylogeny: Nature, v. 574, no. 7776, p. 103–107, doi: 10.1038/s41586-019-1555-y.

Carozzi, A.V., 1964, Lamarck's theory of the Earth: Hydrogeologie: Isis, v. 55, no. 3, p. 293–307, doi: 10.1086/349863.

Carozzi, A.V., 1976, Horace Benedict de Saussure: Geologist or educational reformer?: Journal of Geological Education, v. 24, no. 2, p. 46–49, doi: 10.5408/0022-1368-24.2.46.

Cavallo, E.A., Powell, A., and Becerra, O., 2010, Estimating the direct economic damage of the earthquake in Haiti, IDB Working Paper IBD-WP-163: Washington, D.C., Inter-American Development Bank, Washington, D.C.

Charnay, B., Le Hir, G., Fluteau, F., Forget, F., and Catling, D.C., 2017, A warm or a cold early Earth? New insights from a 3-D climate-carbon model: Earth and Planetary Science Letters, v. 474, p. 97–109, doi: 10.1016/j.epsl.2017.06.029.

CIRES, 2013, Severe flooding on the Colorado Front Range, September 2013, Preliminary Assessment: University of Colorado, 4 p.

Clavel, J., and Morlon, H., 2017. Accelerated body size evolution during cold climatic periods in the Cenozoic: Proceedings of the National Academy of Sciences, v. 114, no. 16, p. 4183–4188, doi: 10.1073/pnas.1606868114.

Cloutier, R., Clement, A.M., Lee, M.S., Noël, R., Béchard, I., Roy, V., and Long, J.A., 2020, Elpistostege and the origin of the vertebrate hand: Nature, v. 579, p. 549–554, doi: 10.1038/s41586-020-2100-8.

Cohen, K.M., Finney, S.C., Gibbard, P.L., and Fan, J.X., 2013 (updated 2020), The ICS international chronostratigraphic chart: Episodes, v. 36, p. 199–204.

Colbert, E.H., 1973, Wandering Lands and Animals: New York, Dutton, 323 p.

Coleman, W., 1962, Lyell and the "reality" of species, 1830–1833: Isis, p. 325–338, doi: 10.1086/349595.

Colorado Water Conservation Board, 2014, CDOT/CWCB hydrology investigation phase one, 2013 peak flow determinations: State of Colorado Technical Memorandum, Denver, 8 p.

Compston, W., and Pidgeon, R.T., 1986, Jack Hills, evidence of more very old detrital zircons in Western Australia: Nature, v. 321, no. 6072, p. 766–769, doi: 10.1038/321766a0.

Conybeare, W.D., 1822, Additional notices on the fossil genera Ichthyosaurus and Plesiosaurus: Transactions of the Geological Society of London, series 2, v. 1, p. 103–123, doi: 10.1144/transgslb.1.1.103.

Conybeare, W.D., 1830, Letter on Mr. Lyell's Principles of Geology: Philosophical Magazine, new series 8, p. 215–219, doi: 10.1080/14786443008675408.

Conybeare, W.D., and Phillips, W., 1822, Outlines of the Geology of England and Wales: London, William Phillips, George Yard, 470 p.

Cope, J.C.W., 2016, Geology of the Dorset Coast (second edition, with contributions from Malcolm Butler; Geologists' Association Guide No. 22): London, Geological Society, 222 p.

Crutzen, P.J., and Stoermer, E.F., 2000, The "Anthropocene": Global Change Newsletter, v. 41, May, p. 17–18, doi: 10.17159/sajs.2019/6428.

Currie, B.S., Colombi, C.E., Tabor, N.J., Shipman, T.C., and Montañez, I.P., 2009, Stratigraphy and architecture of the Upper Triassic Ischigualasto Formation, Ischigualasto Provincial Park, San Juan, Argentina: Journal of South American Earth Sciences, v. 27, p. 74–87, doi: 10.1016/j.jsames.2008.10.004.

Cuvier, G., 1796, Memoire sur les fossiles des environs de Paris (talk): National Institute of France.

Cuvier, G., 1799, Mémoire sur les espèces d'éléphans vivantes et fossiles: Mémoire de l'Academie des Sciences, v. 2, p. 1–32.

Cuvier, G., 1804, Sur les espèces d'animaux dont proviennent les os fossiles: Annales du Muséum Nationale d'Histoire Naturelle, tome troisième, p. 276–289.

Cuvier, G., 1812, Rescarches sur les ossemens fossiles (4 volumes): Paris, Chez Deterville, doi: 10.5962/bhl.title.60807.

Cuvier, G., 1818, Essay on the Theory of the Earth, New York, Kirk & Mercein, 431 p., doi: 10.5962/bhl.title.31662.

Cuvier, G., and Brongniart, A., 1810, Carte geognostique des environs de Paris, scale 1:200,000.

Cuvier, G., and Brongniart, A., 1811, Essai sur la géographie minéralogique des environs de Paris, avec une carte géognostique, et des coupes de terrain (2 volumes): Paris, Baudouin.

Cuvier, G., and Brongniart, A., 1822, Description géologique des environs de Paris: Paris, Chez G. Dufour et E. D'Ocagne. 428 p., doi: 10.5962/bhl.title.149831.

Dahm, R., 2008, Discovering DNA: Friedrich Miescher and the early years of nucleic acid research: Human Genetics, v. 122, p. 565–581, doi: 10.1007/s00439-007-0433-0.

Dana, J.D., 1837, A System of Mineralogy, Including an Extended Treatise on Crystallography, with an Appendix Containing the Application of Mathematics to Crystallographic Investigation, and a Mineralogical Bibliography. 1st edition: New Haven, Durrie & Peck and Herrick and Noyes, 608 p.

Dana, J.D., 1848, The Manual of Mineralogy, Including Observations on Mines, Rocks, Reduction of Ores, and the Application of the Science to the Arts: With 260 Illustrations Designed for the Use of Schools and Colleges: New Haven, Durrie & Peck, 430 p.

Dana, J.D., 1863, Manual of Geology, Treating of the Principles of the Science with Special Reference to American Geological History, for the Use of Colleges, Academies, and Schools of Science: Philadelphia, T. Bliss & Co., 798 p., doi: 10.5962/bhl.title.61162.

Dana, J.D., 1864, A Text-book of Geology: Designed for Schools and Academies: Philadelphia, T. Bliss & Co., 354 p.

Darroch, S.A., Boag, T.H., Racicot, R.A., Tweedt, S., Mason, S.J., Erwin, D.H., and Laflamme, M., 2016, A mixed Ediacaran-metazoan assemblage from the Zaris Sub-basin, Namibia: Palaeogeography, Palaeoclimatology, Palaeoecology, v. 459, p. 198–208, doi: 10.1016/j.palaeo.2016.07.003.

Darwin, C.R., 1835, Letter 282, Darwin Correspondence Project, https://www.darwinproject.ac.uk/letter/DCP-LETT-282.xml, accessed February 25, 2020.

Darwin, C.R., 1859, On the Origin of Species by Means of Natural Selection, or the Preservation of Favoured Races in the Struggle for Life: London, John Murray, 564 p., doi: 10.5962/bhl.title.82303.

Darwin, C.R., 1860, On the Origin of Species by Means of Natural Selection, or the Preservation of Favoured Races in the Struggle for Life: New York, Appleton, 474 p., doi: doi.org/10.5962/bhl.title.162283.

Davidson, J.P., 2002, Bonehead mistakes: The background in scientific literature and illustrations for Edward Drinker Cope's first restoration of Elasmosaurus platyurus: Proceedings of the Academy of Natural Sciences of Philadelphia, v. 152, no. 1, p. 215–240, doi: 10.1635/0097-3157(2002)152[0215:HPOVBM]2.0.CO;2.

Dawson, J.W., 1855. Acadian Geology: An Account of Geological Structure and Mineral Resources of Nova Scotia, and Portions of the Neighboring Provinces of British America: Edinburg, Oliver and Boyd, 388 p.

Davis, L.E., 2009, Mary Anning of Lyme Regis: 19th century pioneer in British palaeontology: Headwaters, Faculty Journal of the College of Saint Benedict and Saint John's University, v. 26, p. 105–106.

Davis, W.M., 1915, Biographical memoir of John Wesley Powell: National Academy of Sciences, Biographical Memoirs, Part of Volume III, Washington, National Academy of the Sciences, v. 83, p. 12, doi: 10.5962/bhl.title.31326.

Dean, D.R., 1975, James Hutton on religion and geology: The unpublished preface to his Theory of the Earth (1788): Annals of Science, v. 32, no. 2, p. 187–193, doi: 10.1080/00033797500200241.

Dean, D.R., 1992, James Hutton and the History of Geology: Ithaca, Cornell University Press, 303 p.

De la Beche, H., 1839, Report on the Geology of Cornwall, Devon, and West Somerset: London, Longman, Orme, Brown, Green, and Longmans, 686 p.

De la Beche, H.T., 1848, Obituary notes: Quarterly Journal of the Geological Society of London, p. xxi–cxx.

Denton, W., 1875, On the asphalt bed near Los Angeles, California: Proceedings of the Boston Society of Natural History, v. 18, p. 185–186.

De Muizon, C., 2001, Walking with whales: Nature, v. 413, no. 6853, p. 259–260, doi: 10.1038/35095137.

Desnoyers, J., 1829, Observations sur un ensemble de dépôts marins plus récents que les terrains tertiaires du bassin de la Seine, et constituant une formation géologique distincte: Précédées d'un aperçu de la nonsimultanéité des bassins tertiaires: Annales Scientifiques Naturelles, v. 16, p. 171–214, 402–419.

De Vries, M.V.W., Bingham, R.G., and Hein, A.S, 2017, A new volcanic province: An inventory of subglacial volcanoes in West Antarctica: Geological Society of London Special Publications, v. 461, no. 1, p. 231–248, doi: 10.1144/SP461.7.

D'Halloy, J.J.O., 1822, Observations sur un essai de carte géologique de Pay-Bas de la France, et de quelques contrées voisinés: Namur, Imprimerie de Madame Huzard, 26 p.

Denton, W., 1875, On the asphalt bed near Los Angeles, California: Proceedings of the Boston Society of Natural History, v. 18, p. 185–186.

Dietz, R.S., 1961, Continent and ocean basin evolution by spreading of the sea floor: Nature, v. 190, p. 854–857, doi: 10.1038/190854a0.

Dietz, R.S., 1968, Reply [to "Arthur Holmes: Originator of Spreading Ocean Floor Hypothesis"]: Journal of Geophysical Research, v. 73, p. 6567, doi: 10.1029/JB073i020p06567.

Dodick, J., and Argamon, S., 2006, Rediscovering the historical methodology of the earth sciences by analyzing scientific communication styles, *in* Manduca, C.A., and Mogk, D.W., eds., Earth and Mind: How Geologists Think and Learn about the Earth: Geological Society of America Special Paper 413, p. 105–120, doi: 10.1130/SPE413.

Donarummo, J., Jr., Ram, M., and Stoermer, E.F., 2003, Possible deposit of soil dust from the 1930's U.S. Dust Bowl identified in Greenland ice: Geophysical Research Letters, v. 30, no. 6, 4 p., doi: 10.1029/2002GL016641.

D'Orbigny, A., 1849–1852, Cours élémentaire de paléontologic et de géologie stratigraphies (2 volumes): Paris, Masson, 1146 p., doi: 10.5962/bhl.title.154975.

Dorset County Council, 2000, Nomination for the Dorset and East Devon Coast for inclusion in the World Heritage List, UNESCO, p. 25–27.

Dott, R.H., 1998, Charles Lyell's debt to North America: His lectures and travels from 1841 to 1853: Geological Society of London Special Publications, v. 143, no. 1, p. 53–69, doi: 10.1144/GSL.SP.1998.143.01.06.

Du Toit, A., 1937, Our Wandering Continents: An Hypothesis on Continental Drifting: Edinburgh, Oliver and Boyd, 366 p.

Eakins, P.R., and Eakins, J.S., 1990, Dawson, Sir John William, *in* Dictionary of Canadian Biography (Vol. 12): Toronto, University of Toronto, p. 1892.

Eiseley, L.C., 1958, Darwin's Century: Evolution and the Men Who Discovered It: Garden City, NY, Doubleday, 378 p.

Eiseley, L.C., 1959, Charles Lyell: Scientific American, v. 201, no. 2, p. 98.

Eldredge, N., and Gould, S.J., 1972, Punctuated equilibria: An alternative to phyletic gradualism, *in* Schopf, T.J.M., ed., Models in Paleobiology: San Francisco, Freeman Cooper, p. 82–115, doi: 10.4319/lo.1974.19.2.0375.

Evans, D., et al., 2018, Eocene greenhouse climate revealed by coupled clumped isotope-Mg/Ca thermometry: Proceedings of the Natural Academy of Science, v. 115, no. 6, p. 1174–1179, doi: 10.1073/pnas.1714744115.

Evans, D.A.D., and Pisarevsky, S.A., 2008, Plate tectonics on early Earth? Weighing the paleomagnetic evidence, *in* Condie, K.C., and Pease, V., eds., When Did Plate Tectonics Begin on Planet Earth?: Geological Society of America Special Paper 440, p. 249–263, doi: 10.1130/SPE440.

Falcon-Lang, H.J., and Calder, J.H., 2005, Sir William Dawson (1820–1899): A very modern paleobotanist: Atlantic Geology, v. 41, p. 103–114.

Federov, A.V., Brierley, C.M., Lawrence, K., Liu, Z., Dekens, P.S., and Ravelo, A.C., 2013, Patterns and mechanisms of early Pliocene warmth: Nature, v. 496, no. 7443, p. 43–49, doi: 10.1038/nature12003.

Field, C.B., Barros, V.R., Dokken, D.J., Mach, K.J. Mastrandrea, M.D., Bilir, T.E., Chatterjee, M., Ebi, K.L. Estrada, Y.O., Genova, R.C., Girma, B., Kissel, E.S., Levy, A.N., MacCracken, S., Mastrandrea, P.R., and White, L.L. (eds.), 2014, Climate change 2014: Impacts, adaptation, and vulnerability, Part A, Global and sectoral aspects, *in* Contribution of Working Group II to the Fifth Assessment Report of the Intergovernmental Panel on Climate Change: Cambridge, Cambridge University Press, 1150 p. www.cambridge.org/9781107641655.

Firestone, R.B., et al., 2007, Evidence for an extraterrestrial impact 12,900 years ago that contributed to the megafauna extinctions and the Younger Dryas cooling: Proceedings of the National Academy of Sciences, v. 104, no. 41, p. 16016–16021, doi: 10.1073/pnas.0706977104.

First, D., 2003, The music of the sphere: An investigation into asymptotic harmonics, brainwave entrainment and the Earth as a giant bell: Leonardo Music Journal, v. 13, p. 31–37, doi: 10.1162/096112104322750755.

Folland, C.K., Karl, T.R., Christy, J.R., Clarke, R.A., Gruza, G.V., Jouzel, J., Mann, M.E., Oerlemans, J., Salinger, M.J., and Wang, S.W., 2001, Observed climate variability and change, *in* Houghton, J.T., Ding, Y., Griggs, D.J., Noguer, M., van der Linden, P.J., Dai, X., Maskell, K., and Johnson, C.A., eds., Climate Change 2001: The Scientific Basis: Cambridge, Cambridge University Press, 881 p.

Forsyth, D., and Uyeda, S., 1975, On the relative importance of the driving forces of plate motion: Geophysical Journal International, v. 43, no. 1, p. 163–200, doi: 10.1111/j.1365-246X.1975.tb00631.x.

Foster, G.L., Lear, C.H., and Rae, J.B.W., 2012, The evolution of pCO2, ice volume and climate during the middle Miocene: Earth and Planetary Science Letters, v. 311–344, August, p. 243–254, doi: 10.1016/j.epsl.2012.06.007.

Fox, D., 2016, What sparked the Cambrian explosion?: Nature, v. 530, no. 7590, p. 268–270, doi: 10.1038/530268a.

Frank, F.C., 1968, Curvature of island arcs: Nature, v. 220, no. 5165, p. 363, doi: 10.1038/220363a0.

Frazer, P., 1888, A short history of the origin and acts of the International Congress of Geologists, and of their American Committee delegation to it: American Geologist, January, v. 1, no. 1, p. 86–100.

Frazer, P., 1890, The American Association for the Advancement of Science, of 1890: American Naturalist, Proceedings of Scientific Societies, v. 24, part 2, p. 987.

Frey, E., and Martill, D.M., 1996, A reappraisal of Arambourgiania (Pterosauria, Pterodactyloidea): One of the world's largest flying animals: Neues Jahrbuch für Geologie und Paläontologie-Abhandlungen, p. 221–247, doi: 10.1127/njgpa/199/1996/221.

Friedlander, G., Kennedy, J.W., and Macias, E.S., 1981, Nuclear and Radiochemistry: New York, Wiley, 704 p.

Froman, N., 1996, Marie and Pierre Curie and the discovery of polonium and radium (originally delivered as a lecture at the Royal Swedish Academy of Sciences, Stockholm, February 28, translated by Marshall-Lundén, N.), https://www.nobelprize.org/nobel_prizes/themes/physics/curie/. https://www.nobelprize.org/nobel_prizes/themes/physics/curie/.

Fuentes, A.J., Clyde, W.C., Weissenburger, K., Bercovici, A., Lyson, T.R., Miller, I.M., Ramezani, J., Isakson, V., Schmitz, M.D., and Johnson, K.R., 2019, Constructing a time scale of biotic recovery across the Cretaceous–Paleogene boundary, Corral Bluffs,

Denver Basin, Colorado, USA: Rocky Mountain Geology, v. 54, no. 2, p. 133–153, doi: 10.24872/rmgjournal.54.2.133.

Garcia-Castellanos, D., Estrada, F., Jiménez-Munt, I., Gorinin, C., Fernàndez, M., Vergés, J., and De Vincente, R., 2009, Catastrophic flood of the Mediterranean after the Messinian salinity crisis: Nature, v. 462, p. 778–782, and supplement, 4 p., doi: 10.1038/nature08555.

Gauger, T., Konhauser, K., and Kappler, A. 2015, Protection of phototrophic iron(II)-oxidizing bacteria from UV irradiation by biogenic iron(III) minerals: Implications for early Archean banded iron formation: Geology, v. 43, no. 12, p. 1067–1070, doi: 10.1130/G37095.1.

Gazin, C., 1941, Paleocene mammals from the Denver Basin, Colorado: Journal of the Washington Academy of Sciences, v. 31, no. 7, p. 289–295.

Gibbons, A., 2002, In search of the first hominids: Science, v. 295, no. 5558, p. 1214–1219, doi: 10.1126/science.295.5558.1214.

Gingerich, P.D., 2006, Environment and evolution through the Paleocene–Eocene thermal maximum: Trends in Ecology & Evolution, v. 21, no. 5, p. 246–253, doi: 10.1016/j.tree.2006.03.006.

Gingerich, P.D., 2012, Evolution of whales from land to sea: Proceedings of the American Philosophical Society, v. 156, September 3, p. 309–323.

Gingerich, P.D., Arif, M., Bhatti, M.A., Anwar, M., and Sanders, W.J., 1997, Basilosaurus drazindai and Basiloterus hussaini, new Archaeoceti (Mammalia, Cetacea) from the middle Eocene Drazinda Formation, with a revised interpretation of ages of whale-bearing strata in the Kirthar Group of the Sulaiman Range, Punjab (Pakistan): Contributions from the Museum of Paleontology, University of Michigan, v. 30, no. 2, p. 55–81.

Gingerich, P.D., von Koenigswald, W., Sanders, W.J., Smith, B.H., and Zalmout, I.S., 2009, New protocetid whale from the middle Eocene of Pakistan: Birth on land, precocial development, and sexual dimorphism: PLOS One, v. 4, no. 2, p. e4366, doi: 10.1371/journal.pone.0004366.

Glassman, S., Bolt, E.A., Jr., and Spamer, E.E., 1993, Joseph Leidy and the "Great Inventory of Nature": Proceedings of the Academy of Natural Sciences of Philadelphia, v. 144, p. 1–19.

Glasspool, I.J., and Scott, A.C., 2010, Phanerozoic concentrations of atmospheric oxygen reconstructed from sedimentary charcoal: Nature Geoscience, v. 3, p. 627–630, doi: 10.1038/ngeo923.

Gomes, R., Levison, H.F., Tsiganis, K., and Morbidelli, A., 2005, Origin of the cataclysmic Late Heavy Bombardment period of the terrestrial planets: Nature, v. 435, no. 7041, p. 466–469, doi: 10.1038/nature03676.

Gordon, R.G., 2000, Diffuse oceanic plate boundaries: Strain rates, vertically averaged rheology, and comparison with narrow plate boundaries and stable plate interiors, *in* Richards, M., Gordon, R.G., and Van Der Hilst, R.D., eds., History and Dynamics of Global Plate Margins: Geophysical Monographs 121, American Geophysical Union, p. 143–159, doi: 10.1029/GM121.

Gornitz, V., Rosenzweig, C., and Hillel, D., 1997, Effects of anthropogenic intervention in the land hydrologic cycle on global sea level rise: Global and Planetary Change, v. 14, nos. 3–4, p. 147–161, doi: 10.1016/S0921-8181(96)00008-2.

Gould, J., 1837, Remarks on a group of ground finches from Mr. Darwin's collection, with characters of the new species: Proceedings of the Zoological Society of London, v. 5, p. 4–7.

Gould, S.J., 1987, Time's Arrow; Time's Cycle: Cambridge, Mass., Harvard University Press, 222 p.
Gould, S.J., 1989, Wonderful Life: The Burgess Shale and the Nature of History: New York, Norton, 352 p.
Gould, S.J., and Eldredge, N., 1977, Punctuated equilibria: The tempo and mode of evolution reconsidered: Paleobiology, v. 3, no. 2, p. 115–151.
Gould, S.J., and Eldredge, N., 1993, Punctuated equilibrium comes of age: Nature, v. 366, no. 6452, p. 223–227, doi: 10.1038/366223a0.
Greene, M.T., 1985, Geology in the Nineteenth Century: Changing Views of a Changing World (Cornell History of Science Series): Ithaca, Cornell University Press, 328 p.
Greene, M.T., 2015, Alfred Wegener: Science, Exploration and the Theory of Continental Drift: Baltimore, Johns Hopkins University Press, p. 327.
Greenough, G.B., 1820, A geological map of England and Wales: London, Longmans, Hurst, Rees, Orme & Brown, scale 5 nautical miles to 1 inch, 4 sheets.
Gumsley, A.P., Chamberlain, K.R., Bleeker, W., Söderlund, U., de Kock, M.O., Larsson, E.R., and Bekker, A., 2017, Timing and tempo of the Great Oxidation Event: Proceedings of the National Academy of Sciences, v. 114, no. 8, p. 1811–1816, doi: 10.1073/pnas.1608824114.
Guntau, M., 2009, The rise of geology as a science in Germany around 1800, *in* Lewis, C.L.E., and Knell, S.J., eds., The Making of the Geological Society of London: Geological Society of London Special Publications, v. 317, no. 1, p. 163–177, doi: 10.1144/SP317.9.
Gupta, S., Collier, J.S., Palmer-Felgate, A., and Potter, G., 2007, Catastrophic flooding origins of the shelf valley system in the English Channel: Nature, v. 448, July 19, p. 342–345, doi: 10.1038/nature06018.
Guthrie, R.D., 2003, Rapid body size decline in Alaskan Pleistocene horses before extinction: Nature, v. 426, no. 6963, p. 169–171, doi: 10.1038/nature02098.
Hague, A. and Emmons, S.F., 1877, United States Geological exploration of the 40th parallel, Vol. 2, Descriptive Geology: U.S. Army Engineering Department Professional Paper No. 18: Washington, Government Printing Office, 890 p., doi: 10.5962/bhl.title.49454.
Hague, J.D., 1870, United States Geological exploration of the 40th parallel, Vol. 3, Mining Industry: U.S. Army Engineering Department Professional Paper No. 18: Washington, Government Printing Office, 647 p., doi: 10.5962/bhl.title.49454.
Haile, N.S., 1987, Time and age in geology: The use of upper/lower, late/early in stratigraphic nomenclature: Marine and Petroleum Geology, v. 4, no. 3, p. 255–257, doi: 10.1016/0264-8172(87)90048-1.
Halpern, J.M., 1951, Thomas Jefferson and the geological sciences: Rocks and Minerals, v. 74, p. 601–602.
Hanks, T.C., and Kanamori, H., 1979, A moment magnitude scale: Journal of Geophysical Research: Solid Earth, v. 84, no. B5, p. 2348–2350, doi: 10.1029/JB084iB05p02348.
Hansen, J.M., 2009, On the origin of natural history: Steno's modern, but forgotten philosophy of science: Bulletin of the Geological Society of Denmark, v. 203, p. 1–24.
Harlan, R., 1834, Notice of fossil bones found in the Tertiary in the State of Louisiana: Transactions of the American Philosophical Society, v. 4, p. 397–403.
Harrell, T.L., Jr., Gibson, M.A., and Langston, W., Jr., 2016, A cervical vertebra of Arambourgiania philadelphiae (Pterosauria, Azhdarchidae) from the late Campanian micaceous facies of the Coon Creek Formation in McNairy County, Tennessee, USA: Bulletin of the Alabama Museum of Natural History, v. 33, no. 2, p. 94–103.

Hatcher, R.P., Jr., 2010, The Appalachian orogeny: A brief summary, *in* Tollo, R.P., Bartholomew, M.J., Hibbard, J.P., and Karabinos, P.M., eds., From Rodinia to Pangea: The Lithotectonic Record of the Appalachian Region: Memoir Volume 206, Geological Society of America Memoire 206, p. 1–20, doi: 10.1130/MEM206.

Heath, R.C., 1998, Basic ground-water hydrology: Water Supply Paper 2220, U.S. Department of the Interior, USGS, 86 p., doi: 10.3133/wsp2220.

Heezen, B.C., and Tharp, M., 1977, World ocean floor panorama (map, painted by Berann, H.): Columbia University, Office of Naval Research.

Hennig, W., 1966, Phylogenetic Systematics: Urbana, University of Illinois Press, 263 p.

Herbert, S., 2005, Charles Darwin, Geologist: Ithaca, Cornell University Press, 480 p.

Herring, S.C., Hoell, A., Hoerling, M.P., Kossin, J.P., Schreck, C.J., III, and Stott, P.A., eds., 2016, Explaining extreme events of 2015 from a climate perspective: Bulletin of the American Meteorological Society, v. 97, no. 12, p. S1–S145, doi: 10.1175/BAMS-ExplainingExtremeEvents2015.1.

Hertel, T.W., and Liu, J., 2016, Implications of water scarcity for economic growth: OECD Environment Working Papers 109, Paris,doi: 10.1787/5jlssl611r32-en.

Hess, H.H., 1962, History of the ocean basins: Petrologic Studies, November, p. 590–620.

Hickey, L.J., and Doyle, J.A., 1977, Early Cretaceous fossil evidence for angiosperm evolution: Botanical Review, v. 43, no. 1, p. 3–4, doi: 10.1007/BF02860849.

Hildebrand, A.R., Penfield, G.T., Kring, D.A., Pilkingotn, M., Camargo, Z.A., Jacobsen, S.B., and Boynton, W.V., 1991, Chicxulub Crater: A possible Cretaceous/Tertiary impact crater in the Yucatán Peninsula, Mexico: Geology, v. 19, p. 867–871, doi: 10.1130/0091-7613(1991)019%3C0867:CCAPCT%3E2.3.CO;2.

Hodgskiss, M.S., Crockford, P.W., Peng, Y., Wing, B.A., and Horner, T.J., 2019, A productivity collapse to end Earth's great oxidation: Proceedings of the National Academy of Sciences, v. 116, no. 35, p. 17207, doi: 10.1073/pnas.1900325116.

Hoegh-Guldberg, O., et al., 2018, Impacts of 1.5ºC Global Warming on Natural and Human Systems, *in* Masson-Delmotte, V., et al., eds., Global Warming of 1.5°C: An IPCC Special Report on the impacts of global warming of 1.5°C above pre-industrial levels and related global greenhouse gas emission pathways, in the context of strengthening the global response to the threat of climate change, sustainable development, and efforts to eradicate poverty, p. 175–311.

Hoffman, P.F., et al., 2017, Snowball Earth climate dynamics and Cryogenian geology-geobiology: Science Advances, v. 3, no. 11, p. e1600983, doi: 10.1126/sciadv.1600983.

Holden, A.R., Koch, J.B., Griswold, T., Erwin, D.M., and Hall, J., 2014, Leafcutter bee nests and pupae from the Rancho La Brea Tar Pits of southern California: Implications for understanding the paleoenvironment of the late Pleistocene: PLOS One, v. 9, no. 4, p. e94724, doi: 10.1371/journal.pone.0094724.

Holmes, A., 1941, Principles of Physical Geology: London, Thomas Nelson, 532 p.

Hörnes, M., 1853, Mittheilung an Professor BRONN gerichtet, Vienna: Neues Jahrbuch fur Mineralogie, Geologie, Geognosie und Petrefaktenkunde, p. 806–810.

House, C.H., 2015, Penciling in details of the Hadean: Proceedings of the National Academy of Sciences, v. 112, no. 47, p. 14410–14411, doi: 10.1073/pnas.1519765112.

House, M.R., 2002, Strength, timing, setting and cause of mid-Palaeozoic extinctions: Palaeogeography, Palaeoclimatology, Palaeoecology, v. 181, p. 21, doi: 10.1016/S0031-0182(01)00471-0.

Hsu, K.J., Montadert, L., Bernoulli, D., Cita, M.B., Erikson, A., Garrison, R.E., Kidd, R.B., Melières, F., Müller, C., and Wright, R., 1977, History of the Mediterranean salinity crisis: Nature, v. 267, p. 399–403, doi: 10.1038/267399a0.

Huang, H.H., Lin, F.C., Schmandt, B., Farrell, J., Smith, R.B., and Tsai, V.C., 2015, The Yellowstone magmatic system from the mantle plume to the upper crust: Science, v. 348, no. 6236, p. 773–776, doi: 10.1016/j.jvolgeores.2009.08.020; 10.1126/science.aaa5648.

Hume, D., 1848 [1777], An Enquiry Concerning Human Understanding, in Philosophical Essays Concerning Human Understanding: London, A. Millan, p. 36–37.

Hume, D., 2000 [1777], An Enquiry Concerning Human Understanding: A Critical Edition (Vol. 3): Oxford, Oxford University Press, 139 p.

Hunter, P., 2013, Molecular fossils probe life's origins: European Molecular Biology Organization Reports, v. 14, no. 11, p. 964–967,doi: 10.1038%2Fembor.2013.162.

Huntington, T., 1998, The great feud: American History, v. 33, no. 3, p. 17–18.

Hutton, J., 1785, Concerning the system of the Earth, its duration and stability, a paper read to the Royal Society of Edinburgh, on the 7th of March and 4th of April 1785: Edinburgh, Royal Society of Edinburgh.

Hutton, J., 1788, Theory of the Earth; or an investigation of the laws observable in the composition, dissolution, and restoration of land upon the globe: Earth and Environmental Science Transactions of the Royal Society of Edinburgh, v. 1, no. 2, p. 209–304.

Hutton, J., 1795, Theory of the Earth with Proofs and Illustrations (2 volumes): Edinburgh, William Creech.

Hutton, J., 1899, Theory of the Earth with Proofs and Illustrations, (Vol. 3 edited by Sir Archibald Geikie): London, Geological Society.

Huxley, J. 1942, Evolution: The Modern Synthesis: London, George Allen & Unwin, 645 p.

Intergovernmental Panel on Climate Change, 2013, Climate change 2013: The physical science basis, *in* Stocker, T.F., Qin, D., Plattner, G.K., Tignor, M., Allen, S.K., Boschung, J., Nauels, A., Xia, Y., Bex, V., and Midgley, P.M., eds., Contribution of Working Group I to the Fifth Assessment Report of the Intergovernmental Panel on Climate Change: Cambridge, Cambridge University Press, doi:10.1017/CBO9781107415324.

Intergovernmental Panel on Climate Change, 2015, Climate change 2014 synthesis report, *in* Pachauri, R.K., Allen, M.R., Barros, V.R., Broome, J., Cramer, W., Christ, R., Church, J.A., Clarke, L., Dahe, Q., Dasgupta, P., and Dubash, N.K., eds., Contribution of Working Groups I, II, and III to the Fifth Assessment Report of the Intergovernmental Panel on Climate Change: Geneva, IPCC, 151 p.

International Atomic Energy Agency, 1992, Statistical treatment of environmental isotope data in precipitation (revised edition), Technical Reports Series No. 331: Vienna, IAEA.

International Human Genome Sequencing Consortium, 2001, Initial sequencing and analysis of the human genome: Nature, v. 409, p. 860–921, doi: 10.1038/35057062.

Irmscher, C., 2013, Louis Agassiz: Creator of American Science: New York, Houghton Mifflin Harcourt, 448 p.

Jamieson, A.J., Malkocs, T., Piertney, S.B., Fujii, T., and Zhang, Z., 2017, Bioaccumulation of persistent organic pollutants in the deepest ocean fauna: Nature Ecology and Evolution, v. 1, article 0051, 1 p., doi: 10.1038/s41559-016-0051.

Javaux, E.J., and Marshall, C.P., 2005, Tracking the record of early life: Carnets de Géologie (M02/05), p. 27–31.

Jiang, D.Y., et al., 2016, A large aberrant stem ichthysauriform indicating early rise and demise of ichthysauromorphs in the wake of the end-Permian extinction: Scientific Reports, v. 6, no. 26232, 9 p., doi: 10.1038/srep26232.

Johannsen W. 1909, Elemente der exakten Erblichkeitslehre [Elements of the Exact Theory of Inheritance]: Jena, Gustav Fischer, 515 p., doi: 10.5962/bhl.title.1060.

Johanson, D.C., and White, T.D., 1979, A systematic assessment of early African hominids: Science, v. 203, no. 4378, p. 321–329, doi: 10.1126/science.104384.

Johnson, T.E., Brown, M., Gardiner, N.J., Kirkland, C.L., and Smithies, R.H., 2017, Earth's first stable continents did not form by subduction: Nature, v. 543, March 9, p. 239–242, doi: 10.1038/nature21383.

Jolley, D., Gilmour, I., Gurov, E., Kelley, S., and Watson, J., 2010, Two large meteorite impacts at the Cretaceous–Paleogene Boundary: Geology, v. 38, no. 9, p. 835–838, doi: 10.1130/G31034.1.

Jones, P.D., Briffa, K.R., Barnett, T.P., and Tett, S.F.B., 1998, High-resolution paleoclimate records for the last millennium: Interpretation, integration and comparison with general circulation model control-run temperatures: Holocene, v. 8, p. 455–471, doi: 0959- 6836(98)HL258XX.

Keller, G., Adatte, T., Pardo, J.A., and Lopez-Oliva, J.G., 2009, New evidence concerning the age and biotic effects of the Chicxulub impact in NE Mexico: Journal of the Geological Society of London, v. 166, no. 3, p. 393–411, doi: 10.1144/0016–76492008–116.

Kennett, J.P., and Stott, L.D., 1991, Abrupt deep-sea warming, palaeoceanographic changes and benthic extinctions at the end of the Palaeocene: Nature, v. 353, no. 6341, p. 225–229, doi: 10.1038/353225a0.

Kerr, R.A., 1984, Making the moon from a big splash: Science, v. 226, p. 1060–1062.

Kettlewell, H.B.D., 1955, Selection experiments on industrial melanism in the Lepidoptera: Heredity, v. 9, p. 323–342.

Kielan-Jaworowska, Z., and Hurum, J.H., 2006, Limb posture in early mammals: Sprawling or parasagittal: Acta Palaeontologica Polonica, v. 51, no. 3, p. 393–406.

Kielan Jaworowska, Z., Hurum, J.H., and Lopatin, A.V., 2005, Skull structure in Catopsbaatar and the zygomatic ridges in multituberculate mammals: Acta Palaeontologica Polonica, v. 50, no. 3, p. 487–512.

King, C., 1874, Mountaineering in the Sierra Nevada (fourth edition): New York, Charles Scribner's Sons, 378 p.

King, C., 1877, Catastrophism and evolution: American Naturalist, v. 11, n. 8, p. 449–470, doi: 10.1086/271929.

King, C., 1878, United States Geological exploration of the 40th parallel, Vol. 1, Systematic Geology: U.S. Army Engineering Department Professional Paper No. 18: Washington, Government Printing Office, 803 p., doi: 10.3133/70038097.

King, P.B., and Beikman, H.M., 1974, Explanatory text to accompany the geologic map of the United States: U.S. Geological Survey Professional Paper 901, Washington, D.C., U.S. Government Printing Office, 40 p., doi: 10.3133/pp901.

King, P.B., Beikman, H.M., and Edmonston, G.J., 1974, Geologic map of the United States (exclusive of Alaska and Hawaii): US Geological Survey, Scale 1:2,500,000, 2 Plates: 40.75 × 52.50 inches and 40.63 × 52.49 inches; Legend, doi: 10.3133/70136641.

Knoll, A.H., Niklas, K.J., and Tiffney, B.H., 1979, Phanerozoic land-plant diversity in North America: Science, v. 206, no. 4425, p. 1400–1402, doi: 10.1126/science.206.4425.1400.

Krause, D.W., et al., 2020, Skeleton of a Cretaceous mammal from Madagascar reflects long-term insularity: Nature, p. 1–7, v. 581, doi: 10.1038/s41586-020-2234-8.

Kresic, N., 2006, Hydrogeology and Groundwater Modeling (second edition): Boca Raton, CRC Press, 828 p.

Krijgsman, W., Hilgen, F.J., Raffi, I., Sierro, F.J., and Wilson, D.S., 1999, Chronology, causes and progression of the Messinian salinity crisis: Nature, v. 400, August 12, p. 652–655, doi: 10.1038/23231.

Krill, A., 2011, Fixists vs. mobilists in the geological contest of the century, 1844–1969: Trondheim, A. Krill, 299 p.

Kunio, K., and Oshima, N., 2017, Site of asteroid impact changed the history of life on Earth: The low probability of mass extinction: Scientific Reports, v. 7, no. 1, p. 14855, doi: 10.1038/s41598-017-14199-x.

Lamarck, J.B., 1801, Système des animaux sans vertèbres . . .: Paris, Chez Deterville, 468 p., doi: 10.5962/bhl.title.14255.

Lamarck, J.B., 1802, Hydrogéologie ou recherches sur l'influence qu'ont les eaux sur la surface du globe terrestre; sur les causes de l'existence du bassin des mers, de son déplacement et de son transport successif sur les différens points de la surface de ce globe, enfin sur les changemens que les corps vivans exercent sur la nature et l'état de cette surface: Paris, An X, 268 p.

Lammer, H., et al., 2018, Origin and evolution of the atmospheres of early Venus, Earth and Mars: Astronomy and Astrophysics Review, v. 26, no. 1, p. 1–72, doi: 0.1007/s00159-018-0108-y.

Lapworth, C., 1879, On the tripartite classification of the lower Paleozoic rocks: Geological Magazine, N series, v. 6, p. 1–15, doi: 10.1017/S0016756800156560.

Lawson, D.A., 1975, Pterosaur from the latest Cretaceous of West Texas: Discovery of the largest flying creature: Science, v. 187, no. 4180, p. 947–948, doi: 10.1126/science.187.4180.947.

Lea, D.W., Pak, D.K., Peterson, L.C., and Hughen, K.A., 2003, Synchroneity of tropical and high-latitude Atlantic temperatures over the last glacial termination: Science, v. 301, no. 5638, p. 1361–1364, doi: 10.1126/science.1088470.

Leakey, M.D., and Hay, R.L., 1979, Pliocene footprints in the Laetolil Beds at Laetoli, northern Tanzania: Nature, v. 278, p. 317–323, doi: 10.1038/278317a0.

Leakey, R.E., and Lewin, R., 1996, The Sixth Extinction: Patterns of Life and the Future of Humankind: New York, Anchor 288 p., doi: 10.1175/2008BAMS2613.1.

Lehmann, J.G., 1756, Versuch einer Geschichte von Flötz-Gebürgen: Berlin, n.p.

Leidy, J., 1847, On the fossil horse of America: Proceedings of the Academy of Natural Sciences of Philadelphia, v. 3, p. 262–266, doi: 10.7135/UPO9780857286512.019.

Leidy, J., 1865, Cretaceous Reptiles of the United States (Smithsonian Contributions to Knowledge 192): New York, Appleton, 208 p., doi: 10.5962/bhl.title.39830.

Leidy, J., 1870, Remarks on Elasmosaurus platyurus: Proceedings of the Academy of Natural Sciences of Philadelphia, v. 22, p. 9–10.

Leidy, J., and Gibbes, R.W., 1847, On the fossil horse of America: Description of new species of Squalides from the Tertiary Beds of South Carolina: Proceedings of the Academy of Natural Sciences of Philadelphia, v. 3, no.11, p. 262–269.

Leonardo da Vinci, 1888, The Notebooks of Leonardo da Vinci (translated by Richter, J.P.): Milan, n.p.

Le Quéré, C., et al., 2020, Temporary reduction in daily global CO2 emissions during the COVID-19 forced confinement: Nature Climate Change, v. 10, p. 1–7, doi: 10.1038/s41558-020-0797-x.

Lihoreau, F., Boisserie, J.R., Manthi, F.K., and Ducrocq, S., 2015, Hippos stem from the longest sequence of terrestrial cetartiodactyl evolution in Africa: Nature Communications, v. 6, no. 1, p. 1–8, doi: 0.1038/ncomms7264.

Lock, S.J., and Stewart, S.T., 2017, The structure of terrestrial bodies: Impact heating, corotation limits, and synestias: Journal of Geophysical Research: Planets, v. 122, no. 5, p. 950–982, doi: 10.1002/2016JE005239.

Lock, S.J., Stewart, S.T., Petaev, M.I., Leinhardt, Z., Mace, M.T., Jacobsen, S.B., and Cuk, M., 2018, The origin of the moon within a terrestrial synestia: Journal of Geophysical Research: Planets, v. 123, no. 4, p. 910–951, doi: 10.1002/2017JE005333.

Lombardo, P.A., 1982, The great chain of being and the limits to the Machiavellian cosmos: Journal of Thought, v. 17, no. 1, p. 37–52.

Lorenz, E., 2000, The butterfly effect: World Scientific Series on Nonlinear Science, Series A, v. 39, p. 91–94.

Lovejoy, A.O., 2001, The Great Chain of Being: A Study of the History of an Idea: Abingdon, Routledge, 382 p.

Lowrie, W., 2007, Fundamentals of Geophysics (third edition): London, Cambridge University Press, 425 p.

Lozovsky, V.R., 2005, Olson's Gap or Olson's Bridge, that is the question, *in* Lucas, S.G., and Zeigler, K.E., eds., The Nonmarine Permian, New Mexico Museum of Natural History and Science Bulletin no. 30, p. 179–184, doi: 10.1130/G32669.1.

Lucas, S.G., 2005, Olsen's Gap or Olsen's Bridge, an answer, *in* Lucas, S.G., and Zeigler, K.E., eds., The Nonmarine Permian: New Mexico Museum of Natural History and Science Bulletin, no. 30, p. 185–186.

Lucas, S.G., 2010, The Triassic time scale: An introduction, *in* Lucas, S.G., ed., The Triassic Time Scale: Geological Society of London Special Publications, v. 334, p. 1–16, doi: 10.1144/SP334.1 0305–8719/10.

Lucas, S.G., 2013, No gap in the middle Permian record of terrestrial vertebrates (forum comment): Geological Society of America, September, p. e293, doi: 10.1130/G33734C.1.

Lyell, C., 1830, Principles of Geology (Vol. 1): London, John Murray, 552 p.

Lyell, C., 1833, Principles of Geology, Being an Attempt to Explain the Former Changes of the Earth's Surface, by Reference to Causes Now in Operation (Vol. 3): London, John Murray, 398 p., doi: 10.5962/bhl.title.50860.

Lyell, C., 1838, Elements of Geology: London, John Murray, 543 p.

Lyell, C., 1863, Geological Evidences of the Antiquity of Man: London, John Murray, 500 p., doi: 10.5962/bhl.title.19191.

MacFadden, B.J., 2005, Fossil horses: Evidence for evolution: Science, v. 308, p. 1728–1730, doi: 10.1126/science.1105458.

MacFadden, B.J., Oviedo, L.H., Seymour, G.M., and Ellis, S., 2012, Fossil horses, orthogenesis, and communicating evolution in museums: Evolution: Education and Outreach, v. 5, p. 29–37, doi: 10.1007/s12052-012-0394-1.

Maclure, W., 1817, Observations on the Geology of the United States of America, with some remarks on the effect produced on the nature and fertility of soils, by the decomposition of the different classes of rocks; and an application to the fertility of every State in the Union, in reference to the accompanying geologic map: American Philosophical Society Transactions, Memoire, 2 plates, 127 p.

Maclure, W., Tanner, H.S., and Lewis, S., 1809, Observations on the geology of the United States, explanatory of a geological map: American Philosophical Society Transactions, v. 6, 411 p. and map.

Majerus, M.E., 1998, Melanism: Evolution in Action: Oxford, Oxford University Press, 364 p.

Mann, M.E., Bradley, R.S., and Hughes, M.K., 1998, Global-scale temperature patterns and climate forcing over the past six centuries: Nature, v. 392, no. 6678, p. 779–787, doi: 10.1038/33859.

Mann, M.E., Bradley, R.S., and Hughes, M.K., 1999, Northern hemisphere temperatures during the last millennium: Inferences, uncertainties, and limitations: Geophysical Research Letters, v. 26, p. 759–762, doi: 10.1029/1999GL900070.

Marsh, O.C., 1875, Odontornithes, or birds with teeth: American Naturalist, v. 9, no. 12, p. 625–631, doi: doi.org/10.1086/271556.

Marsh, O.C., 1876, Notice of new Tertiary mammals, V: American Journal of Science, v. 71, p. 401–404, doi: doi.org/10.2475/ajs.s3-12.71.401.

Marsh, O.C., 1880, United States Geological exploration of the 40th parallel, Vol. 7, Odontornithes, A Monograph on the Extinct Toothed Birds of North America: U.S. Army Engineering Department Professional Paper No. 18: Washington, Government Printing Office, 201 p., doi: 10.5962/bhl.title.49454.

Marsh, O.C., 1881, Principal characters of American Jurassic dinosaurs, Part V: American Journal of Science, series 3, v. 21, May, p. 417–423, doi: 10.2475/ajs.s3-21.125.417.

Marsh, O.C., 1883, Birds with Teeth. 3rd Annual Report of the Secretary of the Interior: Washington, D.C., U.S. Government Printing Office, v. 3, p. 43–88.

Martill, D.M., Vidovic, S.U., Howells, C., and Nudds, J.R., 2016, The oldest Jurassic dinosaur: A basal neotheropod from the Hettangian of Great Britain: PLOS One, v. 11, no. 1, p. e0154352, doi: 10.1371/journal.pone.0154352.

Maslin, M.A., Brierley, C.M., Milner, A.M., Shultz, S., Trauth, M.H., and Wilson, K.E., 2014, East African climate pulses and early human evolution: Quaternary Science Reviews, v. 101, p. 1–17, doi: 10.1016/j.quascirev.2014.06.012.

Mason, R.G., 1958, A magnetic survey off the west coast of the United States between latitudes 32° and 36° N. and longitudes 121° and 128° W.: Geophysical Journal of the Royal Astronomical Society, v. 1, p. 320–329.

Mastin, L.G., Van Eaton, A.R., and Lowenstern, J.B., 2014, Modeling ash fall distribution from a Yellowstone supereruption: Geochemistry, Geophysics, Geosystems, v. 15, no. 8, p. 3459–3475, doi: 10.1002/2014GC005469.

Matthews, S.C., and Missarzhevsky, V., 1975, Small shelly fossils of late Precambrian and early Cambrian age; a review of recent work: Journal of the Geological Society of London, v. 131, p. 289–304, doi: 10.1144/gsjgs.131.3.0289.

Matthews, W.H., III, 1969, The Geologic Story of the Palo Duro Canyon: Austin, Bureau of Economic Geology, https://www.gutenberg.org/files/52179/52179-h/52179-h.htm#fig6.

Mayr, E., 1977, Darwin and natural selection: How Darwin may have discovered his highly unconventional theory: American Scientist, v. 65, no. 3, p. 321.

McCulloch, A.W., 2010, Sir John William Dawson: A profile of a Nova Scotian scientist: Proceedings of the Nova Scotian Institute of Science, v. 45, no. 2, p. 3–4.

McInerney, F.A., and Wing, S.L., 2011, The Paleocene–Eocene thermal maximum: A perturbation of carbon cycle, climate, and biosphere with implications for the future: Annual Review of Earth and Planetary Sciences, v. 39, May, p. 489–516, doi: 10.1146/annurev-earth-040610-133431.

McKerrow, W.S., Mac Niocaill, C., and Dewey, J.F., 2000, The Caledonian orogeny redefined: Journal of the Geological Society of London, v. 157, p. 1149–1154, doi: 10.1144/jgs.157.6.1149.

McLelland, J., Daly, J.S., and McLelland, J.M., 1996, The Grenville orogenic cycle (ca. 1350–1000 Ma): An Adirondack perspective: Tectonophysics, v. 265, nos. 1–2, p. 1–28, doi: 10.1016/S0040-1951(96)00144-8.

McPhee, J., 1980, Basin and Range: New York, Farrar, Strauss & Giroux, 224 p.

Meek, F.B., Hall, J., Whitfield, R.P., and Ridgeway, R., 1877, United States Geological exploration of the 40th parallel, Vol. 4, Palaeontology Part I (Meek), Palaeontology Part II (Hall and Whitfield), Ornithology Part III (Ridgway): U.S. Army Engineering Department Professional Paper No. 18: Washington, Government Printing Office, 699 p., doi: 10.5962/bhl.title.49454.

Menand, L., 2001, Morton, Agassiz, and the origins of scientific racism in the United States: Journal of Blacks in Higher Education, v. 34, p. 110–113: doi: 10.2307/3134139.

Menard, H.W., 1955, Deformation of the northeastern Pacific Basin and the west coast of North America: Geological Society of America Bulletin, v. 66, p. 1149–1198, doi: 10.1130/0016-7606(1955)66[1149:DOTNPB]2.0.CO;2.

Menne, M.J., Williams, C.N., Jr., and Vose, R.S., 2009, The U.S. historical climatology network monthly temperature data, version 2: New bias adjustments reduce uncertainty in temperature trends for the United States: Bulletin of the American Meteorology Society, v. 90, no. 7, p. 993–1007, doi: 10.1175/2008BAMS2613.1.

Messing, C.G., Neumann, A.C., and Lang, J.C., 1990, Biozonation of deep-water lithoherms and associated hardgrounds in the northeastern Straits of Florida: Palaios, v. 5, no. 1, p. 15–33, doi: 10.2307/3514994.

Meyer, H.W., 2003, The Fossils of Florissant: Washington, D.C., Smithsonian Books, 258 p.

Mikhail, S., and Sverjensky, D.A., 2014, Nitrogen speciation in upper mantle fluids and the origin of Earth's nitrogen-rich atmosphere: Nature Geoscience, v. 7, no. 11, p. 816–819, doi: 10.1038/ngeo2271.

Miko, I., 2008, Gregor Mendel and the principles of inheritance: Nature Education, v. 1, no. 1, p. 134–137.

Milankovitch, M., 1930, Mathematische Klimalehre und astronomische Theorie der Klimaschwankungen: Handbuch der Klimatologie 1: Berlin, Borntraeger, 176 p.

Millar, C.D., and Lambert, D.M., 2013, Ancient DNA: Towards a million-year-old genome: Nature, v. 499, no. 7456, p. 34–35, doi: 10.1038/nature12263.

Mitchill, S.L., 1826, Catalogue of Organic Remains Presented to the New York Lyceum of Natural History: New York, Seymour, 42 p., doi: 10.5962/bhl.title.62913.

Moodie, R.L., 1916, The Coal Measures Amphibia of North America (No. 238): Carnegie, Institution of Washington, 296 p.

Morgan, J.V., et al., 2016, The formation of peak rings in large impact craters: Science, v. 354, no. 6314, p. 878–882, doi: 10.1126/science.aah6561.

Morlon, H., Parsons, T.L., and Plotkin, J.B., 2011, Reconciling molecular phylogenies with the fossil record: Proceedings of the National Academy of Sciences, v. 108, no. 39, p. 16327–16332, doi: 10.1073/pnas.1102543108.

Morris, S.C., and Whittington, H.B., 1985, Fossils of the Burgess Shale: A national treasure in Yoho National Park, British Columbia (Vol. 43): Natural Resources Canada, 31 p.

Mosher, S., 1998, Tectonic evolution of the southern Laurentian Grenville orogenic belt: Geological Society of America Bulletin, v. 110, no. 11, p. 1357–1375, doi: 10.1130/0016-7606(1998)110<1357:TEOTSL>2.3.CO;2.

Müller, R.D., Sdrolias, M., Gaina, C., and Roest, W.R., 2008, Age, spreading rates, and spreading asymmetry of the world's ocean crust: Geochemistry, Geophysics, Geosystems, v. 9, no. 4, doi: 10.1029/2007GC001743.

Murchison, R.I., 1839, The Silurian System, Founded on Geological Research, in Two Parts (Part 1): London, John Murray, 656 p.

Murchison, R.I., 1841, First sketch of some of the principal results of a second geological survey of Russia, in a letter to M. Fischer: Philosophical Magazine and Journal of Science, series 3, no. 19, p. 417–422, doi: 10.1080/14786444108650460.

Næraa, T., Scherstén, A., Rosing, M.T., Kemp, A.I.S., Hoffmann, J.E., Kokfelt, T.F., and Whitehouse, M.J., 2012, Hafnium isotope evidence for a transition in the dynamics of continental growth 3.2 Gyr ago: Nature, v. 485, no. 7400, p. 627–630, doi: 10.1038/nature11140.

Nagel, T.J., Hoffmann, J.E., and Münker, C., 2012, Generation of Eoarchean tonalite-trondhjemite-granodiorite series from thickened mafic arc crust: Geology, v. 40, no. 4, p. 375–378, doi: 10.1130/G32729.1.

Narbonne, G.M., and Gehling, J.G., 2003, Life after snowball; the oldest complex Ediacaran fossils: Geology, v. 31, no. 1, p. 27–30, doi: 10.1130/0091-7613(2003)031<0027:LASTOC>2.0.CO;2.

Narendra, B.L., 1979, Benjamin Silliman and the Peabody Museum: Discovery, v. 14, no. 2, p. 1–29.

National Academy of Sciences, 1999, Science and Creationism: A View from the National Academy of Sciences (second edition): Washington, D.C., National Academy Press, 48 p.

National Academy of Sciences, 2016, Correction for Zahid et al., Agriculture, population growth, and statistical analysis of the radiocarbon record: Proceedings of the National Academy of the Sciences, v. 113, no. 8, p. E 2546, doi: 10.1073/pnas.1605181113.

Naumann, C.F., 1866, Lehrbuch der geognosie (Vol. 3): Leipzig, Englemann, 192 p.

Nutman, A.P., Bennett, V.C., Friend, C.R., Van Kranendonk, M.J., and Chivas, A.R., 2016, Rapid emergence of life shown by discovery of 3,700-million-year-old microbial structures: Nature, v. 537, no. 7621, p. 535–538, doi: 10.1038/nature19355.

Ogg, J.G., Ogg, G.M., and Gradstein, F.M., 2016, A concise geologic time scale: Amsterdam, Elsevier, 234 p.

Oldroyd, D.R., 1971, The Vulcanist–Neptunist dispute reconsidered: Journal of Geological Education, v. 19, no. 3, p. 124–129.

Oldroyd, D.R., and McKenna, G., 1995, A note on Andrew Ramsay's unpublished report on the St David's area, recently discovered: Annals of Science, v. 52, no. 2, p. 193–196, doi: 10.1080/00033799500200181.

Olson, E.C., 1991, George Gaylord Simpson: June 16, 1902–October 6, 1984: National Academy of Sciences Biographical Memoirs, v. 60, p. 332, doi: 10.17226/6061.

O'Neill, C., and Debaille, V., 2014, The evolution of Hadean–Eoarchaean geodynamics: Earth and Planetary Science Letters, v. 406, p. 49–58, doi: 10.1016/j.epsl.2014.08.034.

O'Neill, J., Boyet, M, Carleson, R.W., and Paquette, J.L., 2013, Half a billion years of reworking the Hadean mafic crust to produce the Nuvvuagittuq Eoarchean felsic crust: Earth and Planetary Science Letters, v. 379, p. 13–25, doi: 10.1016/j.epsl.2013.07.030.

Oppel, A., 1858–1856, Die Juraformatión Englands, Frankreichs und das Südwestlichen Deutschlands: Stuttgart, Ebner & Serebert, 857 p.

Orcutt, M., 1954, The discovery in 1901 of the La Brea fossil beds: Historical Society of Southern California Quarterly, v. 36, no. 4, p. 338–341, doi: 10.2307/41168510.

Oreskes, N., 1999, The Rejection of Continental Drift: New York, Oxford University Press, 420 p., doi: 10.1093/oso/9780195117325.001.0001.

Oreskes, N., 2003, From continental drift to plate tectonics, *in* Oreskes, N., ed., Plate Tectonics: An Insider's History to the Modern Theory of the Earth: Boulder, Westview Press, p. 3–30.

Ospovat, A., 1960, Abraham Gottlob Werner and his influence on mineralogy and geology [Ph.D. thesis]: Norman, University of Oklahoma Graduate College, 259 p.

Ospovat, A., 1969, Reflections on A.G. Werner's "Kurze Klassifikation," *in* Schneer, C.J., ed., Toward a History of Geology: Cambridge, Massachusetts, MIT Press, p. 242–256.

Owen, R., 1840, A description of a specimen of the Plesiosaurus macrocephalus, Conybeare, in the collection of Viscount Cole, MP, DCL, FGS: Transactions of the Geological Society of London, v. 2, no. 3, p. 515–535, doi: 10.1144/transgslb.5.3.515.

Owen, R., 1841a, Description of the fossil remains of a mammal (Hyracotherium leporinum) and of a bird (Lithornis vulturinus) from the London Clay: Transactions of the Geological Society of London, v. 2, no. 1, p. 203–208, doi: 10.1144/transgslb.6.1.203.

Owen, R., 1841b, Report on British Fossil Reptiles, Part II, Report of the British Association for the Advancement of Science, 11th Meeting: London, Richard and John E. Taylor, 145 p.

Padian, K., Clemens, W.A., and Valentine, J.W., 1985, Terrestrial vertebrate diversity: episodes and insights, *in* Valentine, J.W., ed., Phanerozoic Diversity Patterns: Profiles in Macroevolution: Princeton, Princeton University Press, p. 41–96.

Pander, C., 1856, Monographie der fossilen Fische des Silurischen Systems der russisch–baltischen Gouvernements: St. Petersburg, Akademie der Wissenschaften, 91 p.

Parnreiter, C., 2009, Megacities in the geography of global economic governance: Die Erde, v. 140, no. 4, p. 371–390.

Penny, D., 2011, Darwin's theory of descent with modification, versus the biblical tree of life: PLOS Biology, v. 9, no. 7, p. e1001096, doi: 10.1371/journal.pbio.1001096.

Phillips, J., 1860, Life on Earth: Its Origin and Succession: Cambridge, Macmillan, 224 p., doi: 10.5962/bhl.title.22153.

Picard, M.D., 1989, Harry Hammond Hess and the theory of sea-floor spreading: Journal of Geological Education, v. 37, p. 346–349.

Pickford, S., 2015, "I have no pleasure in collecting for myself alone": Social authorship, networks of knowledge and Etheldred Benett's Catalogue of the Organic Remains of the County of Wiltshire (1831): Journal of Literature and Science, v. 8, no. 1, p. 69–85, doi: 10.12929/jls.08.1.05.

Pigliucci, M., and Muller, G., 2010, Evolution: The Extended Synthesis: Cambridge, Massachusetts, MIT Press, 504 p.

Piñeiro, G., Ferigolo, J., Menechel, M., and Laurin, M., 2012, The oldest known amniotic embryos suggest viviparity in mesosaurs: Historical Biology, v. 24, no. 6, p. 620–630, doi: 10.1080/08912963.2012.662230.

Pingali, P.L., 2012, Green revolution: Impacts, limits and the path ahead: Proceedings of the National Academy of Sciences, v. 109, no. 31, p. 12302–12308, doi: 10.1073/pnas.0912953109.

Pino, M., et al., 2019, Sedimentary record from Patagonia, southern Chile supports cosmic-impact triggering of biomass burning, climate change, and megafaunal extinctions at 12.8 ka: Scientific Reports, v. 9, no. 1, p. 1–27, doi: 10.1038/s41598-018-38089-y.

Piper, J.D., 2013, A planetary perspective on Earth evolution; lid tectonics before plate tectonics. Tectonophysics, v. 589(C), p. 44–56, doi: 10.1016/j.tecto.2012.12.042.

Playfair, J., 1802, Illustrations of the Huttonian Theory of Earth: Edinburgh, Neill & Co. for Caddell and Davis, London, William Creech, 528 p., doi: 10.5962/bhl.title.50752.

Playfair, J., 1805, Biographical account of the late Dr. James Hutton, F.R.S.: Transactions of the Royal Society of Edinburgh, v. V, part III, p. 39–99, published online by Cambridge University Press, January 17, 2013, doi: 10.1017/S0080456800020937.

Pliny the Younger, 2016, Letters of Pliny (Bosanquet, F.C.T., ed.; Melmoth, W., trans.): Project Gutenberg Ebook.

Plumb, K.A., 1991, New Precambrian time scale: Episodes, v. 14, no. 2, p. 139–140, doi: 10.18814/epiiugs/1991/v14i2/005.

Popper, K.R., 1958–1959, Back to the pre-Socratics: Proceedings of the Aristotelian Society, v. 59, no. 1, p. 1–24.

Powell, J.W., 1875, Exploration of the Colorado River in the West and Its Tributaries; Explored in 1869, 1870, 1871 and 1872, under the Direction of the Secretary of the Smithsonian Institution (Monograph): Washington, D.C., U.S. Government Printing Office, 291 p., 2 plates, doi: 10.3133/70039238.

Powell, J.W., 1876, Report on the Geology of the Eastern Portion of the Uinta Mountains and a Region of the Country Adjacent Thereto (Monograph): Washington, D.C., U.S. Government Printing Office, 218 p., 8 atlas sheets, doi: 10.3133/70039913.

Powell, J.W., 1877, Report on the Geological and Geographical Survey of the Rocky Mountain Region (Monograph): Washington, D.C., U.S. Government Printing Office, 19 p., 1 map, doi: 10.3133/70039914.

Powell, J.W., 1895a, Canyons of the Colorado River: Meadville, Flood & Vincent, 127 p.

Powell, J.W., 1895b, The Exploration of the Colorado River and Its Canyons: New York, Dover, 458 p.

Prothero, D.R., 2013, Bringing Fossils to Life: An Introduction to Paleobiology: New York, Columbia University Press, 672 p.

Prothero, D.R., 2016, The Princeton Field Guide to Prehistoric Mammals (Vol. 112): Princeton, Princeton University Press, 240 p.

Quennell A.M., 1958, The structural and geomorphic evolution of the Dead Sea Rift: Journal of the Geological Society of London, v. 114, p. 1–24, doi: 10.1144/gsjgs.114.1.0001.

Rabbitt, M.C., McKee, E.D., Hunt, C.B., and Leopold, L.B., 1969, The Colorado River Region and John Wesley Powell (U.S. Geological Survey Professional Paper 669-A): Washington, D.C., U.S. Government Printing Office, 145 p., doi: 10.3133/pp669A.

Raup, D.M., 1991, Extinction: Bad Genes or Bad Luck?: New York, Norton, 224 p.

Reboul, H.P.I., 1833, Géologie de la période Quaternaire: Paris, F.G. Levrault, 222 p.

Reichow, M.K., et al., 2009, The timing and extent of the eruption of the Siberian Traps large igneous province: Implications for the end-Permian environmental crisis: Earth and Planetary Science Letters, v. 277, nos. 1–2, p. 9, doi: 10.1016/j.epsl.2008.09.030.

Renne, P., and Basu, A.R., 1991, Rapid eruption of the Siberian Traps flood basalts at the Permo-Triassic boundary: Science, v. 253, no. 5016, p. 176–179, doi: 10.1126/science.253.5016.176.

Richter, C.F., 1935, An instrumental earthquake magnitude scale: Bulletin of the Seismological Society of America, v. 25, no. 1, p. 1–32.

Robb, L.J., Knoll, A.H., Plumb, K.A., Shields, G.A., Strauss, H., and Veizer, J., 2004, The Precambrian: The Archean and Proterozoic Eons, *in* Gradstein, F.M. and Ogg, J.G., eds.

A Geologic Time Scale, 2004: Cambridge, Cambridge University Press, p. 129–140, doi: 10.1017/CBO9780511536045.010.

Robert, F., 2001, The origin of water on Earth: Science, v. 293, no. 5532, p. 1056–1058, doi: 10.1126/science.1064051.

Rohde, R.A., and Muller, R.A., 2005, Cycles in fossil diversity: Nature, v. 434, no. 7030, p. 208–210, doi: 10.1038/nature03339.

Röhl, U., Westerhold, T., Bralower, T.J., and Zachos, J.C., 2007, On the duration of the Paleocene-Eocene thermal maximum (PETM): Geochemistry, Geophysics, Geosystems, v. 8, no. 12, p.1 1–13, doi: 10.1029/2007GC001784.

Romm, J., 1994, A new forerunner for continental drift: Nature, v. 367, no. 6462, p. 407–408, doi: 10.1038/367407a0.

Rosenberg, G.D., 2009, Introduction: The revolution in geology from the Renaissance to the Enlightenment, *in* Rosenberg, G.D., ed., The Revolution in Geology from the Renaissance to the Enlightenment: Geological Society of America Memoir 203, p. 1–11, doi: 10.1130/2009.1203(00).

Rosing, M.T., 1999, 13C-depleted carbon microparticles in >3700-Ma sea-floor sedimentary rocks from West Greenland: Science, v. 283, no. 5402, p. 674–676, doi: 10.1126/science.283.5402.674.

Ross, R.J., Jr., 1984, The Ordovician System, progress and problems: Annual Review of Earth and Planetary Science, v. 12, p. 307335, doi: 10.1146/annurev.ea.12.050184.001515.

Rothman, D.H., Fourier, G.P., French, K.L., Alm, E.J., Boyle, E.A., Cao, C., and Summons, R.E., 2014, Methanogenic burst in the end-Permian Carbon Cycle: Proceedings of the National Academy of Sciences, v. 111, no. 15, p. 5462–5467, doi: 10.1073/pnas.1318106111.

Royal Society of Edinburgh, 1826, Transactions, p. 452.

Rudwick, M.J.S., 1988, The Great Devonian Controversy: The Shaping of Scientific Knowledge among Gentlemanly Specialists: Chicago, University of Chicago Press, 528 p.

Rudwick, M.J.S., 2005, Bursting the Limits of Time: The Reconstruction of Geohistory in the Age of Revolution: Chicago, University of Chicago Press, 732 p.

Runcorn, S., 1959, Rock magnetism: The magnetization of ancient rocks bears on the questions of polar wandering and continental drift: Science, v. 129, no. 3355, p. 1002–1012, doi: 10.1126/science.129.3355.1002.

Rutherford, E., 1904, Radio-activity, *in* Neville, F.H., and Whetham, W.C.D., eds., Cambridge Physical Series: London, Cambridge University Press, 399 p., doi: 10.5962/bhl.title.62543.

Rutherford, E., and Soddy, F., 1902, The cause and nature of radioactivity: London, Edinburgh, and Dublin Philosophical Magazine and Journal of Science, v. 4, 6th series, p. 370–396 (Part I), doi: 10.1080/14786440209462856; p. 569–585 (Part II), doi: 10.1080/14786440209462881.

Ryan, W.B.F., 2009, Decoding the Mediterranean salinity crisis: Sedimentology, v. 56, p. 95–136, doi: 10.1111/j.1365-3091.2008.01031.x.

Sagan, C., 1980, Cosmos: New York, Random House, 384 p.

Saitta, D., 2003, Stephen Jay Gould: In memoriam: Rethinking Marxism, v. 15, no. 4, p. 445–449, doi: 10.1080/0893569032000163357.

Salvador, A., ed., 2013, Internationalstratigraphic Guide (second edition): Boulder, International Union of Geological Sciences and Geological Society of America, 214 p., doi: 10.1130/9780813774022.

Sánchez-Villagra, M.R., 2010, Developmental palaeontology in synapsids: The fossil record of ontogeny in mammals and their closest relatives: Proceedings of the Royal Society of Britain, v. 277, no. 1685, p. 1139–1147, doi: 10.1098/rspb.2009.2005.

Sandoval-Castellanos, E., Wutke, S., Gonzalez-Salazar, C., and Ludwig, A., 2017, Coat colour adaptation of post-glacial horses to increasing forest vegetation: Nature Ecology & Evolution, v. 1, p. 1816–1819, doi: 10.1038/s41559-017-0358-5.

Sandweiss, M.A., 2009, Passing Strange: A Gilded Age Tale of Love and Deception across the Color Line: New York, Penguin, 403 p.

Sanford, J.C., Snedden, J.W., and Gulich, S.P.S., 2016, The Cretaceous–Paleogene boundary deposit in the Gulf of Mexico: Large scale oceanic basin response to the Chicxulub impact: Journal of Geophysical Research, Solid Earth, v. 121, no. 3, p. 1240–1261, doi:10.1002/2015JB012615.

Santayana, G., 2013, Life of Reason: Amherst, New York, Prometheus Books, 504 p.

Sarafian, A.R., Nielsen, S.G., Marschall, H.R., McCubbin, F.M., and Monteleone, B.D., 2014, Early accretion of water in the inner solar system from a carbonaceous chondrite-like source: Science, v. 346, no. 6209, p. 623–626, doi: 10.1126/science.1256717.

Sarich, V.M., 1985, Molecular clocks and eutherian phylogeny (paper): Fourth International Theriological Congress, Edmonton, August 13–20.

Schneiderman, J.S., 1997, Rock stars: A life of firsts: Florence Bascom: GSA Today, July, p. 8–9.

Schoene, B., Eddy, M.P., Samperton, K.M., Keller, C.B., Keller, G., Adatte, T., and Khadri, S.F., 2014, U-Pb geochronology of the Deccan Traps and relation to the end-Cretaceous mass extinction: Science, v. 363, no. 6429, p. 862–866, doi: 10.1126/science.aaa0118.

Schuchert, C., 1910, Paleogeography of North America: Geological Survey of America Bulletin, February 5, v. 20, p. 513.

Schuchert, C., 1916, Correlation and chronology in geology on the basis of paleogeography: Geological Society of America Bulletin, September 1, p. 491–514, doi: 10.1130/GSAB-27-491.

Schulte, P., et al., 2010, The Chicxulub asteroid impact and mass extinction at the Cretaceous–Paleogene boundary: Science, v. 327, no. 5970, p. 1214–1218, doi: 10.1126/science.1177265.

Schumann, W.O., 1952, Über die strahlungslosen Eigenschwingungen einer leitenden Kugel, die von einer Luftschicht und einer Ionosphärenhülle umgeben ist: Zeitschrift für Naturforschung A, v. 7, no. 2, p. 149–154, doi: 10.1515/zna-1952-0202.

Scotese, C.R., 2009, Late Proterozoic plate tectonics and palaeogeography: A tale of two supercontinents, Rodinia and Pannotia: Geological Society of London Special Publications, v. 326, no. 1, p. 67–83, doi:10.1144/SP326.4.

Scott, G.R., and Cobban, W.A., 1965, Geologic and biostratigraphic map of the Pierre Shale between Jarre Creek and Loveland, Colorado: U.S. Geological Survey Miscellaneous Geologic Investigation Map I-439, scale 1:48,000, separate text, 4 p., doi: 10.3133/i439.

Scott, R.F., 1913, Scott's Last Expedition (Vol. 1): New York, Dodd, Mead and Co, 443 p., doi: 10.5962/bhl.title.11355.

Stuart, S., 1972, Biography Luis W. Alvarez, *in* Nobel Lectures, Physics, 1963–1970: Amsterdam, Elsevier, p. 291–292.

Sedgwick, A., and Murchison, R.I., 1835, On the Silurian and Cambrian systems, exhibiting the order in which older sedimentary strata succeed each other in England and Wales: British Association for the Advancement of Science, Report 5th Meeting, p. 59–61.

Sedgwick, A., and Murchison, R.I., 1839, On the classification of the older stratified rocks in Devonshire and Cornwall: London and Edinburgh Philosophical Magazine and Journal of Science, series 3, v. 14, no. 89, p. 241–260, doi: 10.1080/14786443908649732.

Seeley, H.G., 1887, On a sacrum apparently indicating a new type of bird, Ornithodesmus cluniculus Seeley: Quarterly Journal of the Geological Society of London, v. 43, p. 206–211, doi: 10.1144/GSL.JGS.1887.043.01-04.19.

Seeley, H.G., 1888, On Thecospondylus daviesi (Seeley), with some remarks on the classification of the Dinosauria: Quarterly Journal of the Geological Society of London, v. 44, p. 79–87, doi: 10.1144/GSL.JGS.1888.044.01-04.11.

Seeley, H.G., 1895, Researches on the structure, organization, and classification of the fossil Reptilia, Part IX, Section 5, On the skeleton in the New Cynodontia from the Karroo Rocks: Philosophical Transactions of the Royal Society of London, B 186, p. 59–148, doi: 10.1098/rstb.1895.0002.

Şengör, A., 2002, On Sir Charles Lyell's alleged distortion of Abraham Gottlob Werner in Principles of Geology and its implications for the nature of the scientific enterprise: Journal of Geology, v. 110, no. 3, p. 355–368, doi: 10.1086/339537.

Sepkoski, J.J., 1984, A kinetic model of Phanerozoic taxonomic diversity, III, Post-Paleozoic families and mass extinctions: Paleobiology, v. 10, no. 2, p. 246–267, doi: 10.1017/S0094837300005972.

Sepkoski, J.J., 1992, A compendium of fossil marine animal families: Contributions in Biology and Geology, v. 83, p. 1–156.

Sereno, P.C., 2013, Basal sauropodomorphs and the vertebrate fossil record of the Ischigualasto Formation (Late Triassic: Carnian-Norian) of Argentina: Journal of Vertebrate Paleontology, v. 32, no. supl, Memoir 12, p. 1–9, doi:10.1080/02724634.2013.819809.

Sereno, P.C., Forster, C.A., Rogers, R.R., and Monetta, A.M., 1993, Primitive dinosaur skeleton from Argentina and the early evolution of Dinosauria: Nature, v. 361, no. 6407, p. 64–66, doi: 10.1038/361064a0.

Seward, A.C., 1914, Antarctic fossil plants: British Museum (Natural History) report, British Antarctic ("Terra Nova") report, 1910, Natural History Report: Geological Studies: London, Printed by order of the Trustees of the British Museum,1914–1964, v. 1, p. 1–49, doi: 10.5962/t.174120.

Sharpe, T., and McCartney, P.J., 1998, The papers of H.T. De la Beche (1796–1855) Geological Series 17 (short summary letter, not complete): National Museum of Wales, Cardiff, p. 69–70.

Shen, S.Z., et al., 2011, Calibrating the end-Permian mass extinction: Science, v. 334, no. 6061, p. 1367–1372, doi: 10.1126/science.1213454.

Shen, S.Z., and Bowring, S.A., 2014, The end-Permian mass extinction: A still-unexplained catastrophe: National Science Review, v. 1, no. 4, p. 492–495, doi: 10.1093/nsr/nwu047.

Shubin, N.H., Daeschler, E.B., and Jenkins, F.A., 2006, The pectoral fin of Tiktaalik roseae and the origin of the tetrapod limb: Nature, v. 440, no. 7085, p. 764–771, doi: 10.1038/nature04637.

Simonetta, A., 1975, The Cambrian non trilobite arthropods from the Burgess Shale of British Columbia: A study of their comparative morphology taxonomy and evolutionary significance: Palaeontographia Italica, v. 69, tabs. I–LXI, p. 1–37.

Simpson, G.G., 1944, Tempo and Mode in Evolution: New York, Columbia University Press, 237 p.

Singer, B.S., Jicha, B.R., Mochizuki, N., and Coe, R.S., 2019, Synchronizing volcanic, sedimentary, and ice core records of Earth's last magnetic polarity reversal: Science Advances, v. 5, no. 8, p. eaaw4621, doi: 10.1126/sciadv.aaw4621.

Singh, V., and Singh, K., 2018, Modern synthesis, *in* Vonk, J., and Shackelford, T.K., eds., Encyclopedia of Animal Cognition and Behavior: New York, Springer, n.p., doi: 10.1007/978-3-319-47829-6_203-1.

Sloan, R.E., and Van Valen, L., 1965. Cretaceous mammals from Montana: Science, v. 148, no. 3667, p. 220–227, doi: 10.1126/science.148.3667.220.

Smith, R.B., Jordan, M., Steinberger, B., Puskas, C.M., Farrell, J., Waite, G.P., Husen, S., Chang, W.L., and O'Connell, R., 2009. Geodynamics of the Yellowstone hotspot and mantle plume: Seismic and GPS imaging, kinematics, and mantle flow: Journal of Volcanology and Geothermal Research, v. 188, p. 26–56, doi: 10.1016/j.jvolgeores.2009.08.020.

Smith, W., 1815. A Delineation of the Strata of England and Wales with Part of Scotland; Exhibiting the Collieries and Mines, the Marshes and Fen Lands Originally Overflowed by the Sea, and the Varieties of Soil According to the Variations in the Substrata, Illustrated by the Most Descriptive Names, British statute miles, 30[= 155 mm: London, J. Carey.

Smith, W., 1816–1819, Strata Identified by Organized Fossils, Containing Prints on Colored Paper of the Most Characteristic Specimens in Each Stratum (in 4 parts): London, Arding, doi: 10.5962/bhl.title.106808.

Smith, W., 1817, Stratigraphical system of organized fossils with reference to the specimens of the original Geologic Collection in the British Museum explaining their state of preservation and their use in identifying British strata: London, E. Williams, 118 p.

Snider-Pellegrini, A., 1858, La création et ses mystères dévoilés; ouvrage où l'on expose clairement la nature de tous les ětres, les éléments dont ils sont composés et leurs rapports avec le globe et les astres, la nature et la situation du feu du soleil, l'origine de l'Amérique, et de ses habitants primitifs, la formation forcée de nouvelles planètes, l'origine des langues et les causes de la variété des physionomies, le compte courant de l'homme avec la terre, etc.: Paris, A. Franck, 487 p., 10 plates.

Soddy, F., 1923, The origins of the conception of isotopes: Scientific Monthly, v. 17, no. 4, p. 305–317, doi: 10.1038/112208a0.

Sohl, L.E., Chandler, M.A., Jonas, J., and Rind, D.H., 2014, Energy and heat transport constraints on tropical climates of the Sturtian Snowball Earth: AGU Fall Meeting Abstracts, PP43C-1487.

Soriano, A., Navarro, E.A., Paul, D.L., Portí, J.A., Morente, J.A., and Craddock, I.J., 2005, Finite difference time domain simulation of the Earth-ionosphere resonant cavity: Schumann resonances: IEEE Transactions on Antennas and Propagation, v. 53, no. 4, p. 1535–1541, doi: 10.1109/TAP.2005.844415.

Spamer, E.E., Brogan, A.E., and Torrens, H.S., 1989, Recovery of the Etheldred Benett collection of fossils mostly from the Jurassic-Cretaceous strata of Wiltshire, England, analysis of the taxonomic nomenclature of Benett (1831), and notes and figures of type specimens contained in the collection: Proceedings of the Academy of Natural Sciences of Philadelphia, v. 141, p. 115–180.

Stanley, S.M., 2016, Estimates of the magnitudes of major marine mass extinctions in Earth history: Proceedings of the National Academy of Sciences, v. 113, no. 42, p. E6325–E6334, doi: 10.1073/pnas.1613094113.

Steele, A., 2012, Looking backwards, looking forwards: A consideration of the foibles of action research within teacher work: The Canadian Journal of Action Research, v. 13, no. 2, p. 17–34, doi: 10.33524/cjar.v13i2.36.

Steno, N., 1667, Elementorum myologiae specimen seu "Musculi Descriptio Geometrica" cui accedunt "Canis carchariae dissectum caput" et "Dissectus Piscis ex Canum Genere": Florence, Stellae, 123 p.

Steno, N., 1669, De solido intra solidum naturaliter contento dissertationis prodromus: Florence, n.p., 78 p., doi: 10.5962/bhl.title.148841.

Steno, N., 1916 [1669], The prodromus of Nicolaus Steno's dissertation, concerning a solid body enclosed by process of nature within a solid (translated by John Garrett Winter, English version with an introduction and explanatory notes): London, Macmillan, 283 p., doi: 10.5962/bhl.title.54340.

Stillwell, J.D., and Long, J.A., 2011, Frozen in Time: Prehistorical Life in Antarctica: Clayton, Australia, CSIRO, 248 p., doi: 10.1111/j.1095-8312.2012.01932.x.

Stose, G.W., and Ljungstedt, O.A., 1932, Geologic map of the United States: U.S. Geological Survey, scale 1:2,500,000.

Subramanian, M., 2019, Anthropocene now: influential panel votes to recognize Earth's new epoch: Nature, v. 21, p. 2019, doi: 10.1038/d41586-019-01641-5.

Sues, H.D., Nesbitt, S.J., Berman, D.S., and Henrici, A.C., 2011, A late-surviving basal theropod dinosaur from the Latest Triassic of North America: Proceedings of the Royal Society, v. B 278, no. 1723, p. 3459–3464, doi: 10.1098/rspb.2011.0410.

Suess, E., 1885, Das antlitz der Erde, Bd. 1: Vienna, F. Tempsky; Leipzig, G. Freytag, 778 p.

Suess, E., 1909, The Face of the Earth (translated by Sollas, H.B.C.): Oxford, Clarendon Press, 673 p.

Switek, B., 2011, Written in Stone: London, Icon Books, 321 p., doi: 10.1016/j.pgeola.2011.12.006.

Szaniawski, H., 2009, The earliest known venomous animals recognized among conodonts: Acta Palaeontologica Polonica, v. 54, no. 4, p. 669–676, doi: 10.4202/app.2009.0045.

Taquet, P., 2009, Geology beyond the channel, *in* Lewis, C.L.E., and Knell, S.J., eds., The Making of the Geological Society of London: Geological Society of London Special Publications, v. 317, no. 1, p. 155–162, doi: 10.1144/SP317.8.

Tarduno, J.A., Cottrell, R.D., Davis, W.J., Nimmo, F., and Bono, R.K., 2015, A Hadean to Paleoarchean geodynamo recorded by single zircon crystals: Science, v. 349, no. 6247, p. 521–524, doi: 10.1126/science.aaa9114.

Tashiro, T., Ishida, A., Hori, M., Igisu, M., Koike, M., Méjean, P., Takahata, N., Sano, Y., and Komiya, T., 2017, Early trace of life from 3.95 Ga sedimentary rocks in Labrador, Canada: Nature, v. 549, no. 7673, p. 516–518, doi: 10.1038/nature24019.

Thackery, J.C., 1976, The Murchison–Sedgwick controversy: Journal of the Geological Society, v. 132, p. 367–372, doi: 10.1144/gsjgs.132.4.0367.

Theberge, A.E., 2012, The myth of the telegraphic plateau: Hydro International, May 10, 2 p., https://www.hydro-international.com/content/article/the-myth-of-the-telegraphic-plateau.

Thomas, L., 2013, Coal Geology: Hoboken, New Jersey, Wiley-Blackwell, 454 p.

Torrens, H.S., 1995, Mary Anning (1799–1847) of Lyme: "The greatest fossilist the world ever knew": British Journal for the History of Science, v. 28, p. 257–284, doi: 10.1017/S0007087400033161.

Torrens, H.S., 2001, Timeless order: William Smith (1769–1839) and the search for raw materials 1800–1820: Geological Society of London Special Publications, v. 190, no. 1, p. 61–83, doi: 10.1144/GSL.SP.2001.190.01.06.

Torrens, H.S., 2015, Rock stars: William "Strata" Smith: GSA Today, September, p. 38–40.

Torrens, H.S., Benamy, E., Daescher, E.B., Spamer, E.E., and Bogan, A.E., 2000, Etheldred Benett of Wiltshire, England, the first lady geologist: Her fossil collection in the Academy of Natural Sciences of Philadelphia, and the rediscovery of lost specimens of Jurassic Trogoniidae (Mullusco: Bivalva) with their soft parts preserved: Proceedings of the Academy of Natural Sciences, v. 150, April 14, p. 59–123.

Torsvik, T.H., Doubrovine, P.V., Steinberger, B., Gaina, C., Spakman, W., and Domeier, M., 2017, Pacific plate motion change caused the Hawaiian-Emperor Bend: Nature Communications, v. 8, no. 15660, 12 p., doi: 10.1038/ncomms15660.

Trenberth, K.E., 2011, Changes in precipitation with climate change: Climate Research, v. 47, nos. 1–2, p. 123–138, doi: 10.3354/cr00953.

Turner, S., Burek, C.V., and Moody, R.T.J., 2010, Forgotten women in an extinct saurian (man's) world, *in* Moody, R.T.J., Buffetetaut, E., Naish, D., and Martill, D.M., eds., Dinosaurs and Other Extinct Saurians: A Historical Perspective: London, Geological Survey, p. 111–153, doi: 10.1144/SP343.7.

Twitchett, R.J., 2006, The palaeoclimatology, palaeoecology and palaeoenvironmental analysis of mass extinction events: Palaeogeography, Palaeoclimatology, Palaeoecology, v. 232, no. 2–4, p. 190–213, doi: 10.1016/j.palaeo.2005.05.019.

Ulrich, E.O., and Bassler, R.S., 1926, A classification of the toothlike fossils, conodonts, with descriptions of American Devonian and Mississippian species: U.S. National Museum Proceedings, v. 68, no. 2613, article 12, 63 p., 11 plates, doi: 10.5479/si.00963801.68-2613.1.

UNESCO, 2015, Tsunami warning and mitigation systems to protect coastal communities, Indian Ocean Tsunami Warning and Mitigation System (IOTWS) 2005–2015, Fact Sheet, May: Paris, Intergovernmental Oceanographic Commission.

United Nations, Department of Economic and Social Affairs, Population Division, 2017, World population prospects: The 2017 revision, key findings and advance tables, Working Paper ESA/P/WP/248: New York, United Nations.

Vacarri, E., 2006, The "classification" of mountains in eighteenth century Italy and the lithostratigraphic theory of Giovanni Arduino (1714–1795), *in* Vai, G.B., Glen, W., and Caldwell, E., eds., The Origins of Geology in Italy: Geological Society of America Special Paper 411, p. 157–177, doi:10.1130/2006.2411(10).

Valley, J.W., et al., 2014, Hadean age for a post-magma-ocean zircon confirmed by atom-probe tomography: Nature Geoscience, v. 7, p. 219–223, doi: 10.1038/ngeo2075.

Valley, J.W., Peck, W.H., King, E.M., and Wilde, S.A., 2002, A cool early Earth: Geology, v. 30, no. 4, p. 351–354, doi: 10.1130/0091-7613(2002)030<0351:ACEE>2.0.CO;2.

Van Valen, L., and Sloan, R.E., 1966, The extinction of the multituberculates: Systematic Zoology, v. 15, no. 4, p. 261–278, doi: 10.2307/2411985.

Vanity Fair, 1861, Grand ball given by the whales: Vanity Fair, April 20, 1861, v. 3, no. 1861, p. 186.

Venter, O., et al., 2016, Sixteen years of change in the global terrestrial footprint and implications for biodiversity conservation: Nature Communications, v. 7, no. 1, p. 1–11, doi: 10.1038/ncomms12558.

Vigil, J.F., n.d., This dynamic planet (wall map): U.S. Geological Survey, Smithsonian Institution, and U.S. Naval Research Laboratory. https://pubs.usgs.gov/gip/earthq1/plate.html.

Vine, F.J., and Matthews, D.H., 1963, Magnetic anomalies over oceanic ridges: Nature, v. 199, p. 947–949, doi: 10.1038/199947a0.

Von Herzen, R.P., 1959, Heat-flow values from the South-Eastern Pacific: Nature, v. 183, p. 882–883, doi: 10.1038/183882a0.

Von Herzen, R.P., and Uyeda, S., 1963, Heat flow through the Eastern Pacific ocean floor: Journal of Geophysical Research, v. 68, no. 14, p. 4219–4450, doi: 10.1029/JZ068i014p04219.

Walker, J.D., Geissman, J.W., Bowring, S.A., and Babcock, L.E., 2013, The Geological Society of America geologic time scale, GSA Bulletin, v. 125, no. 3–4, p. 259–272, doi: 10.1130/B30712.1.

Ward, P.D., Garrison, G.H., Haggart, J.W., Kring, D.A., and Beattie, M.J., 2004, Isotopic evidence bearing on the late Triassic extinction events, Queen Charlotte Islands, British Columbia, and implications for the duration and cause of the Triassic–Jurassic mass extinction: Earth and Planetary Science Letters, v. 224, p. 589–600 doi: 10.1016/j.epsl.2004.04.034.

Warren, S.G., 2015, Can human population be stabilized?: Earth's Future, v. 3, p. 82–94, doi: 10.1002/2014EF000275.

Watson, J.D., and Crick, F.H.C., 1953a, Molecular structure of nucleic acids: Nature, v. 171, no. 4356, p. 737–738, doi: 10.1038/171737a0.

Watson, J.D., and Crick, F.H.C., 1953b, The structure of DNA: Cold Spring Harbor Symposia on Quantitative Biology, v. 18, p. 123–131, doi: 10.1101/sqb.1953.018.01.020.

Watson, S., 1871, United States Geological exploration of the 40th parallel, Vol. 5, Botany: U.S. Army Engineering Department Professional Paper No. 18: Washington, Government Printing Office, 525 p., doi: 10.5962/bhl.title.49454.

Wegener, A., 1912, Die Herausbildung der Grossformen der Erdrinde (Kontinente und Ozeane), auf geophysikalischer Grundlage: Petermanns Geographische Mitteilungen, v. 63, p. 185–195.

Wegener, A., 1966 [1915], The Origins of the Continents and Oceans (translated from the fourth revised German edition by Biram, J.): New York, Dover, 274 p.

Werner, A.G., 1787, Kurze Klassification und Beschreibung der verschiedenen Gebirgsarten: Dresden, Waltherischen Hofbuchhandlung, 28 p.

Werner, A.G., 1971 [1786], A Short Classification and Description of the Various Rocks (translated with an introduction and notes by Ospovat, A.M.): New York, Hafner, 194 p.

Werner, A.G., 1805 [1774], A Treatise on the External Characters of Fossils (translated by Weaver, T.): Dublin, M. N. Mahon, 312 p.

Wertenbaker, W., 2000, Rock stars: William Maurice Ewing: Pioneer explorer of the ocean floor and architect of Lamont: GSA Today, October, p. 28–29.

Westerling, A.L., Hidalgo, H.G., Cayan, D.R., and Swetnam, T.W., 2006, Warming and earlier spring increase western U.S. forest wildfire activity: Science, August 18, p. 940–943, doi: 10.1126/science.1128834.

Whittington, H.B., and Briggs, D.E.G., 1985, The largest Cambrian animal, Anomalocaris, Burgess Shale, British-Columbia: Philosophical Transactions of the Royal Society of London B, Biological Sciences, v. 309, no. 1141, p. 569–609, doi: 10.1098/rstb.1985.0096.

Wilkins, J.S., 2009, Species: A History of the Idea: Berkeley, University of California Press, 320 p.

Williams, H.S., 1891, Correlation Papers: Devonian and Carboniferous (U.S. Geological Survey Bulletin 80): Washington, U.S. Government Printing Office, 288 p., doi: 10.5962/bhl.title.52384.

Williams, H.S., 1893, Studies for students, the elements of the geologic time scale: Journal of Geology, v. 1, p. 283–295, doi: 10.1086/606184.

Willis, B., 1912, Index to the stratigraphy of North America: U.S. Geological Survey Professional Paper No. 71, 894 p., with Map 1:5,000,000 by Willis, B., and Stose, G.

Williston, S.W., 1925, The Osteology of the Reptiles (Society for the Study of Amphibians and Reptiles), W.K. Gregory, ed.: Cambridge, Massachusetts, Harvard University Press, 324 p., doi: 10.5962/bhl.title.6573.

Wilson, J.T., 1963, A possible origin of the Hawaiian Islands: Canadian Journal of Physics, v. 41, no. 6, p. 863–870, doi: 10.1139/p63-094.

Wilson, J.T., 1965, A new class of faults and their bearing on continental drift: Nature, v. 207, p. 343–347, doi: 10.1038/207343a0.

Wilson, J.T., 1968, Static or mobile earth: The current scientific revolution: Proceedings of the American Philosophical Society, v. 112, no. 5, p. 309–320, doi: 10.1016/0040-1951(69)90033-X.

Wilson, L.G., 1998, Lyell, the man and his times, *in* Blundell, D.F., and Scott, A.C., eds., Lyell: The Past Is the Key to the Present: London, Geological Society, p. 21–37, doi: 10.1144/GSL.SP.1998.143.01.04.

Williston, S.W., 1925, Osteology of the Reptiles: Society for the Study of Amphibians and Reptiles: Cambridge, Harvard University Press, 324 p.

Winchester, S., 2001, The Map That Changed the World: New York, HarperCollins, 368 p.

Winsor, M.P., 1979, Louis Agassiz and the species question: Studies in the History of Biology, v. 3, p. 89–117.

Woodbridge, W.C., 1838, A System of Universal Geography on the Principles of Comparison and Classification: Hartford, Connecticut, J. Beach, 336 p.

Woodford, A.O., 1971, Catastrophism and evolution: Journal of Geological Education, v. 19, no. 5, p. 229–231, doi: 10.5408/0022-1368-XIX.5.229.

Woodward, H.B., 1908, The History of the Geological Society of London: London, Longmans, Green, 336 p.

Wool, D., 2001, Charles Lyell—"the father of geology"—as a forerunner of modern ecology: Oikos, v. 94, no. 3, p. 385–391, doi: 10.1034/j.1600-0706.2001.940301.x.

Worster, D., 1979, The Dust Bowl: The Southern Plains in the 1930s: New York, Oxford University Press, 304 p.

Xing Xu, Zhonghe Zhou, and Xiaolin Wang, 2000, The smallest known non-avian theropod dinosaur: Nature, v. 408, no. 7, p. 705–708, doi: 10.1038/35047056.

Yochelson, E.L., 1996, Discovery, collection, and description of the Middle Cambrian Burgess Shale biota by Charles Doolittle Walcott: Proceedings of the American Philosophical Society, v. 140, no. 4, p. 469–545.

Yochum, S.E., 2015, Colorado Front Range flood of 2013: Peak flows and flood frequencies: Proceedings of the 3rd Joint Federal Interagency Conference on Sedimentation and Hydrologic Modeling, April 19–23, Reno, p. 537–548, doi: 10.13140/RG.2.1.3439.1520.

Zahid, H.T., Robinson, E., and Kelly, R.L., 2016, Agriculture, population growth, and statistical analysis of the radiocarbon record: Proceedings of the National Academy of Sciences, v. 114, no. 4, p. 931–935, doi: 10.1073/pnas.1517650112.

Zalasiewicz, J.A., Taylor, L., Rushton, A.W.A., Loydell, D.K., Rickards, R.B., and Williams, M., 2009, Graptolites in British stratigraphy: Geological Magazine, v. 146, no. 6, p. 785–850, doi: 10.1017/S0016756809990434.f.

Zemp, M., et al., 2015, Historically unprecedented global glacial decline in the early 21st century: Journal of Glaciology, v. 61, no. 228, p. 745–762, doi: 10.3189/2015JoG15J017.

Zirkel, F., 1876, United States Geological exploration of the 40th parallel, Vol. 6, Microscopic Petrography: U.S. Army Engineering Department Professional Paper No. 18: Washington, Government Printing Office, 297 p., doi: 10.5962/bhl.title.49454.

Index

For the benefit of digital users, indexed terms that span two pages (e.g., 52–53) may, on occasion, appear on only one of those pages.

Tables and figures are indicated by an italic *t* or *f* following the page number.